T0289228

Structural Safety

Structural Safety

Theory and Practice

Allan Mann

BSc PHD C Eng FIStructE FREng

Whittles Publishing

Published by
Whittles Publishing Ltd.,
Dunbeath,
Caithness, KW6 6EG,
Scotland, UK

www.whittlespublishing.com

ISBN 978-184995-152-4

Printed and bound by CPI Group (UK) Ltd, Croydon, CR0 4YY

To my beloved family for their continuous support and to all my engineering colleagues for all I have learned from them.

Contents

Preface

It is the constant case that events pertaining to structural safety are continually evolving in ways affecting design practice even at the point of this book's publication. In the aftermath of the Grenfell Fire (2017) the UK government is introducing a Building Safety Act relating to high rise dwellings. Special emphasis is being placed on structural and fire safety and the Act will require engineers to provide positive demonstrations of building safety via a new demand for safety cases which will certainly challenge the professions. Alongside this development, there has been no let-up in news items highlighting issues relevant to structural safety. Even in the current year, there have been some major failures; dramatic fires; terrible earthquake damage in Turkey/ Syria; and the dramatic heat of the summer brought with it extensive wild fires; intense rain and terrible flooding. Hence, the need for us all to understand how to cope with safety challenges remains, indeed the demand is growing.

Allan Mann

Acknowledgements

After the infamous 1968 collapse of the Ronan Point tower block in London, the Institution of Civil Engineers, the Institution of Structural Engineers and the Health and Safety Executive (HSE) set up the SCOSS, the Standing Committee on Structural Safety. Ever since then, for over 50 years, the SCOSS has maintained a watchful eye on developments and failures, issuing feedback and periodic warnings of trends, always promoting a safety agenda. Since 2005, the SCOSS has run a confidential reporting system (CROSS), which issues quarterly bulletins plus separate reports pertaining to construction safety. Both the SCOSS and the CROSS are referred to frequently throughout this book. However, all the opinions expressed throughout are mine alone and any errors are mine alone. I am indebted though to my friend and colleague John Carpenter, who has contributed Chapter 12, and for his other valuable comments. We are both greatly indebted to Jenny Macgregor who has patiently edited our text and added much valuable clarification. Many failures reported herein are not referenced directly, but these days, the background to incidents can easily be found via a trawl on the Web. Mostly, I have just assembled publicly available information into a pattern and a digestible format and in the main this text is a personal view ranging over the subject matter.

Prologue

Engineers are often quizzed about whether their designs are safe. For the most part, we rely on code compliance to presume acceptability and make no direct justification. Yet many failures and disasters suggest that some wider understanding of 'safety' would be prudent. Unfortunately, 'safety', as a skill, has not generally been taught or considered a separate topic within the curriculum. Furthermore, answering questions such as 'what is safe?' or 'what is safe enough?', and demonstrating safety, are not at all easy. Given the crucial challenges of making infrastructure safe, and the corporate and personal consequences of 'getting it wrong', this is profoundly unsatisfactory.

Any review of the contemporary technical press will reveal that danger lurks within the construction industry: there are plenty of incidents to show that all that is being built is not necessarily safe and a few failures are truly spectacular. Hence an ability to respond to questions on 'safety' really is a key skill that all engineers should acquire.

The responsibilities of those who build are great. Society depends on lifeline systems, i.e. essential systems that form our infrastructure, and everyone needs those to function safely and reliably. Moreover, one direct purpose of much infrastructure is to protect populations from the significant range of man-made and natural hazards which threaten us. It only takes a glance at the news to appreciate the challenges. Year on year, large populations are at risk from potential industrial disasters and from the extremes of nature. Storms, floods and earthquakes bring misery, devastation and death, and cost us vast amounts.

Engineering communities across the world have expended considerable effort over the years to improve their design and construction methodologies to make them ever more reliable. And by and large, if proper procedures are in place and if knowledge is deployed, all will be well. Yet there is absolutely no reason for complacency. The HSE annually publish statistics on work-related harms and those remain non-trivial and only maintained at their current level by constant vigilance. Allied to that, in any year it is easy to pick out significant construction failures making headlines. Recent times have been no exception, with several bridge and building collapses. There can be few incidents more horrific than the burn out of the Grenfell Tower in London (2017) or of the instantaneous collapse of the Morandi motorway bridge outside Genoa (2018) or the 2020 blast in Beirut. The public are rightly asking how these events could happen in our supposedly well-regulated society.

Further afield, these last few years have seen the usual crop of natural disasters with annual reports from the World Economic Forum making gloomy reading. Under their section headed 'Our planet on the brink', the WEF describe how environmental risks have grown in prominence over the report's 13-year history. Furthermore, the numbers and concentration of populations at risk from such extremes continue to rise with world population increase and as the trend from rural to urban dwelling intensifies and as cities around the world expand into mega-centres.

The WEF report draws attention to the connection existing between these environmental hazards and other risks of immediate implication: 'And as the impact of Hurricane Maria on Puerto Rico has starkly illustrated, environmental risks can also lead to serious disruption of critical infrastructure.' That last statement defines a challenge for the engineering community. All this is set against a background of fear that the underlying causes of climate change (which is thought to be driving extreme and erratic weather) are not being tackled. By 2021, the language being used had shifted from Climate Change to Climate Emergency and the 2021 Glasgow COP conference sounded alarms as never before. Pressing challenges exist in the form of extreme weather events with many statistics for the years 2017 to 2022 breaking records.

Against this background, efforts are required from all those engaged in construction to minimise the risk of failures and to design and construct safely. Not only is this a moral obligation, it is a task strongly linked to self-interest. There are few experiences more traumatic than dealing with structural failure: careers can be blighted; the costs may be horrendous and the consequences punitive. If just one incident can be avoided, this book will have been worthwhile.

The overall task is formidable since learning about 'safety' is a major study area and a very much expanded text could be produced about any of the topics discussed within this book. Nevertheless, there is much each of us can do. One starting point is to systematically learn from mistakes that others have had the misfortune to make. This book describes a number of such mistakes, and it is invariably the case that a pattern exists; that what goes wrong has happened before and could have been prevented if lessons had been learned. And it is possible to learn, albeit that for those just entering the profession there is a bewildering amount to learn and not all of it is clear without practical experience of design and construction. A key message is therefore to impress the importance of acquiring a grounding and then building on that by continuing observation of events that life throws up. For sure there will be some events recurring on a regular basis.

The book is targeted at civil and structural engineers. The first chapters (1 to 5) address the tricky problem of what it is that makes structures safe or unsafe. But in nearly all failures a human being has got something wrong, so Chapter 6 discusses the role of human error. One group of failures can be categorised as those related to material or products, with an obvious factor being degradation over time: Chapter 7 describes examples falling within this group. Chapter 8 is devoted to failures under the heading of 'design and construction' and many incidents are covered; alas there is no shortage of others that could be added. The Grenfell fire was mentioned earlier as being a seminal event. But in truth, that fire was just one more in a long history of similar disasters. Fire prevention should not be a bolt-on attribute of building safety, it should be an integral part of building design. Chapter 9 discusses this. Some of the hardest challenges we face are exemplified by the tremendous loss of assets and life occurring in disasters due to natural or man-made events: these are covered in Chapters 10 and 11. Apart from designing structures to be safe in service, we all have obligations to protect those who undertake the construction task itself. In years gone by there was a fatalistic attitude, it being accepted that construction was inherently dangerous, and some workers would inevitably be maimed or killed. That is unacceptable, and Chapter 12 sets out current obligations that designers need to be aware of. Finally, Chapter 13 addresses the topic everyone should be interested in: how might failure be avoided?

1 Introduction

1.1 Introduction

Despite all our knowledge and advanced understanding of materials and design, construction disasters still recur. Hence society continues to endure many structural failures along with the economic and social problems those bring. Often there is human injury and death. Even in the last few years, there have been spectacular failures of major structures and engineering systems when it might have been supposed the design and construct teams employed were from the most highly skilled and experienced staff available. It is therefore puzzling why these events happened and why we do not seem to learn, since there is plenty of evidence for failure repetition. There is no unique answer. Partly the response is that all humans make mistakes; partly that there is a vast amount to learn and it is a formidable task for each generation of engineers to absorb all available knowledge. Work forces roll over, as new recruits join, and older workers retire, making it self-evident that the process of training and passing on knowledge can never end. As for learning, it has been observed that organisations very rarely have corporate memory; only individuals have memory and if individuals have not been informed, they will not have learned.

Yet another factor is that the systematic study of failures and the drawing of lessons from them, plus the discipline of safety, is relatively recent. It seems hardly ever to be taught. Hereto it has been presumed that engineers aim to design safely but too often that just means engineers designing individual components to a margin of safety but neglecting to look at the whole: this is especially likely when the whole encompasses many different technical disciplines. Furthermore, any major engineering project is the synthesis of design, manufacture and construction (and often in-service maintenance); an immensely complex task and weaknesses exist along all the interfaces. Many structural failures occur in the construction stage for reasons often explainable by lack of communication between the design and construction teams and from a lack of appreciation of each other's requirements.

Apart from headline incidents such as major collapses, the topic of safety includes aspects which affect the health and safety of those who construct. Not so long ago the industry was regarded as inherently hazardous; injuries and deaths were commonplace and long-term health problems were rife. Significant improvements have been made and the statistics for bodily harm are nowadays much reduced, but maintaining that trend requires constant vigilance. Those entering the industry need to be proactively trained to ensure they understand the activities that bring risks to occupational health and safety.

1.2 Failure cost

Engineering failures can be immensely costly; the bill for the 2010 Deepwater Horizon oil blowout in the Gulf of Mexico has been put at $40 billion (Figure 1.1). The cost of reconstruction after the 2011 tsunami and earthquake in northern Japan has been put at $300 billion. And those costs hide other tragedies; 11 workers were killed in the Gulf of Mexico blowout and there was a hugely destructive environmental impact. There were about 20,000 dead in Japan, but even that number is dwarfed by the loss of life in the earlier 2010 Haitian earthquake of maybe 315,000. Alongside the dead and injured, huge numbers were made homeless and suffered disrupted lives. In the last decade, storm flooding worldwide has affected major cities, displaced thousands of residents and cost immense amounts. The same could be said of regional fires. On top of that, for Europeans (who have disasters of their own) these disasters are not just something that happened to others in far-away lands. Because we live in a globalised economy, disasters affect us all, either by insurance claims, by fluctuations in commodity prices, by industrial supply disruption or via social/political problems. The future of nuclear

Figure 1.1 Deepwater Horizon rig on fire. (Dreamstime 18174231)

power was put in jeopardy by the 2011 failures of the Fukushima plant. Flood damage in central China in 2011 was extensive enough to push up world food prices. And food price instability around the world, consequent on drought or rain (which engineering can mitigate) has been responsible for severe population unrest.

1.3 Risk and uncertainty

Over recent years the approach to engineering safety has become more generalised. The construction community has embraced the concepts of risk and uncertainty proactively. Rather than just considering a project or a design as 'risky' with adverse connotations, the modern approach is to recognise there are risks in everything we do; thereafter proper management involves identification and control of those risks (Carpenter, 2008). Communities cannot avoid all risk; that is impossible. What is required is to identify and address risks, positively seeking to contain them at tolerable levels. Some of these risks are posed by the environment; others by the nature of the project. Others just arise because something might not work out as expected: witness the world financial crisis of 2007–2008 or the COVID-19 crisis that started in 2020. On a day-to-day basis, apart from self and professional interest in minimising risk, all those engaged in construction now have legal duties with regard to risk control. Hence it is imperative that engineers fully understand what those duties are (Chapters 2, 4 and 12). Various techniques are available to assist in the hazard/risk identification process (Chapter 13).

1.4 Man-made and environmental hazards

Whilst there are many examples of individual structural failures from design or construction faults, equally there are many examples of widespread regional damage to engineered structures wreaked by man-made or environmental hazards. And the consequences endured by populations at risk are often dreadful. The Deepwater Horizon blowout is an example of a man-made catastrophe. There is uncertainty in natural hazards but only to the extent of which one will strike and where. For sure, every year there will be potentially damaging events such as earthquakes, floods, storms and so on. Civil and structural engineers try to design infrastructure to be resilient against the forces of nature but do not always succeed. There is great difficulty in forecasting what might happen and great difficulty in providing affordable protection.

1.5 Mitigating consequences

Whilst we cannot avoid natural disasters, there is a lot we can do to mitigate their consequences. To confirm this, just compare damage and loss of life between countries subjected to comparable storms or earthquakes where one country has a properly engineered infrastructure against others that do not. Such differences reveal that engineers and scientists can do much to predict the likely magnitudes of environmental hazards and prepare for their aftermath.

Most engineers are not directly involved with extreme environmental events. But all are involved somehow in construction projects great and small. A regular review of the press will reveal a trickle of failures with occasional headline events. These nearly always involve commercial loss and may involve loss of life or serious injury. Periodically there is a significant disaster: the New York Twin Towers attack of 2001 was one; the 2017 London Grenfell Tower fire was another and the Miami, Surfside condominium collapse in 2021 yet another. So, in design for all projects, it pays dividends for designers to consider what might go wrong and try and configure their designs accordingly to contain the consequences of failure to something less than disaster. Building robustness and resilience into structures and projects has been shown to be a wise strategy in the cause of promoting safety.

1.6 Robustness and resilience

Everyone desires that engineering products should be safe, but we live in an increasingly complex and integrated world where what might seem a relatively trivial lapse or failure can have widespread consequences. There is a very old children's nursery rhyme that sums this up rather well:

> For want of a nail the shoe was lost.
> For want of a shoe the horse was lost.
> For want of a horse the rider was lost.
> For want of a rider the battle was lost.
> For want of a battle the kingdom was lost.
> And all for the want of a horseshoe nail.

Engineers need to look for that nail whose loss will precipitate disaster. This is achieved by following the mental process of considering what might go wrong and trying to make sure the fault is avoided. In turn, that process is significantly informed by observing what has gone wrong elsewhere, and there are plenty of precedents to learn from.

The failure of the blowout preventer in the Gulf of Mexico led to horrendous consequences and so is one example of a lack of robustness. But we should not think examples are restricted to such major projects on which few of us work. If we recall the possibility of lack of maintenance on a single point system on the railway, the consequent derailment of a train, the consequent closure of the railway network followed by chaos on the roads and a significant loss of economic output, we can see the general idea (examples of such incidents are given in Chapter 11). We might also think of this in terms of the infamous Ronan Point collapse (Chapter 8): one faulty gas fitting followed by an explosion in one flat which triggered off total collapse of a significant part of the building. Very frequently fires fall into the category of a small initiating event producing a conflagration (witness the fire cause in Grenfell being a single faulty electrical kitchen device) or the fire which apparently set off the giant blast in the Port of Beirut in 2020. The possibility of small events triggering much larger consequences are been a study area for the UK HSE (CIRIA/HSE, 2011). Chapters 4 and 5 discuss desirable attributes of safety such as robustness/resilience and how they should be incorporated into engineering systems.

1.7 Teaching

All failures are traumatic, not just for the victims but also for the parties involved. Costs are large, careers are blighted, and the mental scars are great. Yet very rarely have structural failures been caused by a deliberate act or even significant negligence. Frequently there are a myriad of causes, all contributing. Against that background, it is in some ways surprising that we neglect to teach failures as object lessons in undergraduate courses. In other ways it is not surprising, since at the time of their initial learning experience, students probably lack the technical background to appreciate all the implications. Lessons are therefore more likely to be beneficial to those engineers who have enough experience to grasp the essential points and who have an appreciation of how the failures could have come about.

It is the purpose of this book to describe various failures/disasters as a basis for learning how not to repeat the mistakes. Later chapters will focus on that objective. But initially it is useful to review the development of engineering and safety systems to see how we have arrived at the current status.

1.8 The beginning

Engineering, in whatever branch, is about making products for people to use. Perhaps the oldest branch is related to construction for protection, since shelter is surely one of our most basic needs. For millennia, engineers have also tried to control nature or harness its power: in hot countries by constructing irrigation schemes, in wet countries by constructing flood defences. In ancient times, all the great civilisations were characterised by their ability to organise human effort into creating infrastructure. Very early on, artisans must have grasped the notion of nature's variability and the forces acting on structures and tried to evolve methodologies of control, not least since their clients were apt to seek stronger retribution for failure than mere monetary compensation. Those early societies were certainly aware of the power of fire and earthquakes to destroy cities or the danger of storms and floods. Biblical and Mesopotamian renditions of the Flood story are testimony to the traumatic consequences of such events. There is no shortage of evidence of tremendous destruction by fire: it is only necessary to recall the Great Fire of London in 1666 to grasp the havoc caused. Whilst societies could see the potential for control of fires and some could see the potential for flood control, it was not until comparatively recently that disasters were thought of as other than symbols of divine retribution. The Great Lisbon Earthquake in 1755 caused near destruction of the city and had profound political consequences across Europe, yet it also sparked off the scientific study of disasters and created a shift in attitudes. Nonetheless, the effects of natural disasters have been so huge that little could be done to mitigate them until well into the 20th century.

1.9 Early failures

For much of history the design of engineering projects had little science applied. It was only late in the 19th century that professional civil engineering really emerged, driven by the enormous demands of the Industrial Revolution. In earlier times, designs were guided by intuition, empiricism and historical precedent. Spectacular failures were not unknown, as in 1284 when the choir vaulting in Beauvais Cathedral collapsed. It is hard not to admire the nerve of those mediaeval cathedral builders when recalling the later collapse of the same cathedral's 153 metre central tower (1573). Who would risk building such an edifice these days, reliant solely on intuition? Beauvais was not alone and reveals that our forefathers were not blessed with infallibility. In 1972, major underpinning was carried out under York Minster's central tower for it was then considered close to failure. Of course, everyone knows of the foundation failure at the tower of Pisa and the gradual sinking of Venice (the Venice campanile (~ 100 metres tall), dating from the 16th century, collapsed spectacularly in 1902, which just shows what can happen even up to modern times).

Yet failures of this kind must have been learning opportunities and many structures survive extraordinarily well. Ancient engineers were certainly learning and, for a long time, attempts have been made to minimise the risk of flooding by appropriate siting and to contain the risk of fire by various means. There are many instances of switching building materials from timber to stone to control the risk and it was standard practice to locate bakehouses strategically to minimise the fire hazard (recall that the Great Fire of London started from a bakehouse). Occasionally citizens must have applied an early form of risk assessment. Venice is clearly at risk of flood, but its siting offered commercial and defensive benefits which seem to have outweighed the inconvenience of occasional inundation.

1.10 Development of the profession: learning from failures

1.10.1 Mechanical failures

As the profession of engineering grew in the late 19th century, engineers were learning continually, often from mistakes. As transport systems developed, trains and bridges were required, and both suffered failures. This was partly because the demands engineers placed on their inventions were continually growing. Steam-raising boilers had been used since the Newcomen engine of 1712 (improved by Watt in about 1763). But the need to work under higher pressures made ever-increasing demands on boiler materials, on structural form and on joints, such that dramatic failures were common. Early bad incidents included the blowing-up of several steamboats in America in the 1860s. In 1865, boilers on the steamboat Sultana exploded, killing about 1,700 passengers.

Boiler failures offer good examples of whole systems that can be reviewed to illustrate generic factors and terminology in the abstract study of safety: hazards, failure modes, consequences and mitigation. One inherent hazard for boilers arises from use of high-pressure steam. Secondly, there are multiple modes and causes of failure under internal pressure and thirdly, the failure consequences can range from simple pressure loss to catastrophic explosion with perhaps collateral damage from flying debris. Against these eventualities, mitigation steps can be deployed via quality control, via installed safety devices to limit pressure, via in-service monitoring and instrumentation, and by maintenance and inspection. Human error is another potential cause of failure, as is corrosion, and the corrosion hazard highlights that levels of safety vary with time with the value generally degrading as materials decay. These generic principles are of interest to all engineers. In the early days of the British Industrial Revolution, boiler failures were so common that William Fairbairn set up the Manchester Steam Users' Association (1854) to raise standards, whilst in the US, the American Society of Mechanical Engineers (ASME) developed their Boiler and Pressure Vessel Code. Hewison (1983) gives a comprehensive account of British boiler explosions, listing 137 between 1815 and 1962. Whilst 122 of these were in the 19th century, there were only 15 in the 20th century, so things appear to have improved as better technology was deployed.

Although boiler failures lie within the province of mechanical engineering, there is commonality with civil structures; moreover, boiler failures illustrate other crucial points in safety. Primarily, there is the need to think of overall systems and understand the hazards applicable to the whole, not least since civil structures frequently need designing against hazards consequent on mechanical failure. During the Fukushima nuclear plant failures in 2011 (consequent on the earthquake and tsunami) (Chapter 10), TV viewers watched live as released hot gas exploded and blew the cladding off the plant buildings. The world also watched intensely, knowing that the only protection against meltdown and intense radioactive release was being provided by the civil engineering containment.

1.10.2 Structural failures

Running parallel with mechanical failures and their resolution, there were many early structural failures associated with the railway system. Engineers were still searching for a better understanding of materials and the technology of the time was inadequate to produce metal with reliable properties. Robert Stephenson himself suffered a severe setback in 1847 when the cast iron girder bridge he had designed for the River Dee near Chester collapsed as a train crossed over; the collapse perhaps precipitated by fracture of the girder's bottom flange. This failure was not a one-off. In 1860, part of the Bull Bridge (Derbyshire) collapsed, again whilst a train was crossing. The failure cause was fracture of a cast iron girder emanating from a defect. The report of the inspectorate revealed that the fracture surface was rusted, so cracks had obviously been growing for some time. This early failure shows two critical errors that should be avoided in modern practice. First, the need for adequate quality control to show that a structure goes into service in its intended state and secondly, that in-service inspection is required to detect the onset of degradation before defects progress to dangerous levels. The use of cast iron itself can be said to violate one key safety attribute – that materials should possess ductility – a topic that will be discussed in Chapter 4.

In the following year, 1861, another cast iron bridge failed: the Wooten Bridge (Warwickshire) followed by others. At Wooten, five cast iron girders supporting the base of the wooden bridge all fractured near their centre. These days it is to be hoped that the thread of commonality exhibited by using cast iron in girders to support moving loads might have caused earlier concern. However, steel was not generally available, so material substitution did not occur till the end of the century. Any safety survey seeks particularly to determine if there is some fundamental pattern in successive failures.

After this last catastrophe, use of cast iron was phased out, with the Norwood Junction (London) railway crash of 1891 being the final blow; the cast iron bridge there collapsed by fracture, once again propagating from a large flange blowhole. After the subsequent inquiry, the Board of Trade recommended that all cast iron bridges be replaced.

One of the more infamous collapses had occurred earlier in 1879, when the Tay Bridge in Scotland failed during a December storm. The whole centre section collapsed whilst a train was running on its single track and 75 passengers were killed. The bridge was about two miles long and consisted of a series of lattice girder spans supported on piers. The girders were fabricated from a mixture of wrought and cast iron. Following the failure, there was an enormous public outcry and an official inquiry was set up. The inquiry found use of inferior material allied with poor quality construction, and so collectively, poor quality control and evidence of decidedly dangerous practice. Notoriously, the design had made no direct allowance for wind loading and the bridge layout clearly made it vulnerable. Interestingly, failure was precipitated by fracture of some lugs that were supposed to attach bracing onto columns, perhaps another example of one small error leading on to catastrophe. The designer, Sir Thomas Bouch, shouldered much of the blame for it was concluded that the bridge was 'badly designed, badly built and badly maintained, and that its downfall was due to inherent defects in the structure, which must sooner or later have brought it down'. Alongside that, there was evidence of missed opportunities in maintenance, for it was reported that tie bars had been vibrating for some time. Overall, the Tay Bridge disaster provides several generic lessons.

1.11 Generic lessons

Firstly, there is the question of why the event became so infamous? That might seem obvious with 75 deaths, yet on the other hand, in those days, loss of life in construction was common and scarcely raised comment. But the drama of the failure provides the lesson that some types of disasters cause more concern than others; probably here the horror of a whole train speeding at night into icy waters. Even today, the public are far more alarmed by deaths on trains than they are by deaths on roads. Likewise, it is a real concern for engineers to deal with public perceptions of nuclear safety which, rational or not, are a reality and a hindrance to the construction of nuclear power projects. The public are far more tolerant of deaths via 'natural' disasters (even if these could have been mitigated) than they are over deaths from 'man-made' disasters, even if these were pure accidents. In 2017 there was tremendous outrage after the Grenfell Tower fire, which could be seen live on TV screens and resulted in 71 deaths, leaving the iconic backdrop of a burned-out tower block. On the other hand, it has been a contemporary struggle to persuade construction workers to take elementary precautions for their own health and safety, for many have been accustomed to risks and so fail to take hazards seriously. All of this leads up to the difficult questions of what makes structures 'safe', what justifies them being 'safe enough', and what is society's tolerance of risk?

Secondly, we could take comfort that our technical understanding has moved on enormously since 1879, certainly to the extent that ignoring wind loading is barely conceivable in the absence of gross negligence. Furthermore, our understanding of materials and our ability to produce reliable materials have moved on enormously. Thus, we could probably rule out a similar contemporary accident occurring from either of those two causes. Nevertheless, the possibility of human error remains; the possibility of poor maintenance remains, and the possibility of poor detailing remains. The possibility of ignoring warning signs remains. Moreover, even over the last 25 years, there have been some spectacular bridge failures, so there is no room for complacency.

Although dramatic failures occurred in the 19th century, the profession of engineering was keen to learn from them and develop knowledge. Robert Stephenson wrote a century ago:

> Nothing was so instructive to the younger members of the profession, as records of accidents in large works, and of the means employed in repairing the damage.

And so ever since, the professions have taken an interest in safety and tried to promote safety thinking. The fact that there is still much to do can again be shown by reference to the Tay Bridge. The inquiry was chaired by the then President of the Institution of Civil Engineers, William Barlow, but almost a century later in 1967 the same institution produced a report on contemporary conditions which included in its preface: 'The cost to the construction industry in lives, injuries and money over the past years has been so great that action was clearly an urgent matter' (see Chapter 2; and COMSAFCIVENG and Shirley Smith, 1969).

We may gain similar insight from the Institution of Structural Engineers, founded in 1908, for in only the second edition of its journal proper, it featured a paper on 'The failure of the Knickerbocker Theatre' (Godfrey, 1923). This was a theatre in Washington DC (USA), the roof of which collapsed in 1922, killing 100 people. The collapse was said by some to have been triggered by snow, but the paper's author was scathing in describing the snow as merely the trigger. He put the real cause down to: 'miserable design and construction'. He described every part of this construction as weak, 'including the roof slabs, the roof beams, the roof trusses, the columns, and the walls'. But he supposed that the failure was precipitated by movement of one of the main trusses off its supporting wall. He cites two errors: first, poor vertical capacity and second, lack of anchorage such that thermal expansion and contraction gradually caused the truss to move off its supports as well as pushing the wall outwards. Note the failure cause does not involve overstress as such, which is often taken to be the key indicator of safety; rather the failure was caused by instability and lack of overall robustness (see Chapters 4 and 5).

The fact that failures continued to occur and needed some oversight was recognised by both institutions and by the HSE when, following the 1968 Ronan Point collapse, they jointly set up a 'Standing Committee on Structural Safety' (SCOSS) in 1976. More of that later.

1.12 Industrial society and occupational health

Britain had the first industrial society in the world and for those employed, working and living conditions were notoriously bad. In 1844 Frederic Engels wrote of Manchester (Engels, 1892):

> Such is the Old Town of Manchester, and on re-reading my description, I am forced to admit that instead of being exaggerated, it is far from black enough to convey a true impression of the filth, ruin, and uninhabitableness, the defiance of all considerations of cleanliness, ventilation, and health which characterise the construction of this single district, containing at least twenty to thirty thousand inhabitants. And such a district exists in the heart of the second city

> of England, the first manufacturing city of the world. If any one wishes to see in how little space a human being can move, how little air – and such air! – he can breathe, how little of civilisation he may share and yet live, it is only necessary to travel hither. True, this is the Old Town, and the people of Manchester emphasise the fact whenever any one mentions to them the frightful condition of this Hell upon Earth; but what does that prove? Everything which here arouses horror and indignation is of recent origin, belongs to the industrial epoch.

Conditions in the mills around Manchester were truly dreadful: long working hours combined with poor working conditions and inadequately protected machinery gave rise to poor health and many industrial injuries. Conditions in construction were no better and examples can be seen from the development of railways, at its peak in the era of Railway Mania in the 1840s. In his book 'The Railway Navvies' (Coleman, 1965), Terry Coleman writes that the first recorded death took place in 1827 at Edgehill in Liverpool (the worker was crushed under collapsing clay).

> But this labourer was the first of many thousands, the work was inherently dangerous. Gangs of men half disciplined if at all, hacked away at great banks of earth and sheer cliffs of rock, or blasted out tunnels hundreds of feet below ground. Deaths were expected, and the navvies increased the ever-present hazard by their own recklessness. It was a sort of bravado with many of them not to take care. A contractor said he knew one fellow who would smoke a pipe near an open barrel of gunpowder.

Elsewhere Coleman records:

> Even the most apparently humane of engineers resigned themselves – or rather the navvies – to the risk of work. Brunel was once shown a list of 131 navvies on the Great Western, not including those slightly injured, who had been taken to Bath Hospital from 30th September 1839 to 24 June 1841. 'I think it is a small list' he replied: 'Considering the very heavy works and the immense amount of powder used, and some of the heaviest and most difficult works. I am afraid it does not show the whole extent of accident incurred in that district.'

Brunel himself, it will be recalled, nearly died when his Rotherhithe Tunnel collapsed in 1828.

To put alongside the loss of life incurred in building railways, there were structural disasters causing deaths. Quoting again from Coleman, he recorded an incident in 1845 at Ashton under Lyne, where the Stalybridge to Ashton line was carried over the River Tame by a nine-arch viaduct:

> At about quarter past three on 19th April a Saturday afternoon an old man called William Kemp was sitting on the battlement of the arches watching a group of labourers putting finishing touches to the viaduct which was due to be opened soon. He saw a crack, stuck his stick into it, and then went down into the valley below to have another look from there. He saw what looked like a steady stream of black mortar falling from one arch. At the same time Henry Morton, a navvy, up on the viaduct, saw the crack too, and pointed it out to his mates. 'It's nothing', they said and laughed at it. Beneath them, Kemp heard the man laughing, looked up again, and was knocked over by a baulk of timber falling from the top. This was the beginning. The arches fell one after another, quickly until they came to the centre of the viaduct; then there was a pause for a second or so, before the arches toppled from the farther bank in orderly succession. Of the twenty or so men on the arches, five jumped or were thrown clear. Henry Morton among them: the rest were buried in the collapsing rubble. It was twenty past three.

Perhaps this is an early record of progressive collapse? And perhaps this time the lesson was not lost, since an inspection of the 1,320 ft-long Ribblehead Viaduct (built 1870–1874), (Figure 1.2) shows intermittent piers being much bulkier than the others. It has been said this was a deliberate feature to limit the risk of a domino-type failure

Figure 1.2 Ribblehead viaduct with intermittent stocky piers. (Dreamstime 25916843)

all from one side. Although structural collapse was avoided at Ribblehead, around 100 men are said to have died in its construction and the toll was so great that the railway company paid for an expansion of the local graveyard.

1.13 Workers' safety

The prime objective of structural safety is avoidance of loss of life or injury, be that from structures failing in service or from hazards posed during their construction. Earlier in the 20th century, loss of life in construction projects was high. Much closer to the present day, workers have been subject to huge risks. Most of us have seen daredevil pictures of New York steelworkers on skyscrapers, practices that should never be tolerated. Within current working lives, workers' site protection from PPE (helmets and protective clothing) was minimal, while site and factory welfare conditions were often dreadful. Worker injury, site deaths and premature death from occupational health causes remained at unacceptable levels (even in present times there remains a legacy of about 200 deaths a year from asbestosis).

Around the middle of the 19th century, society began to try and improve matters by passing the Factory Acts. The first of these were passed in 1833 to protect children working in textile mills. Thereafter the Acts were gradually extended to cover all workplaces to try and restrict the numbers of accidents. Mines were covered in 1843 following a Royal Commission report, which exposed dangerous working conditions, many accidents and a high incidence of disease related to mining conditions (Chapter 12).

However, it was not until 1974 that groundbreaking legislation was enacted. This was the 'Health and Safety at Work etc. Act 1974', which marked a departure from the framework of prescribed and detailed regulations in place at the time. The Act introduced a new system based on less-prescriptive and more goal-based regulations, supported by guidance and codes of practice. The Health and Safety at Work Act established the Health and Safety Commission (HSC) for the purpose of proposing new regulations, providing information and advice, and conducting research. HSC's operating arm, the Health and Safety Executive, was formed shortly afterwards to enforce health and safety law, a duty shared with local authorities.

The Health and Safety Commission (HSC) had objectives set out as

> taking appropriate steps to secure the health, safety and welfare of people at work, to protect the public generally against risks to health and safety arising out of the work situation, to give general direction to the Health and Safety Executive (HSE) and guidance to Local Authorities on the enforcement provisions of the Act, to assist and encourage persons with duties under the Act and to make suitable arrangements for research and the provision of information.

Some of the key health and safety hazards with which HSC was concerned in its first few months included asbestos, construction dusts, genetic manipulation, ionising radiation, lead, noise and vinyl chloride.

The HSE publish annual statistics on injuries in the workplace (HSE, 2018). The figures for deaths in construction have shown a steady decline since the passing of the HSW Act, both in absolute numbers and when defined as 'fatalities /100,000 workers'. That latter rate is currently about 25% of the value in the 1980s, so forms a tribute to efforts put in by industry as whole. The HSE statistics also show that the UK is one of the safest employment areas in the EU, having a fatal accident rate only about 15% of that of the worst performers. Whilst that figure is encouraging, it also shows what might happen if vigilance is relaxed. As the HSE concludes: 'Every casualty is a tragedy and has both a social cost and a personal cost to those directly affected.' The statistics also reveal that construction, as a sector, has more fatalities in total than any other sector, albeit the rate of fatal injuries (per 100,000 workers) is far less than for workers employed in agriculture or waste and recycling; nevertheless the rate in the construction sector remains about four times the average for all sectors of industry. Chapter 12 deals with occupational health and safety.

1.14 Current status

The construction sector forms a key part of the UK economy (about 7%), contributing about £100 billion p.a. and generating around 3 million jobs (10% of the UK total). The sector covers everything from routine house repairs to design and construction of the most complex infrastructure projects. At the time of writing, Crossrail in London is Europe's biggest construction project with a workforce of around 10,000. The industry has significantly improved its attitudes to safety, and nowadays most members are anxious to avoid anything that causes harm and indeed seek positive action to promote safe working practices. Moreover, they are conscious of their legal obligations which apply to all parties: clients, designers, and constructors. We have a highly skilled and trained workforce and the professional standards of designers are overseen by institutions of established repute. Yet we know that failures still take place, some of which cause injury, some of which could have caused injury, and many of which cost significant amounts of money.

It is too simple to regard all these incidents as springing from negligence or even human error and to be fatalistically accepted as such. The reality is that the construction process is extremely complex and one lesson from failures is that what seems obvious in hindsight is easily overlooked in foresight. It is therefore incumbent on the industry to try and set down a history of failures in a manner that allows us all to learn from them. The following chapters in this book seek to do that. Every designer presumes that what he intends to do is produce a safe structure. Yet defining how to go about that is hard and defining what makes a safe structure is quite hard too, so Chapters 2 to 5 discuss safety as a generalised topic.

1.15 Chapter summary

All civilisations and historical eras have been partly characterised by their construction activity. Since records began there have been failures, sometimes through ignorance, sometimes by natural hazard, sometimes by decay. But it is only relatively recently, in the last century and a half, that technical advances have allowed engineers to predict engineering performance scientifically and with enough accuracy to reduce failure risks reliably. That process has accelerated such that progress in the last 50 years (one working lifetime) has been phenomenal. Even so, that half-century has been marked by innumerable disasters. The thought processes of trying to think about safety as a distinct discipline have been developing in parallel with the emergence of the engineering professions but are still not widely disseminated in any formal way. If the professions are to reduce failures and their frightful consequences, they need to assimilate and dispense lessons from failures as part of engineering education. This is especially important since construction activity is high and the sophistication of what we construct is ever increasing, posing greater challenges than ever before.

The worst aspect of failure is loss of life. But many lives are lost and blighted not just by failures but in the construction process itself. A century ago, conditions were dreadful. They remained totally unacceptable only half a century ago and formal legislation creating a sea change has only been introduced within the working lifetime of today's engineers. Much remains to be done. So, the topic of health and safety (at work and by design) is also one covered in this book.

References

Carpenter, J. (2008), 'Safety risk and failure: the management of uncertainty', *The Structural Engineer*, **86**(21).

CIRIA/HSE (2011), *Guidance on Catastrophic Events in Construction*, Report C699, CIRIA.

Coleman, Terry (1965), *The Railway Navvies*, Pelican Books.

COMSAFCIVENG and H. Shirley Smith (1969), 'Report of the Committee on Safety in Civil Engineering', *Proceedings of the Institution of Civil Engineers*, **42**(1), 143–152.

Engels, F. (1892), *The Condition of the Working-Class in England in 1844*, Swan Sonnenschein & Co., 45, 48–53.

Godfrey, E. (1923), 'The failure of the Knickerbocker Theatre', *Journal of the Institution of Structural Engineers*, **1**(2).

Hewison, C. H. (1983), *Locomotive Boiler Explosions*, David and Charles.

HSE (2018), 'Workplace fatal injuries in Great Britain 2018', HSE.

2 General aspects of structural safety

2.1 Aspects of history

2.1.1 Introduction

Everyone expects buildings and infrastructure to be 'safe'. That includes the general public, who don't give the concept much detailed thought until things go wrong; it includes clients and indeed designers. No one involved in construction expects structures to be 'unsafe'. Yet the intellectual concepts of safety do not seem to have been formalised historically. Very early on, structural forms and sizing were based just on what was known to work via rules of thumb, although patently such processes were not foolproof. Early structures were extremely solid and bulky; later ones were somewhat less so. With the developing skills of medieval masons, ever more daring cathedrals of lighter form evolved and not surprisingly some collapsed, as described in Chapter 1.

It was not until the 18th and 19th centuries, with the growth of science and structural mechanics when notions of calculations and stress were developed, that it became even theoretically possible to provide defined margins of structural capacity against presumed demand. After all, Navier's theory of bending was only introduced in the 1820s and the concept of stress itself had only emerged via Cauchy (1789–1857). Thereafter, for a very long time, the only test of safety was via some numerical safety factor defined as a margin between computed demand and actual strength. It was only gradually that wider concepts of safety emerged.

Although we nowadays take calculations for granted and use them to create some sort of safety margin definition, there has always been conflict between the theory and practice of engineering design. 'Practical men' have often been deeply suspicious of theory and it took a long time before calculation methods were accepted as a sound basis for member sizing. For example, the distinguished English engineer, Thomas Tredgold (1788–1829), was able to argue that 'the stability of a building is inversely proportional to the science of the builder'. Even today, many in the industry are suspicious of 'the Academic'. Although this may seem Luddite, there are in fact good reasons for looking much wider than a single numerical margin for a valid test of a structure's safety. There are good reasons for trying to develop more general concepts of safety and for not being too reliant on numerical output as a sole arbiter. There are many variables that affect design, not all highly predictable, and not all susceptible to mathematical definition or mathematical modelling. Moreover, our understanding is still evolving. A review of many branches of civil engineering will show that enormous strides have been made just in the last half-century. Huge progress has been made in our ability to analyse structures using mathematical models and computers. Even so, we should not be deluded into believing these must be 'accurate'. The accuracy of computer predictions can be quite spurious, and it is wise to remember the adage of 'garbage in = garbage out'. Any prediction is only as good as the model it evaluates, and this is discussed further in Chapters 3 and 13. A study of failures also shows that human behaviour is a major factor to be reckoned with, so maybe Mr Tredgold had a point.

Alongside concern about modern theoretical predictions, there has been a growing concern over the years with structures becoming ever lighter and more 'optimised'. Whilst perhaps economically admirable and apparently less wasteful, the oft-expressed unease is that hitherto built-in safety margins are being eroded to a degree not fully grasped. Thus, experience of what has proven safe is not necessarily a guide for future structures. Sometimes this concern is expressed as a need to ensure that as well as meeting stress limits, structures must also be 'robust'. Robustness is a quality often ascribed to our inherited Victorian infrastructure; that it was built to last and could withstand a good deal of punishment before failing. A practical problem is that whilst experienced designers can intuitively make structures 'robust', it is a quality not easily defined and thus hard to implement formally, certainly via equations, even if legislation demands it.

2.1.2 Learning from failures: hazards and consequences

A review of failures across different engineering disciplines will easily reveal that linking safety just to numerical safety factors will be unsatisfactory. In 1912, the newly launched Titanic struck an iceberg on her maiden voyage from Southampton to New York. It remains one of the largest maritime disasters of all time. The ship sank partly through human error and partly because the structural response was not what the designers expected (more than one watertight compartment was punctured). Thereafter the mode of failure was not as anticipated: capsize hindered the launch of what lifeboats there were and the rate of failure was too rapid. On 6 May 1937 the Hindenburg Zeppelin exploded over New Jersey, USA. Caught on film, the Hindenburg explosion, like the sinking of the Titanic, is one of the best-known disasters. The Zeppelin's air frame did not fail structurally from overload, it failed because the hazard of running with hydrogen was too great. Thus, both the Titanic and Hindenburg failures were precipitated by hazards that had not been properly considered. The Hindenburg disaster also reveals some other features. First, one of complacency; many other hydrogen airships had been destroyed by fire previously. So why did the designers of the Hindenburg presume their airship would be immune? Second, the danger of statements on safety that rely on evidence of successful use over time without accident: we are all aware of the phrase 'an accident waiting to happen'. You might strike lucky, you might not. The fact that Zeppelins themselves had been in use for a long time without mishap before the explosion did not provide any firm evidence as to their future safety. We have many modern parallels with these two. The Piper Alpha disaster (1988, Chapter 11) will be one; the sinking of the Costa Concordia (2012, Chapter 6) will be another. The car park fire in the Liverpool Echo Arena (December 2017, Chapter 9) is another.

The Titanic and Hindenburg disasters cited above could have been prevented if the modern approach of looking at potential hazards and modes of failure had been considered. In each of these incidents, there was significant commercial loss plus loss of life. More than 1,500 people died on the Titanic and 97 on the Hindenburg, highlighting another factor to consider, which is: the consequence of failure. Engineers have to judge appropriate levels of safety and different approaches can be taken if the loss is commercial as opposed to cases when failures involve injury or death. Even when losses are just commercial, it pays to think in terms of consequences when responding to a question such as 'What is safe enough?' If a dam or large bridge collapses, the societal disruption costs may far exceed the direct failure capital losses. On the other hand, there is one case recorded of a large cooling tower that developed a weakening crack through its upper rim. The tower was old and due to be scrapped within five years, but predictions suggested it would not withstand a severe wind storm. If the tower failed there would be a loss of output. But if the tower were replaced early, the plant would have to stop anyway and the capital replacement costs would be considerable. The client took the view that, very probably, the tower would survive for five years, and if it did fail, there was no risk of loss of life and he would just accept the commercial risk. In other words, for the time at risk, the structure was safe enough.

It is frequently the case that civil structures support some plant or activity and the robustness of the structure needs to be looked at in that light. The failure of the Emley Moor TV mast in 1969 (Chapter 8) resulted in a complete TV blackout and it might be argued that such a tall structure was not readily repairable or replaceable within a short period. Failure within parts of the North American electricity grid under heavy ice and snow storms over recent years resulted in days of power loss at times when clearly power need was at a premium. In summer 2012, parts of the eastern US were struck by severe storms which left vast numbers of households without power for several days in sweltering temperatures. Many heat-related deaths resulted. After the winter of 2013/2014, a UK parliamentary committee castigated UK utilities over significant domestic power losses with delayed restoration. These are all examples of system failures where the amount of reliability built in (admittedly at a cost) should have been a design consideration.

Hence for engineers, there are cases where increased reliability might well be justified, and the engineering design process should consider this. Conversely, structural codes allow reductions in safety factors for minor structures such as farm sheds simply because the consequences of their failure are that much more tolerable and where, in most cases, repairs are neither urgent nor difficult. In making such decisions, it is inescapable that we should consider cost. However distasteful it is, more safety does frequently come with a bigger price tag and society must consider where to draw the line. This is not at all easy. We have tornadoes and earthquakes in the UK, but we do not specifically design structures for either. There is very little recorded damage from earthquakes and what there is, is restricted to the odd chimney pot falling or some cosmetic wall cracking, so why design for them? Conversely there are plenty of records of tornado damage. In 2005, in Birmingham, near enough a whole street was badly damaged. Nonetheless, taken overall, it must be cheaper for society to accept the risk of localised tornado damage and pay for it via insurance rather than spend enormous amounts of money trying to protect every house from the remote possibility of tornado damage.

We might say the opposite applied to the risks that were perceived from about the 1980s in respect of cancer initiated by ground-released radon. The risks exist and about 1,000 deaths a year in the UK are estimated to be due to radon exposure, some of it from within the home. The cost of reducing that risk in new-build (via a foundation membrane) is fairly trivial, so it was deemed a worthwhile thing to do. We may summarise this by saying that in assessing the demands of safety for any project, one aspect is to look at the consequences of failure and to assess if those are dire enough to justify more expenditure to eliminate the hazard or to mitigate it, or alternatively to increase robustness or increase the structural reliability. We might argue that if the consequences are not that serious, or if repairs are easily made, then there is justification for accepting a safety factor lower than normal. There is no case for saying 'one size fits all'. Even so, it is not sensible engineering to save minor amounts of money when the consequences of failure may be quite disproportionate. Overall, there are occasions when the sensible design approach is to increase a safety margin to give more robustness or equally, acceptance of a lower than normal margin can be quite rational. Either way, the professional engineering approach is to make a conscious judgement. What engineers need not do is fall into the intellectual trap of responding to a question, 'Is the structure safe or not?' That question poses false alternatives, as there is no absolute answer: structures are only safe to a certain level of probability.

2.2 Evolution of safety thinking

2.2.1 Introduction

Having made all the points above, it is worth pointing out that simply increasing a safety factor does not always increase 'safety' (perversely it can have a negative effect).

Achieving safety is more complicated and there are more rational ways of looking at the problem. Some engineering disciplines came to this conclusion quite early on. In the early days of the Industrial Revolution, mechanical engineers were much concerned with boiler failures as there were some quite catastrophic incidents. Several occurred on American steamboats and several on early locomotives (Chapter 1). The causes were numerous: inadequate design, over-pressurisation, corrosion and (in that era) unknown degradation mechanisms. Stress corrosion at joints was one of these, exacerbated by chemical reaction from the water used. Bi-metallic corrosion was another. Learning from these failures enabled engineers to improve overall boiler safety. Boilers had features such as safety valves and fusible links added. The former detects over-pressurisation and the latter, excessive temperature. Many boiler failures arose as a result of water levels being too low: sometimes linked to faulty water gauges, sometimes to human error. In modern times, any hazard analysis ought to have queried the safety of a system consequent on reliance on a single instrument reading. Redundancy is called for, along with inspections to ensure that devices are working properly. It may be recalled that the 2005 Buncefield catastrophe (Chapter 11) was partly caused by signals from faulty instrument readings and there have been other instances, particularly in aircraft crashes (see Chapter 6 on human error). Another cause of early boiler failure was from internal corrosion concentrated at the lap seams. What was not at first realised was the potential for fatigue damage linked to stress perturbations at the junction between single plates and the double plate of overlapping. Cracks developed, more rapidly propagated via corrosion (aerated water can enhance corrosion rates; Chapter 7 concentrates on material failures). Some of this risk was later eliminated by substituting butt joints and by keeping vulnerable joints above the water line and by improved maintenance/inspection and routine pressure testing, i.e. safety was improved by better design. Another feature was that some parts (stays holding the firebox) could corrode undetected since inspection was impossible. To limit that risk, longitudinal telltale holes can be drilled which create detectable water leak before the stays break. As a further refinement, some boilers are so designed to have heater tubes that are weaker than the boiler shell, so they always fail first and warn. Adopting measures such as these created significant reductions in the instances of boiler failures and where failures did occur, in the consequences of those failures.

2.2.2 Generic concepts

Although these stories of boiler failure can seem unconnected from structural engineering, the safety principles are universal. These might be:

- understanding the modes of failure, a process promoted by studying failure cause;
- designing systems to be fail-safe;

- using concepts such as controlling the mode of failure and 'leak before break' and the concepts of redundancy and diversity;
- making structures inspectable; understanding the role of maintenance;
- understanding degradation mechanisms;
- understanding the safety implications of reliance on single systems;
- noting the role of human error.

The potential role of faulty gauges was highlighted above. Another infamous example of engineering failure consequent on a false reading occurred when the Thetis submarine sank in 1938. A torpedo tube has flap valves at both ends; the inner one to load the torpedo and the outer one to let it out. Obviously, it is vital that the outer flap is closed before the inner one is opened. Therefore, to ensure there is no water in the tube before opening the inner flap, a test cock is included. On the Thetis this was opened for verification but it was actually blocked with paint, so no water flowed. Prickers to clear the test cocks had been provided but were not used. Unfortunately, on this occasion when the inner door was opened, the bow cap was also open and water poured in, sinking the submarine and leading to the loss of life of 99 of the men on board. The error seems to have been compounded by the layout of the tubes in that the 'shut' position for the opened tube differed from all the others, creating confusion. In Chapter 6 other examples of disaster due to human error are given to provide the lesson that ideally, by design, it should be almost impossible for a human operator to get it wrong.

2.2.3 Modes of failure

An early example of considering modes of failure and then trying to mitigate them relates to the invention of the safety lift by Mr Otis in 1852. He introduced safety brakes. If the lift hoisting cable fails, the consequences are clearly disastrous. In lifts fitted with Mr Otis's invention, if the car falls too quickly, side brakes automatically spring on and prevent the lift from falling. As a generality, all electro-mechanical engineering systems should envisage the possibility of parts failing and hence include provisions to mitigate the risks and consequences. We are familiar with this just through exposure to small-scale domestic devices. All electrical services include a fuse (or circuit-breaker) to protect the main system from faults. Sensitive equipment is further protected by surge devices; kettles include an automatic cut-out to close their heating coil off once the water is boiling, and further overload protection in case the kettle boils dry. It should not be possible to open a fuse box unless opening the cover automatically isolates the exposed inner circuits. Likewise, opening the door of a microwave oven automatically isolates the magnetron. A washing machine door will not open while spinning is in process. Overall, these are examples of interlocking, which is a safety process for preventing undesired states and prevents the system from harming the operator or damaging itself. In terms of design, for each device the designer has had to consider how the system may fault, how it might be damaged or how it may harm a user and the mindset of carrying out that task positively and making active consideration of safety is something that has evolved over time, often consequent on reported events. Nowadays, we are all familiar with an accompanying set of safety instructions with every purchase. Alas, there are numerous examples of people overriding such instructions and so causing harm.

The processes described above can equally apply to civil and structural engineering. The task is to identify all potential modes of failure and consider them. Some might be merely inconvenient, some disastrous. For certain structures under certain hazard loadings, it is desirable to ensure that the mode of failure is a preferable one. A good example of this occurs in earthquake engineering where principles of weak beam / strong column are followed and members are detailed to ensure bending rather than shear failures. In modern safety analysis, structured ways of carrying out safety assessments have been evolved; these are described in Chapter 13.

2.2.4 Development of safety thinking

The development of such safety thinking can be illustrated by tracking the design evolution of many products we are familiar with; trains and cars are good examples. Civil and structural engineers can learn from this history. People obviously thought about vehicle safety right from the beginning. When self-propelled vehicles were first introduced, there was concern about their safety and the Locomotive Act 1865 (Red Flag Act) set speed limits of 4 mph in the country (2 mph in towns) and required each vehicle to have a crew of three, one of whom had to carry a red flag and walk 60 yards ahead. Effectively this reduced speeds to walking pace. The Act wasn't fully repealed until 1896. Early passenger train use got off to a bad start when, on the opening of the Liverpool to Manchester railway in 1830, the local MP, William Huskisson, was hit by the train ('The Rocket') injuring his legs severely. He was transported on the Rocket to Manchester for treatment but died the same day. Initially trains travelled at about 17 miles per hour, since it was found that excessive speeds could force the running rails apart. Initially also, passengers travelled in open-topped carriages which, apart from being uncomfortable, left them vulnerable in case of a crash.

One of the first crashes occurred on the same Manchester to Liverpool railway in 1831, when an employee failed to change the points because he was asleep, and there were incidences of drivers falling asleep as well; all examples of human error. As a reaction to the possibility of this hazard of 'lack of human control', the 'dead man's handle' device was deployed, especially when trains were operated by single drivers. In principle, the handle is a fail-safe device, the idea being that the device will not operate unless the handle is depressed; the train's default position being brakes 'on' unless power is supplied to get them 'off'. The system is not foolproof, but is a significant benefit (see the Moorgate and Croydon tram incidents in Chapter 6). Other early train crashes were caused by conflict between trains and items on the track: sometimes vehicles, sometimes landslips. The latter are an example of the conflict between civil engineering works and the transport systems they support. Other classic civil engineering interactions were the Dee Bridge disaster of 1847 and the Tay Bridge disaster of 1879. In both cases, the supporting structure failed completely whilst trains were crossing. A similar modern incident is that at Gerrards Cross in 2005, when a tunnel lining being built over the track collapsed (Chapter 8). Fortunately, this collapse was spotted by the driver, so no lives were lost. Alas, in 2020 after heavy rain, a train in Scotland collided with debris swept onto the track; derailment occurred and three died. Several train crashes occurred early on through signal faults or through signals being obscured by fog or snow, failures that are echoed in the problems of the Watford incident in 1996, the Southall rail crash of 1997 and at Ladbroke Grove in 1999; all these were occasions when signals were passed at danger partly because they could not be seen easily (note the link to the Thetis sinking). Although these are primarily examples of failures within mechanical systems, there is frequently interaction with civil engineering infrastructure and lessons may be drawn. First, in any system design there is a need to consider the generic hazard of human error. Second, engineering systems frequently involve a combination of disciplines: here electrical (power), mechanical and civil engineering and in terms of safety, it is essential to look at the system 'as a whole' and to have regard to the discipline interactions within it. More widely, there is a whole discipline involved in making sure instruments, signs and signals can be read in all conditions. Every car driver will be familiar with the difficulties of seeing traffic signals in adverse sun alignments.

2.2.5 Crashes and modes of failure

Mechanical failure of some kind has been the cause of many crashes. In some early events it was observed that the mode of train carriage failure contributed to the severity of the consequences (a value of accident observation feedback). Early on, open-topped carriages were avoided, as were timber carriages. Passengers have a far better chance of surviving inside a modern metal carriage, even if it crashes at high speed, since they are better contained. But in some modern lightweight aluminium carriages, crashes caused the seams to burst open and so lessen the containment value of the carriage. Timber carriages were soon replaced to avoid injuries from splintered wood. These points illustrate how a generic safety principle might evolve. In this case, for any vehicle, the generic principle is that passengers should be safely contained. That principle is wide-ranging, encompassing everything from aircraft to roller coasters. The principle also introduces the concept of balance of risk. It is essential to keep passengers safely contained but the logic must extend to how to get passengers free if an accident happens and there is a balance between keeping people in and getting them out. The consequences of this conflict are examined in Chapter 13.

2.2.6 Mechanical failure, maintenance and the ability to warn

Incidents of missing sections of train track or track failure through fatigue exist and have caused terrible derailments (e.g. as at Hatfield in 2000 or Grayrigg in 2007 (Chapter 11)). In terms of track breakage, the safety concern is how might this be known about in advance with enough warning to prevent disaster? The introduction of modern welded track since the 1960s has offered a big safety improvement in limiting the risk since a major problem of jointed track is the risk of fatigue cracking around the bolt holes at connecting junctions. This was the cause of the Hither Green track failure in 1967 (Chapter 11). Other forms of contact fatigue are known about, so the problem has not been eliminated (Chapter 7). But using continuous track also offers the possibility of automatic detection of line breakage, achieved by passing an electrical current down the track which fails (and so warns) if the line breaks. Degradation also plays its part; one of the problems of non-continuous track is the need for constant maintenance. Lack of maintenance, or maintenance carried out badly, is a safety issue and was cited as the cause of the accident at Grayrigg in 2007, a crash in which one person was killed. The safety principles introduced here are the need to ensure safety through life by maintenance and inspection and with inspection, the ability to detect the onset of failure before it becomes too late. Lack of this capability has been a major cause of disaster in many structural failures (Chapters 7, 10 and 13). The cited failures also introduce the role of active and automatic monitoring which is a future possibility (Chapter 13). A general safety principle in structural engineering is of the need to warn before collapse (see Chapter 5).

2.2.7 Failures due to hazards or human error

The potential for conflict on train lines is a clear hazard. Deaths still occur at level crossings (in the UK about ten per year, often linked to drivers taking a chance) with the only certain way to avoid conflict being to eliminate the crossings. However, that is not always economically justifiable on normal track, so the question arises of what is safe enough? Another conflict that might have been anticipated from human error occurred at Great Heck in 2001, when a car veered off a road and rolled onto a railway track nearby. The driver was asleep. Ten people were killed and 82 injured. At the time this was said to be a freak occurrence – i.e. of low probability. However, if we look at the number of places where roads run close to railway track, the number of vehicles that use those roads, and the known incidences of drivers falling asleep, it is not difficult to imagine that although the probability of an incident at any place at any one time is low, the probability of a driver falling asleep and coming off the road near any one of UK's many railway tracks at some point is not that low. Given the geometry of the approach at Great Heck, the probability of any driver swerving onto the verge and ending up on the track close to the road was then very high and, given the number of trains on the track, the probability of a hit was then certain. The risk of the incident could have been reduced by better barriers. The incident at Great Heck is an introduction to the role of probability in safety thinking (Chapter 13). The problem of acceptability can be couched in statistical terms of likelihood and consequences and rational decisions can then be made over tolerability.

Of course, the potential conflict of multiple trains on the same lines is also a significant problem and illustrates another difficult choice in safety. In the UK, it is just not possible to have dedicated lines for each train; somehow multiple trains must be sent down the same line at close intervals and ways must be found of avoiding conflict. As speeds increase, this becomes ever harder. So, for really fast trains, dedicated tracks are used, as in the Japanese Shinkansen which has a strong safety record (this will be an issue for the proposed British high-speed rail link). But even then, accidents occur. A particularly bad one occurred on new high-speed track in China in 2011 (Wenzhou) when two trains collided on a viaduct and four carriages fell off. The incident was said to have occurred due to faulty signal systems which failed to warn a following train of a stationary train on the track ahead, but railway officials were also blamed. Some reports say the first train was stopped by a lightning strike. However, the major point is that all tracks depend on a control system to detect where a train is on the track at any time and to make sure that two trains cannot collide. Such systems have been developing ever since railways first started. For whatever reason in Wenzhou, there was a fault in the logic of the control system on this track insofar as the system failed to stop the second train crashing into the first. In design it should always have been presumed that a forward train could stop (for a myriad of reasons); the logic would then be that a following train could impact with frightful consequences, so the reliability demand made on the control system is maximal. Techniques exist within control system design practice to reflect such demands. The Wenzhou incident and the UK incidents of the 1990s are critical in highlighting enormous social and political repercussions stemming from lack of safety in public systems. Societal changes are such that lack of safety now can have very wide implications, not least in costs, as BP found out after the Deepwater Horizon incident of 2010 (Chapter 11).

In Wenzhou, the rear train could not stop in time to avoid the crash and a generic safety issue on all trains, and indeed all moving structures, is concern over the ability to maintain motion control, irrespective of circumstances. A very early train crash at Armagh in 1889 illustrates this. The incident occurred as a Sunday School excursion train tried to climb a steep gradient but the engine stalled. To try and alleviate the problem, the crew divided the train, but the rear section was inadequately braked and so ran backwards down the incline, colliding with a train behind. Eighty people were killed and 260 injured. The incident led to safety measures becoming a legal requirement. This was not an isolated incident; the famous photograph of an engine 'not stopping' at Montparnasse Station in Paris in 1895 shows what happens (Figure

Figure 2.1 Train crash at Paris Montparnasse Station in 1895

2.1). There are many other records of this form of failure. At London's Battersea funfair in 1972, a car on the big dipper went out of control on the incline after becoming detached from its haulage rope and rolled backwards, again colliding with a following car. Five children were killed and 13 injured. Consequently, all roller coasters now have anti-rollback devices on the upwards slope which are toothed racks into which a hinged 'dog' fits as the car moves along: this is responsible for the typical clackety-clack noise heard as the car is pulled up the hill. The Battersea incident also led to the introduction of formal checking of ride safety in the UK ('rides' may also be thought of as civil/structural/electro/mechanical systems with an overall control system overlaid). In making a safety assessment, it ought to be considered that any part of this overall system might develop a fault. Human error is also a factor. (The tragic role of human error in a 2016 'Smiler' roller coaster incident is described in Chapter 6.)

At Tebay in Cumbria in 2004, a laden wagon ran out of control down a slope at night. Unseen, the wagon reached high speed, and after travelling 3 miles it hit four maintenance workers, killing them. The incident happened because the wagon brakes were faulty and wooden chocks holding the wagon in place became dislodged. In effect there was no control over the wagon's potential motion. The persons responsible for the incident were jailed. Perhaps one of the worst cases of loss of control occurred at Moorgate Underground Station in 1975 (Chapter 6). The cause has never been fully explained but for some reason, the tube driver did not stop at the platform and instead drove his train at full speed into a dead end. As a result, 43 died and because of the crash location, rescue took many hours. The train did have a dead man's handle, but the driver had not released it. Moreover, the standard buffers at the end of the track were quite inadequate to stop the train. A consequence of this incident is the observation that for total safety, reliance cannot be made on human beings. All underground trains now have automatic systems for stopping trains at dead ends even if drivers fail to operate the brakes. Despite this, even these systems are not foolproof and great effort is required to devise an absolutely secure control system. The track in Croydon where a tram derailed in 2016 did not have an automatic excessive-speed detection system (Chapter 6).

2.2.8 Role of control

Control may be exercised by humans, by brakes, perhaps supplemented by fail-safe devices, and by electronic control systems. All mechanical systems also normally include emergency stop devices. The question, always, is whether there is adequate control of anything that is moving, or more widely of parts that are being lifted. An illustration of where control was inadequate arises in incidents with gantries. There have been several of these, for example at the Vasco de Gama Bridge (Portugal) in 1978, one at the Severn Bridge (UK), and one at the Avonmouth Bridge (UK). The Avonmouth Bridge incident was in 1994 when the gantry carrying bridge sections keeled over into the river. Four workmen were killed. The Severn Bridge incident, in 1990, concerned a maintenance gantry. Gantries were hung under the bridge to facilitate routine maintenance, including painting. On the occasion of the accident, the gantry was blown out of control along the track and fell at a section where there was no rail. Pins from the fail-safe system were missing. Two workmen were killed (one other miraculously survived the fall) (see also Chapter 8, Section 8.6.2). Lack of control appears to have been the cause of dropping the central section of a new lifting bridge at Barton on the M62 around Manchester in 2016.

2.2.9 Generic lessons

Although the preceding paragraphs have been primarily about trains, the safety issues are generic and pertinent to many areas of civil/structural engineering, particularly moving structures. The general lessons are that in abstract terms these are all engineering systems (which may be a mixture of disciplines). Within such systems, any part can fail, perhaps in more than one manner. Equally, any part of the control system can fail. As a result of these failures there will be consequences that need to be considered and mitigated. Formal methods of examining systems to help understand causes and effects are described in Chapter 13. Where safety is critically dependent on a control system, the design of that system can be tailored to give increased reliability, a further example of the need to consider consequences and then build in more or less safety accordingly. Pertinent lessons are:

- the need to consider failure modes, with some being less desirable than others (as in the form of carriage failure or in forms of structural failure);
- the need to consider the balance of risk;
- notwithstanding the need to design safely, the desirability of having advance warning of something going wrong;
- the role of maintenance and inspection, observing that all structures deteriorate, and their margins of safety generally diminish with time;

- the need to define all hazards (say from human error to lightning);
- an introduction to the concepts of probability (as at Great Heck);
- in designing a system as whole, the need to cooperate over discipline interfaces;
- to understand the role of a control system in ensuring safety.

2.2.10 Evolution of safety from observation of failure

Looking back over the last century of engineering product development, it is apparent that learning from fault and learning from failures has been essential in the evolution of product safety. This has been necessary for legal and moral reasons but nowadays is also a commercial imperative; perceived safety in cars is a major selling point. The fact that the new Boeing Dreamliner was grounded in 2013 over 'safety concerns' must have been very expensive, as equally must have been many mass car recalls. Even Apple had issues (in 2014) with allegations that its new iPhone could bend. The costs to BP of the Deepwater Horizon disaster (2010) have been enormous. The direct costs to Boeing of the 737 MAX crashes of 2018 have been severe and the ongoing costs attributed to lack of confidence in the plane's reliability must have been very high.

If we look at cars, for example, a modern car has an engine, four wheels and passenger seats just as Henry Ford's Model T did in 1908. However, modern car design has been transformed by safety studies and shows a vast improvement over the Model T. Dual braking, safety-glass windscreens, better tyres, better instrument control, seat belts, airbags, child safety locks, etc. have all been introduced, mostly as a response to perceived weakness. As a theme to all engineering, modern design considers not just what goes right but what could go wrong in adverse circumstances and should draw lessons from incidents. One example of that from car design is the story of the Ford Pinto. In 1977, there were allegations that the car's structural design allowed the fuel tank neck to break off in rear-end collisions (an undesirable form of failure) with resulting fuel spillage and fire and there were legal suits about this. Motorists are already being offering more possibilities of automatic detection of other vehicle proximity with the imposition of avoidance car control.

Passenger train carriage design has benefited from observations of how the carriages respond under crash conditions, all with the objective of changing design so that passengers are contained as safely as possible in adverse circumstances (lack of safe containment was a feature of the 2016 Croydon tram crash; see Chapter 6). Indeed, safety concerns are now seen as so important that the rail, motor and aircraft industries all have confidential reporting systems over matters that might be of concern. A confidential reporting system has also been established for matters to do with structural safety and several reports in this book have been drawn from the CROSS database (CROSS: Collaborative Reporting for Safer Structures: www.cross-safety.org/uk).

2.2.11 Accident prevention

Active concern over accidents and their prevention has been around for about 100 years. The Royal Society for the Prevention of Accidents was founded in 1916, reflecting rising public concern over traffic accidents. Since that time, the Society has run campaigns on cycling; on safety in the home; to reduce industrial hazards and to improve occupational health and safety. The history of health and safety management in construction is the history of adverse public reaction to events; a testament to the work of certain enlightened individuals and organisations, and a gradual promotion of increased standards by government. To counter this, there has been an equally long history of resistance: often framed as objections to 'excessive paperwork' and to the perceived expense of protective measures. Those debates continue today, with the pendulum of what is deemed acceptable always in motion, not least currently against a political climate asserting 'excessive regulation'.

2.2.12 Summary

We all have legal and moral duties to create infrastructure that is safe to build, safe in use and safe to maintain. But what is it we need to know to achieve those objectives? Put simply, buildings and infrastructure being safe means them having an improbably low possibility of failure throughout life. But that statement begs the question of what failure actually means, a matter discussed in Chapter 3. Throughout life raises the possibility of degradation, which has been an important feature in many failures; a topic discussed in detail in Chapter 7. Putting degradation to one side, civil engineering structures are subject to loadings which are complicated and ensuring that their stress states under load are acceptable is not straightforward. It involves defining modes of failure and ensuring these are selectively controlled (Chapters 4 and 5). Throughout life also includes the concept of safety during construction, and many failures have taken place in the construction and demolition stages during which structural forms and stability may differ radically from corresponding conditions in service. Chapter 8 discusses these. Apart from structures physically failing during construction or in service, modern concepts of safety include a need to consider the health and safety of constructors and people who later occupy the building or use the infrastructure. This is a topic of increasing importance and is discussed in Chapter 12. Finally, what does an improbably low possibility of failure mean? That in turn depends on the probability of the hazards acting against the structure and it encompasses concepts of the consequences of failure, public attitudes to risk and to affordable costs, all matters discussed in Chapter 13. Chapter 13 also suggests how safety might be assessed and how a demonstration of safety can be made.

In general, for any piece of infrastructure we need to define the functional demands which govern its design, hazard loads that might apply, the risk of failures and the consequences of such failures. The risk of catastrophic failure in any engineering system can be constrained by deploying some of the concepts given in this section, for there is commonality between mechanical systems and structural systems. But if we want to avoid failure in the future, the best possible starting point is to have experience of what has gone wrong in the past.

2.3 History of structural safety

2.3.1 Introduction

Despite comments made earlier in this chapter about the complications of safety, for most new structures, it is impractical to make individual decisions on safety values. If for no other reason than having a level commercial playing field, we need codified rules of what is deemed acceptable to aid design teams, clients and regulators. In effect, a design code is a formal statement of how to design a structure via a set of rules (often, but not always, couched in stress terms) which will achieve an adequate level of safety. Until recently, each country had its own set of national standards but in Europe these national codes have been superseded by the Eurocodes, albeit national annexes are permitted, which give slightly different interpretations. The extent that one national standard deviates from another, or from the corresponding Eurocodes, is an illustration that safety is not some absolute criterion but is a judgement by the particular code committee. Thus, we might say that a structure which meets a British code but does, or does not, meet a Eurocode, is not by that deviation alone either safe or unsafe. All that can be said is that it meets, or does not meet, the code.

Equally, we all understand that standards have been routinely revised; sometimes to tighten up rules, sometimes to relax them. Thus, we have plenty of structures that were built to the standards of their time, and have survived without mishap, but may not meet modern standards. That does not necessarily mean they are unsafe. Each case must be looked at on its merits. It would not be economically possible to require upgrade automatically, or even reappraise every building in a country when design rules change to modify a safety factor or safety attribute. Circumstances must govern. But it is certainly the case in wider considerations of safety, for example in fire engineering (Chapter 9), that evolution of understanding has required extensive modifications (at one time using asbestos was considered safe but we now know better). After the Ronan Point collapse (Chapter 8), an extensive reappraisal of vulnerable building stock was carried out. After the HAC (high alumina cement) scare (Chapter 7), extensive reappraisal of existing stock was considered essential. No doubt a similar reappraisal should follow the Grenfell fire incident of 2017. Even now that debate is framed by the dilemma of whether or not to retrofit the UK high-rise housing stock with sprinklers.

2.3.2 Historical development

Returning to the pure topic of safety factors, early engineers worked in cast iron. They were conscious of material flaws plus dimensional variation in components. Those engineers were also aware that material strength in compression was a more reliable commodity than strength in tension. To counter such uncertainties, the need for a safety margin was appreciated which practitioners defined either against the material's UTS (ultimate tensile strength) or (for steel) its yield point. Margins of safety in use were around four or five. Given that both four and five are convenient round numbers, it is obvious there was no great science to their selection; the numbers were just deemed appropriate values by experienced practitioners.

As work expanded on the infrastructure, the need for both safety and standardisation became apparent. Dramatic failures of cast iron structures such as the Dee Bridge near Chester in 1847, the Tay Bridge in 1879 and the Norwood Junction Bridge in 1891 (Chapter 7) highlighted concerns over the use of brittle materials. About the turn of the 19th century, reinforced concrete was becoming more popular and design rules were required. Strangely enough, the demand for rules came from architects who were especially focused on fire resistance, not engineers focused on strength. Fire had always been a safety problem and one of the original strong imperatives for bringing in iron as a structural material was to try and make 'fireproof' mills. So, although structural safety might not have been thought about in modern terms, engineers were conscious of the need, recognising the risk of fire, the risk of uncertain materials and being aware of some dramatic failures. As we have seen, in the world of mechanical engineering, the ever-present threat of boiler explosions drove a need for active safety considerations.

2.3.3 Learned Institutions

Throughout the 19th century there was huge interest in all matters scientific and technical and several societies grew up to try and develop engineering knowledge and to try and promote engineering education. there was a growth of mechanics' institutes and many of their fine buildings still stand. Mechanics' institutes were largely founded to promote adult education, especially in technical subjects when these disciplines were mostly ignored by traditional universities. One of the oldest societies is the Manchester Association of Engineers (still in being), dating from 1856 (the Leeds Association of Engineers dates from 1865). The Manchester Association was formally established at a time when Manchester was recognised worldwide as a leading centre of engineering, production and technical excellence. The Association's aims were stated as:

> The bringing together of those engaged in the design, direction or superintendence of engineering works and operations, for mutual

> improvement and assistance, increase of acquaintance and for promoting frequent exchange of opinion on interesting questions constantly arising from the progressive nature of engineering.

Over time, many of these groups became regulatory bodies. At first the study of engineering was broad, with the boundaries between, say, mechanical and civil engineering remaining fluid. The Institution of Civil Engineers (ICE) began in 1818, with Thomas Telford being president in 1820. The Institution of Mechanical Engineers (I Mech E) began in 1847 under the presidency of George Stephenson, with the first meeting being held in the Queens Hotel, Birmingham, although that same year an idea had previously been floated in Manchester to establish 'a highly respectable Mechanics Institution', its purpose being 'to increase knowledge and give an impulse to inventions likely to be useful to the world'. The Institution of Structural Engineers sprang from the Concrete Institute (first meeting held in the Ritz Hotel, London in 1908), which had been established to further knowledge of concrete's fire resistance and to standardise usage rules. Soon after 1908 it was appreciated that a more general approach was required, not least to bring an increasing number of steel structures into the fold, and the Institution of Structural Engineers (ISE) proper dates from 1912. The British Standards Institute (BSI) was established in 1901.

We thus arrive at the beginning of the 20th century with sets of codified rules for our main structural materials which all included guidance on analysing structures and which provided numerical values of 'permissible stresses' or equivalent and which gave general guidance on 'stability'. Over the period since then, codes have been evolving and safety factors modified, and more sophistication introduced, yet even current values can be traced back to their early origins. In truth, then as now, the justification for safety margins has been very largely empirical in the sense that structures designed to the rules have largely proven adequate in service and therefore been deemed 'safe'. Nevertheless, various events have occurred to suggest that all was not as well as it seemed.

2.3.4 More recent times

In 1968 the Institution of Civil Engineers published a report on 'Safety in Civil Engineering' (COMSAFCIVENG and Shirley Smith, 1969). The impetus for the ICE report was summarised in its introduction as: 'the cost to the construction industry in lives, injuries and money over the past years has been so great that action was clearly an urgent matter'. The report drew on the contents of papers published in the ICE's Proceedings and it further drew on information provided by a Mr Brueton, following his study of a staggering total of 2,000 failures.

The report cited possible causes of failure as:

a. lack of proper regard by designers for feasibility of erection;
b. inadequate site investigation;

c. lack of proper communication at all levels, e.g. between engineers and contractor, office, site, supplier, and manufacturer;
d. an unsafe erection or demolition scheme and site procedures; failure to analyse the effect on the structure and temporary works at all stages;
e. lack of supervision, awareness, foresight, or judgement;
f. inadequate engineering knowledge;
g. the use of the wrong material;
h. lack of adequate support, anchorage, or lateral restraint;
i. overloading;
j. the reverse loading of structural members;
k. the adverse influence of restrictive regulations on erection and demolition methods.

1968 is now some half a century ago, yet today's engineers will recognise many contemporary failures recurring under these same headings.

From the list above it is thus clear that the ICE realised there was much more to safety than numerical analysis and that the concepts of safety needed expansion to cover the whole range of construction activity. Then in May 1968, by terrible coincidence, came the infamous collapse of the Ronan Point block of flats (Chapter 8). Investigations into Ronan Point discovered shoddy workmanship and that reinforced the message yet again that safety as a concept cannot be achieved by provision of numerical margins alone. Rather it must also be achieved by sound concepts, sound detailing and by assurance of construction quality. Chapter 13 reiterates that in construction, one aspect of demonstrating safety is to be able to show that what was intended to be built was actually built.

In the aftermath of the Ronan Point failure, the major construction institutions (ICE and ISE), along with the HSE, set up SCOSS (the Standing Committee on Structural Safety) with a mission to maintain a continuing review of building and civil engineering matters affecting the safety of structures. SCOSS aims to identify in advance those trends and developments which might contribute an increasing risk to structural safety. SCOSS produces a biennial report (and these are available on their website: www.cross-safety.org/uk). In 2005, CROSS was introduced (www.cross-safety.org/uk), which is a confidential reporting system on matters to do with structural safety. Newsletters are published quarterly and there are now international affiliations.

Another insight into structural concepts of safety can be gained by studying the development of soil mechanics and foundation engineering. Burland (2008) describes the rise of the art through the first 100 years of the Institution of Structural Engineers. He quotes a paper from 1915:

> In 1915 Wentworth Shield read a paper to the Concrete Institute on the stability of quay walls on earth foundations. He opened with the

> following memorable statement: 'In spite of the large amount of experience which has been gained in the construction of quay walls, it is still one of the most difficult problems in engineering to design a wall on an earth foundation with confidence that it will be stable when completed ... Even then if the designer of the wall is assured that it will stand, he cannot with any confidence tell you what factor of safety it possesses.

In his paper, Burland dates the proper birth of scientific foundation engineering from about the 1930s and he devotes space to the truly pioneering work of Karl Terzaghi. It is worth quoting what Terzaghi himself wrote in 1936:

> In pure science a very sharp distinction is made between hypothesis, theory and laws. The difference between the three categories resides exclusively in the weight of sustaining evidence. On the one hand, in foundation and earthquake engineering, everything is called a theory after it appears in print. And if the theory finds its way into a text book, many readers are inclined to consider it a law.

That quote should be recalled when considering many of the failures described in this book. One thing stands out: the causes of failure are numerous and complex and very rarely as simple as a single inadequacy in a numerical safety margin. Returning to Burland:

> [I]n 1934 Terzaghi delivered a lecture before I Struct E in London with the title 'The actual factor of safety in foundations'. He illustrated his lecture with a large number of case histories of measured distributions of settlement across buildings and their variation with time. He was able to explain the broad features of behaviour using basic principles of soil mechanics and foundation analysis demonstrating how vital it is to establish the soil profile with depth and across the plan area of the building. Even so he showed that local variations in soil properties and stratification make it impossible to predict the settlement patterns with any precision.

We are thus led to the conclusion, certainly in soil mechanics, that whatever we do there is a good deal of uncertainty that has to be coped with. Moreover, an understanding of foundation performance begs the question of just what is failure? Has the Leaning Tower of Pisa failed?

That question of what constitutes failure leads onto many modern notions of safety as required for structures which contain hazardous substances or are required to provide certain standards of life protection under extremes of loading. The normal modern approach is to consider that any structure is subject to a range of hazards (subdivided perhaps into different levels of probability) and for each there are various limit states and for each of those states, an acceptance criterion must be defined. (A limit state defines the condition for a structure beyond which it no longer fulfils its relevant criteria: the normal ones are ultimate limit state (ULS) (strength) and serviceability limit state (SLS), such as deflection.) That criterion is not absolute; it can vary according to the probability of the hazard and according to the consequences of failure as defined by some overall safety case. For example, under normal loading, deflection might be restricted whilst under severe earthquake loading it might not. Under normal loading, the structure might be required to achieve a safety margin greater than one against the applied loading, whilst under extreme loading it might only be required not to collapse. Setting the acceptance criterion requires some understanding of the safety demands for the system as a whole (Chapters 5 and 13). As a complement to the above approach, if the relevant hazards are defined, a hierarchy of risk reduction can be made (see Chapter 5).

2.4 Chapter summary

Engineering professions have existed for about century and throughout that time they have been active in developing technical knowledge. The civil and structural engineering institutions have collated information on the performance of structures and infrastructure over time. Following from that, there has been a gradual improvement in our capabilities of predicting how these entities will perform in service. Distinguished engineers and code producers have been mindful of the uncertainties inherent in materials, loading, and methods of analysis and have debated appropriate margins to ensure that, in service, systems are unlikely to fail. However, the process has necessarily been one of continuous improvement as problems with durability, unanticipated modes of failure and so on have arisen. The professions have also been very instrumental in promoting competence in practising engineers which has a role in lessening the risk of failures from human error. The professions have been less successful in developing overall philosophies of safety and in developing a comprehensive framework for teaching 'safety'. That aspect has come to the fore more recently as society needs assurance of safety for ever more sophisticated structures, against hazardous plant and for security against severe environmental loading.

Exactly what 'failure' is and how it might be affected by loadings and material uncertainty and structural performance is discussed in the following chapter, Chapter 3, whilst Chapters 4 and 5 discuss further attributes of safety.

References

Burland, J. B. (2008), 'Foundation engineering', *The Structural Engineer*, **86**(14).

COMSAFCIVENG and H. Shirley Smith (1969), 'Report of the Committee on Safety in Civil Engineering', *Proceedings of the Institution of Civil Engineers*, **42**(1), 143–152.

3 Definition of failures and uncertainties

3.1 Definitions of failure/limit states

Chapter 2 posed the question, 'Has the Leaning Tower of Pisa failed?' Obviously in some sense it has. On the other hand, the Pisan tourism industry would be loathe to restore the tower to verticality. Every structure endures settlement to some degree but the point at which the change constitutes 'failure' remains a matter for debate. When does a tilted old building change from being quaint to unserviceable?

In discussing the question of 'What is safe?' it may actually be easier to consider the reverse and ask, 'What is unsafe?' To respond to that question, engineers need to be able to define what failure is and to investigate how close the structure is to achieving any particular limit state. Such states are scheduled in codes; for example, Eurocode 0, 'Basis of structural design', establishes principles and requirements for the safety, serviceability and durability of structures (CEN, 2002). However, the intellectual dilemma is how to ensure that all potential failure states have been defined at the outset. Any review of failures will show that many of them arose simply because the original designer overlooked a critical design case or potential failure mode (Chapter 8 gives examples).

Some limit states such as strength and displacement always apply; others such as fatigue or vibration might or might not be relevant. In general, the starting point should be to assess all the functional demands on the structure since it is essential to thoroughly understand the role the structure will play in any engineering system of which it is a part. Beyond that, it is also essential to identify all hazards that might apply and thence be accurate in defining what might constitute 'failure' under these hazard loadings, always taking account of 'consequences'.

A requirement for structures to be strong enough seems obvious. But delving deeper, there are complications. Structures can be strong, but unstable. Moreover, some aspects of strength are more important than others and for truly safe structures, the mode of failure under overload should be controlled since some failure modes are less preferable than others (Chapter 2). Ductile failure modes are always preferable to brittle ones. Under certain hazard loadings the objective is not necessarily to make a structure strong enough to resist loads; rather it is to ensure by design that both the type and rate of failure are controlled to absorb energy such that failure becomes a controlled displacement. Assessing strength by checking stresses is routine, yet even that can be confusing, a topic enlarged on in Section 3.5.

For certain limit states such as fatigue or durability, acceptability may be defined by a life rather than a capacity. For such limit states, it follows that the attribute of safety is not a constant but a factor that reduces over time. Failure can be defined as not achieving the target life.

Defining what the word 'failure' means is not always easy. For many structures a total collapse is obviously unacceptable, but short of that, the design intent is that there must be an adequate margin against a significant change in structural state, taking account of all the applicable uncertainties. That definition begs several questions:

a. What is an 'adequate margin'?
b. What are all the 'applicable uncertainties'?
c. What is meant by 'change of state'?

In normal circumstances, for new-build, the judgement of adequate margin is the definition of safety factor or load factor in the applicable code, always provided other conditions such as stability are met. But that factor might be predicated on a presumption of material quality, a presumption of loading uncertainty and an assumption of life all as discussed earlier. Consequently, there are circumstances when legitimate consideration of uncertainties can, and should, be made perhaps either to increase or decrease codified margins. It does nothing to the cause of safety just to try and comply artificially with a code. When dealing with low probability events, the safety margin is often altered to unity since the only design objective might be a demonstration of structure survival just short of collapse. Academically there seems little point in defining some extreme event with all the uncertainty that entails and then seeking a trivial safety margin on top.

There is significant difficulty in understanding what all the applicable circumstances are. A review of failures described in this book will show that many arose simply through circumstances that the design team failed to envisage. Failure by an inadequate code numerical safety margin almost never occurs, except by gross error. In Chapter 13, techniques are described to try and narrow down the risk of structural failure caused by non-consideration of all applicable uncertainties.

Having a sudden change in state is most undesirable, which is why failure by instability needs to be avoided with high confidence. The ideal is always to have warning before collapse and for collapse (if it occurs) to be localised well before the whole structure can fail. Usually there is no sudden change in state if a permissible stress is exceeded. In fact, the reverse argument might apply that unless the structure can sustain a local overstress without significant change in state, then there is something wrong with it. Safe structures are ones that do not fail in a brittle manner. Safe structures are ones where gross instability such as overturning or flotation cannot occur and where excessive deformation occurs before true member failure. Safe structures are ones whose condition is insensitive to the accuracy of design assumptions. Modern codes go some way to creating a distinction in classifying design rules under either serviceability limit states or strength limit states. The former, such as excessive deflection, might be a nuisance but they are probably not dangerous per se. Where limit states are manifest in deformation (such as by creep or corrosion or settlement), there is the possibility of deploying a structural monitoring regime so that rates of change can be tracked, making it possible to judge the state of safety at any point in time.

The response of structures under hazard loadings such as earthquake or blast presents a good example of the choice offered in terms of 'failure'. A standard requirement is that for low return period earthquakes the structure should survive 'as good as new'. But for moderate and severe earthquakes, the forces are so high that the design philosophy needs to consider significant displacement and energy absorption. Under moderate earthquakes the structure should be repairable whilst under severe earthquakes it might just have to survive. The definition of 'failure' is often therefore one of a permitted (excess) deflection rather than a limit on stress, it being assumed that the material limit stress (yield in steel or ultimate strength of concrete) will be exceeded and plastic hinges will develop. Likewise, in dealing with robustness, we are prepared to accept gross distortion and even partial collapse short of gross failure. The principle is that damage will be accepted (a commercial loss) so long as lives are protected.

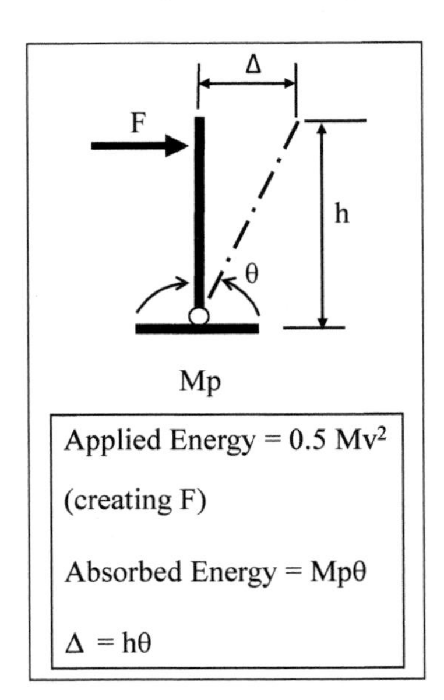

Figure 3.1 Performance of a car crash barrier

All this is no different in principle to the design of a common vehicle crash barrier (Figure 3.1). The function of the barrier is to make sure errant vehicles do not leave the road or car park edge. As the vehicle is moving, the design problem is best couched in terms of energy. The applied energy being the mass of the vehicle × $(\text{velocity})^2$ and the structural response ought to be one whereby applied energy is absorbed by controlled barrier deformation. The failure criterion here is then one of tolerable displacement. For this philosophy to work, it must be ensured that under overload, a cantilevered barrier post will fail by root bending rather than by shearing or by fracturing or pulling out of holding-down bolts. The foundation has to be stronger than the (actual) post plastic hinge capacity. This is therefore one example of how a designer might envisage several failure modes and seek to detail the structure such that the mode of failure is constrained to a desirable one. This example also includes a concept of sensitivity. The imposed energy is a function of $(\text{velocity})^2$ so minor changes in the assumed velocity (which cannot be known accurately) alter the computed kinetic energy significantly. In safety terms, this variability is accommodated by observing that the variation just results in a modified amount of imposed plastic deformation where the safety function (stopping the car) is not strongly linked to precise control of that deformation magnitude (Δ in Figure 3.1).

Generally the decision on what failure means for any civil structure cannot be taken in isolation from its function within a system. A water-retaining structure which cracks sufficiently to allow excessive leakage has for all practical purposes failed even if it retains adequate strength. Under extreme loading conditions, it might be essential for water-retaining structures to retain capacity but not necessarily prevent leakage to the same degree (see the discussion on the Fukushima storage ponds in Chapter 11). This contrasts with other common circumstances where cracking might be just unsightly but will not constitute a safety problem. Lateral sliding of a structure should normally be avoided, but might be tolerated under extreme loading, but not if such displacement permits fracture of interconnected hazardous services (again see Chapters 10 and 11). And sliding is always preferable to overturning. If any postulated failure just represents a financial loss, it is likely to be more tolerable than if a failure threatens injury.

3.1.1 Summary

Many different limit states apply to structures. It is not always easy to spot which ones are relevant and one cause of real-life failures is simply the human error of not considering applicable limit states. For those that are relevant, the acceptance criterion for strength might be achievement of some defined safety margin which might be high where there are many uncertainties, lower where there is much more confidence about governing conditions or it might be as low as unity against some hazard loading. The margin's magnitude should generally be tailored to the degree of uncertainty of an individual form of loading or to the probability or improbability of various load combinations. In other circumstances, the acceptance criterion might be a finite displacement or a finite life or a specified amount of energy absorption.

For safety, it is also usual that several unrelated attributes are met. Strength and stability are obvious ones that are not necessarily related. Others are discussed later in Chapters 4 and 5 and these include design efforts to ensure one mode of failure in preference to another. In all circumstances, it is advantageous to understand structural performance under postulated overload. (Carpenter (2008) discusses safety, risk and failure generally.)

3.2 Materials

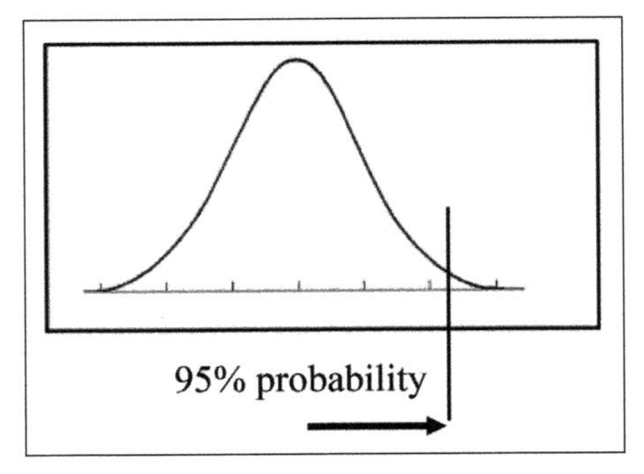

Figure 3.2 Gaussian distribution curve

Obviously a major factor in the overall safety of any structure is having confidence in material quality. In routine design calculations we take it for granted that our structural materials have a certain strength, indeed a standard design specifies accordingly. In reality materials have variable strength and a typical distribution from any tested batch is that shown in Figure 3.2, which follows a Gaussian bell-shaped curve. Academically the target is very low probability of overlap between high loads and low strength at the tail ends of the relative a distribution curves. It has become common practice either to specify a material having guaranteed minimum capacity or a 95% confidence level: this is the characteristic strength.

It is common to specify materials against a standard which will incorporate strength requirements (and other properties) and most materials are now supplied with certification showing how they comply with standards. Thus, day to day, designers are notionally relieved of any responsibility for assuring quality and material strength. However, CROSS has reports of material certificates being falsified (CROSS database) so care ought to be taken as to provenance and spot checks might be in order (the 'Dieselgate' emissions scandal of 2015 onwards against Volkswagen is a cautionary tale). A number of falsification allegations have been made during the Grenfell Tower fire inquiry.

For certain activities, supply certification cannot exist. These normally involve an element of variability linked to workmanship. Thus concrete, welding and brickwork all depend on proper workmanship for their assured strength. Processes to assure quality usually revolve around quality assurance (QA) procedures supplemented by production testing, i.e. quality control (QC) to guarantee that the quality presumed for design has been achieved. Major structural commodities often have national QA schemes (for example, ready-mixed concrete and rebar).

Designers need to be alert when any design aspect relies explicitly or implicitly on workmanship quality and should in those circumstances ensure procedures are in place to assure capacity. Examples could be epoxy-bedded fixings, in situ anchors, piles and so on. QA procedures such as method statements, supplemented by inspection and tests, will frequently be prudent.

The normal presumption is that materials must achieve a minimum strength. Paradoxically, stronger material can sometimes lead to less-safe structures. This applies when excessive strength can force a change in the desired mode of failure. An over-strength crash-barrier post could force an (undesirable) connection/foundation failure rather than the aimed-for impact energy absorption by cantilever root bending. In earthquake design, where energy absorption is a key resistance strategy, codes may be explicit in seeking both maximum and minimum material strengths. Nevertheless, excess strength is rarely detrimental and in fact may well be useful when assessing existing structures for change of use or to verify their continuing safety (which is an important reason for retaining as-built records).

Material strength may vary with time. Concrete strength can increase up to a limit but can also decrease. A case in point relates to trouble in the past with HAC concrete, which converted and quite drastically reduced in strength (Chapter 7). Basic steel strength remains constant, but because steel is supplied with a guaranteed minimum strength, it is likely that actual strengths will be greater than presumed. Checks on material mill certificates can be made. Design presumptions on strength are normally made at the time of construction, but designs need to consider all potential degradation/erosion mechanisms. Self-evidently, a product's strength percentage reduction will be highest for thin sections. In marine steel structures, it is common to add on a compensatory corrosion allowance. In standard buildings, the ingress of water is a frequent cause of rot or material weakening and many cases of failure are related to it (Chapter 7). Water and frost damage in brickwork is a perennial problem either from inadequate material selection, from poor detailing or poor construction. Moisture is an obvious potential cause of longer-term weakness in timber (Hewlett, 2021).

In terms of safety, the focus is mostly on a material's strength, but time-related deformations can also be influential. The 'E' value of steel is constant but some high-strength steels (and cables) creep and this may lead to load redistribution. Timber deformations are time related, especially linked to the applied duration of loading. In concrete, shrinkage and creep can play significant roles in time-changing deformation. For safety, a feature that needs to be watched is the influence of time-related movement on the forces to be sustained by fixings. Time-related deformations, if restrained, can induce high enough fixing loads to cause failure. A good example of such a time-related problem is that related to creep within epoxy-bedded anchorages, which has allegedly been responsible for some catastrophic failures (Rail Accident Investigation Branch, 2013; SCOSS Alert, 2014).

3.2.1 Summary

Materials vary in strength, so safety demands that precautions are taken to verify design assumptions. Such assumptions may only be true at a point in time, often at the time of initial construction and many weakening degradation processes are possible thereafter. Case studies are presented in Chapter 7. When assessing the safety of existing structures, account should be taken of likely material strength, which may be lower, or probably greater, than presumed at the design stage.

3.3 Loading

3.3.1 Introduction

Like material strength, loading is clearly a key component in the assessment of stress and safety. There are three main groups: static, dynamic and strain controlled and each has its own characteristics. Broadly, for static loading, the issues are of uncertainty in the magnitude, disposition and coincidence of different types. Dynamic loading is also characterised by uncertainty and not least since peak magnitudes are a function of interaction between the applied motion and structure stiffness (i.e. dynamic interaction). In addition, the stresses produced should not be looked upon as directly equivalent to static stresses because they are of short duration. Strain-controlled loads and stresses are those due to effects from imposed finite displacement such as occur in thermal changes, shrinkage or as residual stresses. They should not be dismissed out of hand: sometimes they matter, sometimes not. Whilst some loading values might be thought of as absolute, many of practical significance, such as wind or wave loading, can only be defined statistically. It has become accepted to define values as the maximum that might occur over a 50-year period. But for some structures which society wants to last longer (such as bridges) or for which greater reliability is required, loadings from a 1 in 100-year return period or an even longer 1 in 1,000-year return period can be used (values can be derived via extreme value statistical techniques).

3.3.2 Static loading: dead and live

The broad groups of static loading are those from gravity subdivided into dead and live (imposed) loading. Engineers are familiar with the concepts where dead is normally the self-weight of the structure and live is some sort of applied loading such as people or traffic. In routine work, live load is presumed more variable than dead load partly because it can be uncertain in magnitude (just how many people will stand on a floor?), partly uncertain in disposition (where will they stand?), and partly because there may be dynamic effects (will the people jump up and down?). All these variables are routinely wrapped up in a larger partial load factor for live loads than for dead load as the latter ought to be computed with more certainly.

Although the general presumption is that dead load can be computed reasonably accurately, care is needed, guided by experience. In one known example, a concrete floor slab occupied an area of 18 metres × 30 metres. The slab was supported on trusses which had a fabrication tolerance of 1 in 500. Since the floor was 'large' the absolute values of tolerances was important. In this case, the tolerance alone permitted an extra 'sag' of (18 + 30)1000/500 at panel midpoint = 96 mm. Since the concrete floor was only nominally 150 mm thick, its real thickness at the panel centre could be substantially more (or less) than the presumed 150 mm if the floor were laid with its upper surface set to finished level rather than thickness. Moreover, the extra concrete that was actually poured caused more elastic sag than predicted, so increasing the laid thickness even more. Altogether it is not hard to see that there can be considerable uncertainty, even in dead loads.

Magnitudes of live loading are assigned from codes based on surveys which have a long history. Two points follow. First, it is possible that the surveys become out of date. Certainly, office usage and loading have changed over time (Alexander, 2002) and traffic loading on bridges has increased significantly, partly from the introduction of much heavier lorries, plus the probabilities of coincident heavy loading have changed with increasing traffic density. Hence in terms of absolute safety (and when appraising existing structures) it is appropriate to look at the source of loading and make judgements about the potential for variation of either dead or live load and thence assess the adequacy of reserve margins accordingly.

There can be confusion over what is dead and what is live. Water load would normally be classified as 'live' yet if a tank is filled with almost zero freeboard and is protected via an overflow system, then the tank load will be known with a very high confidence both in magnitude and position. Conversely, on a flat roof, water ponding offers potential for significant extra loading but is of uncertain magnitude (it is covered by a standard live load allowance). The greatest risk from roof water occurs during rare, violent rain storms during which the water has no time to drain off the roof. So, is water loading always 'live' or could it sometimes be assumed 'dead'? Hydrostatic pressures against a tank wall are known with high certainty, but granular material pressures in the same tank would not be so certain.

3.3.3 Load factors

The use of partial factors overcomes the intellectual problem linked to using a single load factor when there is a mixture of dead and live loading. It seems intuitively wrong to have a uniform load factor when the bulk of loading can be dead (presumed known with high certainty in magnitude and disposition, perhaps with an inability to change) as opposed to live (which might be uncertain in magnitude and position). Any test of what live load is relates to knowledge of its magnitude and disposition certainty and is not an absolute requirement related to the application. For example, with control of the number of people in a space, it may be possible to presume that the consequent loading is known accurately enough to justify the dead load factor. This is particularly so on the basis that standard imposed loads from people are virtually impossible to create in other than artificial conditions. Many studies show that the imposed loads on floors are likely far less than those demanded by codes whereas the dead load allowance is virtually certain to be realised (Mann, 2019).

3.3.4 Load combinations

In normal building work, the summated loads are often dead + live + wind. It is common for this total combination to attract a lower overall factor from the improbability of peak combinations of all three. This is rational. But designs need to consider cases where the proportions of the three components differ. It would not be right to have a low factor overall when dead + live + wind were in combination where the first two components were extremely low, and the wind load was dominant. This follows not least because the assessment of wind loading is uncertain and subject to potential amplified error if wind speed varies and further because there are potential dynamic enhancement effects.

The comments of the paragraphs above put in context the need, or not, to consider the accuracy of component stress predictions. 'Safety' is not served by excessive concern over minor variations in stress accuracy taken out of context of the variability of the entire applied loading.

In normal design, application of loading, particularly live loading, is straightforward. We might apply it all over a floor slab or in various patterns to create 'worst cases'. Often that is an entirely sensible approach, but safety is not much compromised if the 'critical' combination is one extremely unlikely to occur. Even standard pattern loading of live on alternate spans is improbable: it should be just looked on as a notional design procedure. However, consider the suspended ceiling shown in Figure 3.3. Here the load on the ceiling is due to the accumulated ducts and equipment in the ceiling void. The load on the hangers might be assessed by computing the unit weight of the ducts in the local area. In reality, the actual load at any hanger point is a function of the supported duct stiffness and the alignment of the duct relative to the ceiling supports which may well be governed by the ceiling and duct construction tolerances. Thus, it is not difficult to appreciate that the dead/live load on any single hanger might be almost zero or it might be some integer multiple of what was presumed. Given there is no certainty in being able to compute such loads, and in this sort of task it is not commercially viable to do anything complicated, the design strategy needs to be commensurate. The strategy really needs to fall back on ductility (see also Chapter 4, Section 4.3). Provided the hangers can

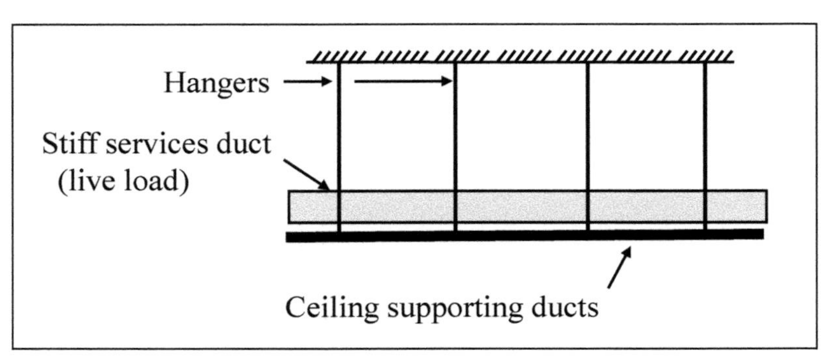

Figure 3.3 Loading on hanging structures

stretch and redistribute the load, all should be well. But that strategy presupposes that the end fixing will not pull out of the concrete slab above and that supposition justifies a large reserve and a large safety factor against fixing failure. (Examples are known of where fixings have pulled out, so initiating a progressive cascade failure of the entire structure as successive hangers became overloaded (Mann, 2019). Many comparable situations arise in structures supporting industrial plant where heavy imposed loads from objects might span several beams. The distribution of load/beam is then a function of the stiffness of the imposed load source and the stiffness of the supporting beams: there is uncertainty.

3.3.5 Dynamic loading

Dynamic loading is that which varies with time. Computing magnitudes can be complex. Essentially, if the frequency of the applied loading closely matches the natural frequency of the supporting structure and loading cycles persist, then structural deformations (and hence corresponding stresses) will build up over successive oscillation cycles and can become very large.

Confusingly, dynamically induced stress does not necessarily have the same significance as gravity induced stress. Figure 3.4 shows two identical cantilevers, each with the same bending stress at their support. One case is due to gravity loading and the other is from dynamic loading. Despite equality of stress, the two states are not directly comparable. The reason is that if, say, the stress under gravity at the cantilever root were above yield, the cantilever would displace indefinitely since there is enough energy in the applied gravity loading to achieve that. Conversely, under short duration,

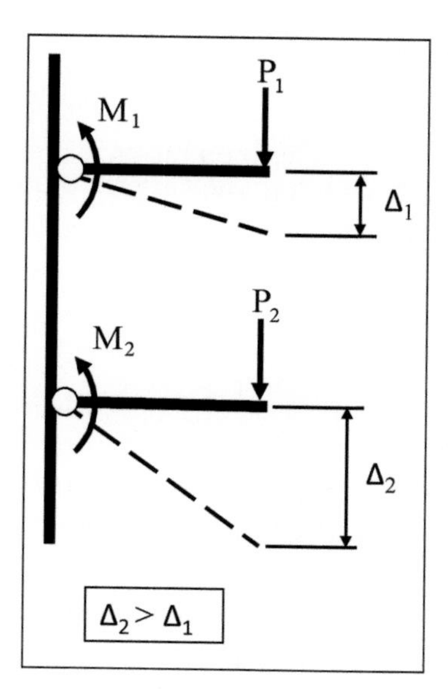

In this diagram, P_1 and P_2 are of equal magnitudes
So $M_1 = M_2$
But P_1 is a short term dynamic load whilst
P_2 is a long term gravity load

So Δ_1 can be less than Δ_2

Figure 3.4 Identical cantilevers with differing performances

dynamic loading (such as blast and earthquake-induced inertial loads) the imposed energy is finite, and the mode of failure can be finite displacement of the cantilever tip. Even if the root stresses are equal, the deformations can differ. For this reason, it is not necessarily fruitful to assess the safety of structures under dynamic loading by stress alone. The real test is to assess the finite displacement under the loading and consider safety on that basis. It is more appropriate to assess the total energy input into the structure and assess displacement response rather than look at forces and stresses. This concept can be grasped quite easily by looking at the response of a car crash barrier under vehicle loading as described in relation to Figure 3.1. The instantaneous stress in the post on impact may be very large indeed but is not of much interest as compared to the deformation of the post after impact.

Note: if the load application is very short, as in impact, then the normal yield strength is not of great significance since it is much enhanced under fast strain conditions.

Dynamic effects are a good example of engineers needing to be cautious when extrapolating from experience. In the last decades of design there has been a demand for longer span floors. There is no great difficulty in designing for such floors based on strength. Alas, the arithmetic so works out that in moving from spans of, say, 6 metres up to spans of, say, 8 metres and above, beam natural frequencies slip downwards into ranges that are readily excited by footfall traffic. Consequently, a whole range of dynamic problems from nuisance vibration has been created by moving to longer span floors. Normally these are not of safety significance, but in the special case of stadia, where flexible cantilever stands can be excited (especially by crowd-imposed regular rhythm as in pop concerts), stand response has become a safety issue. Dynamic loading is a key safety issue for earthquake-resistant structures.

3.3.6 Wind loading

Our technical knowledge of wind loading has been increased enormously: we currently have a much better understanding of the range of wind speeds that might apply related to location and surface topography, and we have a much better understanding of pressures generated on structures. We have a much better understanding of the effects of highly localised pressures. But these advances have come at a price: the issuing of longer, more complex codes with scope for confusion. There remain uncertainty and disputes on interpretation. Wind-related failures still occur.

In relation to speed, it should be recalled that the wind pressure on any surface is a function of V^2. Thus, any error in wind speed prediction can have a significant effect on predicted surface pressures. The fact that wind forces are related to V^2 is an example of the safety attribute of sensitivity which will be discussed later in Chapter 4. Just from arithmetic, variations from assumed speed are likely to be of greater significance at small speeds. Practically this means that temporary structures, which are customarily designed for lower wind speeds than normal buildings, are much more vulnerable to sudden squalls (see Chapter 8 for examples of temporary structures that have failed).

The high effects of wind suction still seem to be commonly overlooked, especially on domestic structures: roofs are sucked off and gable ends sucked out. Many failures occur during construction when elements that would be either sheltered or otherwise stabilised in the finished structure are exposed to wind. (CROSS has examples of internal blockwork walls being blown down during the construction period.)

Perhaps the greatest cause of failure relates to wind dynamics and several incidents point to a lack of appreciation of potential failure modes related either to the fluctuating nature of wind loading or the generation of periodic loading via wind flow across circular and/or slender shapes. The phenomena of vortex shedding and galloping deserve to be better appreciated. The displacement and allied forces created by such periodic loading can be very large, causing failure by that fact alone. If the applied stress created by the fluctuations is low, but repeated many times, fatigue failure is possible (see Chapters 7 and 8).

3.3.7 Construction loading

Many types of loading occur during the construction phase and such loading is often uncontrolled and quite variable (Temporary Works Toolkit, 2017). Loading may also be applied in a manner unanticipated by the designer of the permanent works. Examples include the applied load from wet concrete. The high pressure this exerts on vertical and horizontal formwork should be understood, but formwork collapses are not unknown. Reference to Figure 3.5 will also show that the case of partial loading can be critical. When load is applied to only half the span during a pour, the central shear will be more than in the permanent condition plus the frame will tend to sway sideways. That latter aspect of behaviour illustrates two points: first, that partial loading causes frame stresses

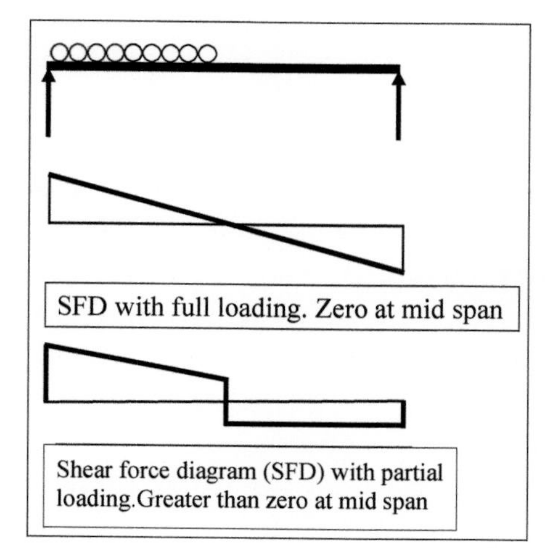

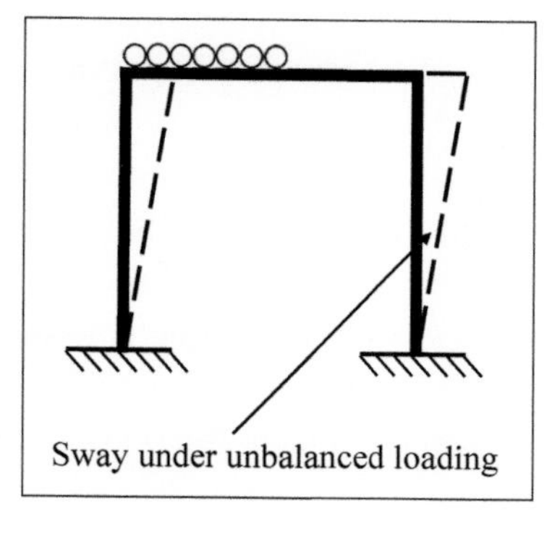

Figure 3.5 Effects of partial loading

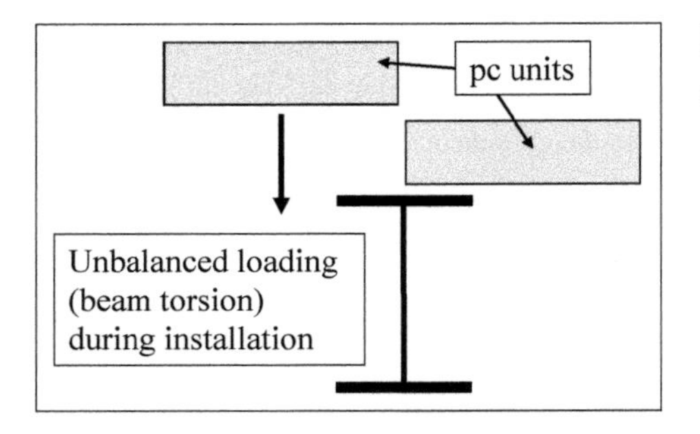

Figure 3.6 Unbalanced loading during construction

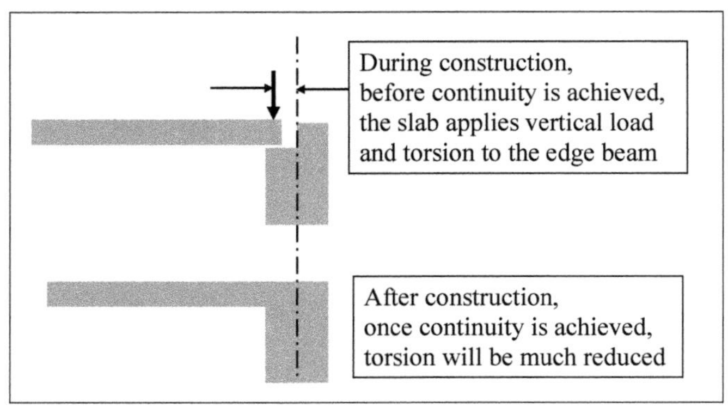

which differ from the permanent case and second, that the diaphragm action which may exist in the permanent case (once concrete has set) is absent from the temporary case so stability is an issue. As a generality, during construction, building elements will not be restrained as completely as in the permanent condition.

Unbalanced loading is quite an issue during construction when assembling precast concrete units and Figure 3.6 will illustrate the potential for introducing significant torsion plus potential instability during the assembly process.

During construction, floors being cast may be propped off floors below (so adding load to them) and once cast, all manner of construction materials may be stacked off them, often in concentrated locations. Chapter 8, Figure 8.16 shows an example and a glance at such pictures will show that real loading is far removed from idealised images of a uniformly applied UDL (uniformly distributed/design load). In short, loads applied during construction are fundamentally uncertain (see, for example, Temporary Works Toolkit, 2019), but for safety need to be controlled. Apart from excess loads due to uncontrolled storage, floor construction loading may peak under activities such as sustaining wheel loads from MEWPS (mobile elevated working platforms).

3.3.8 Accidental loading

Accidental loading is one of the most difficult topics to deal with. Ronan Point collapsed under accidental loading conditions and all manner of car impacts on roadside structures might be termed accidental, with impact energies uncertain. Some dramatic effects of accidental impact loading have been those from forklift trucks initiating progressive collapses in warehouse storage systems (Taylor, 2010). Other examples are given in Chapters 8 and 11. The best defence is to make structures 'robust' (Institution of Structural Engineers, 2010, 2013 and Chapter 4 (Section 4.4)).

3.3.9 Strain-controlled loads

Strain-controlled loads can cause great confusion and it is not always easy to decide if they matter or not. Typical causes are differential settlement, forced fit, shrinkage and thermal effects. If the stresses from differential settlement (or other causes), are calculated they can be huge, but are typically ignored on the basis that if displacements are relatively small, they cannot cause true structural collapse but only result in some cracking/distortion. The justification for this stems from an understanding of the upper and lower bound theorems (Box 3.1).

It follows from these theorems that in a ductile system, collapse cannot come about from any pre-existing set of stresses, i.e. the plastic collapse load of any frame is the same irrespective of whether that frame has been forced together or incurred an imposed strain, say by differential settlement. However, as with many statements, this proposition is not quite true. The argument is true if the only criterion governing frame strength is bending but if capacity is governed by axial capacity and buckling, then strain-controlled stresses do matter (see Section 3.5.5 and residual stresses).

Strain-controlled effects might also affect structural serviceability. Differential settlement may well cause excessive cracking. Similarly, shrinkage stresses can create large stresses but might be relieved by finite cracking. Differential thermal effects can also create huge stress and may or may not be important (Section 3.5.5). The test of acceptability will normally be the tolerability of the cracking/distortion amount.

Some strain-induced stress can be relieved by finite displacement; concrete cracking is a good example (albeit the embedded rebar might remain stressed). Rebar itself offers

Box 3.1 Upper and lower bound theorems

One estimate of a structure's strength may be obtained by postulating a collapse mechanism and then equating the internal work done within the structure on the plastic hinges to the external work done by the load(s) moving through a kinematically consistent set of displacements. The capacity thus evaluated is always an upper bound to true capacity.

Alternatively, if capacity is based on the maximum stress assessed in a structure (less than yield, and the boundary conditions are satisfied) then this is always a lower bound to true capacity.

A discussion on these theorems is contained in MacLeod, 2005.

a good illustration of strain-induced effects. Before it is even installed in concrete, bar is bent well beyond its yield point and goes into service in a strained and stressed state. But that does not detract from its ability to carry more load thereafter: what matters under externally applied load is how far the bar strains from the state of zero external loading up to failure.

Earth pressure against retaining walls is also a value affected by displacement. With massive walls, displacement limited, the pressures will be much higher than active pressures since these require some wall displacement to be realised. This performance provides a common example of trying to understand what 'failure' means and the difference between concentrating on force and concentrating on displacement. In a retaining wall, the key question is whether the movements required to reduce the acting pressures ('at rest') to active are tolerable or not. If they are not, in terms of serviceability, the wall may still be 'safe' in the sense that the wall might not be able to overturn and might only slide a small distance. Moreover, whilst its stem might be overstressed, the section in bending might only rotate a finite amount before the pressure forcing displacement reduces. How much displacement might be tolerable is critically affected by the flexibility/ductility available within the system.

3.3.10 Summary

The loading applied to structures might seem relatively straightforward and often is. But loading is a key variable affecting the capacity of structures overall and hence their safety. Dead loads can be easily computed, allied with the normal presumption that predictions will be accurate. But that is not always true either in magnitude or distribution. Live load intensities are most often defined in standards and are presumed more variable in magnitude and distribution than dead loads, yet typical values are frequently upper bounds. In some cases, live loads can be very accurate, in other cases, uncertain, and in yet other cases, controlled. Where explicit safety checks are required, it pays to consider the inherent uncertainty with the specific loading and consider how that uncertainty might be accommodated (other than by load factor). Especial care is required when assessing any load related to a $(\text{parameter})^2$. Accidental loads are hard to deal with and have been the cause of many failures. Dynamic loading has been the cause of many failures when the possibility of its arising has been overlooked.

Strain-controlled loading can be very confusing as to importance; tackling such loading requires a focus on what 'failure' actually means. The essential problem which strain-controlled effects reveal is that whilst a structure might have to sustain stresses of equal magnitude (generated by different effects) some such stresses matter more than others. In all ductile materials, the quality defining failure is most often displacement and most often strain-controlled stresses have little practical effect. Conversely, in brittle materials, strain-controlled stresses would matter since stress, from whatever source, determines cracking and once a brittle material has cracked, it is generally useless for strength.

3.4 Analysis

3.4.1 Introduction

Following definition of materials and loading, analysis is the next stage in a typical structural design, allowing prediction of stress. As with materials and loadings, there is uncertainty associated with the process of modelling and of assessing the output significance. Nearly all analyses are approximations of reality. Chapters 8 and 13 give examples of failure consequent on inadequate analysis.

Before starting any analysis, it is very important to consider the degree of accuracy required. For years the vast majority of structures were analysed using simplified hand methods which entailed various approximations and hence inaccuracies. Nowadays, of course, it is readily possible to have much more advanced analysis either of the classic kind or of finite element analysis (FEA) to examine stress distributions in detail. Computerised methods may be thought of as just efficient, cheap, tools or they may be thought of as bringing accuracy. But an impression of accuracy is not necessarily relevant in terms of structural safety and may indeed seduce users away from a more essential focus. The prediction of high stress concentrations by FEA (which were cheerfully ignored using hand methodology) might cause unnecessary alarm. The key, as always, is to understand the meaning of failure. If localised stress concentrations can be accommodated by strain limited local yielding or cracking, then their effects on global strength and global displacement are negligible (Section 3.5.5). On the other hand, if a structure's safe state is governed by fatigue, then the true elastic stresses matter and efforts are required to model the structure accordingly.

Discounting specialist applications, most analyses, either by hand or by computer, are of the linear elastic type. For these it is presumed that elastic displacements are relatively low. But under extreme loading, imposed P-Δ effects from sway can become significant and where this is the case, the form of analysis needs to include second-order effects.

3.4.2 Load application

The first variable is that related to application of load onto the structure. In Section 3.3.4 above, it was stated that load distribution between beams might well be governed by the stiffness of the object applying the load, but rarely is that modelled. In most structures, linear dimensions are quite fixed, and the effect of tolerance on length, say, and thence developed bending, is quite small. Nevertheless, it is instructive to be alert to cases where small variations might be significant and to consider the sensitivity of performance to any assumptions. As an example, Figure 3.7 shows a short cantilever. The root bending moment (BM) is P × L. But if L is short, and of the same order as a position tolerance (say 20 mm) then the BM can be either P × 20 or P × 40 – a 100% difference. Conversely,

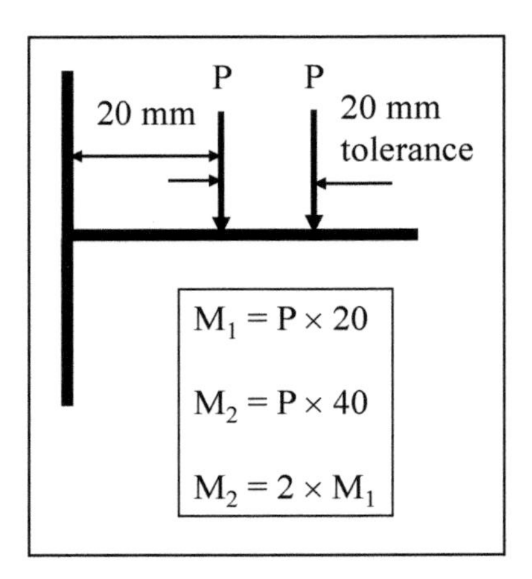

Figure 3.7 Effects of tolerance on bending moment

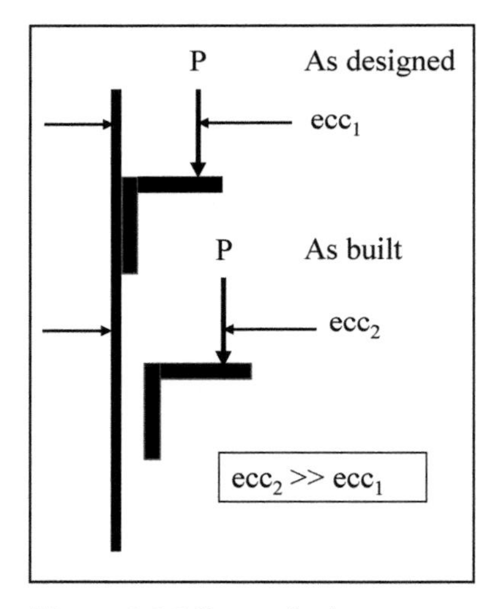

Figure 3.8 Effects of tolerance on bending moment

if L were, say, 1 metre, the BM would be either P × 1000 or P × 1020 (only a 2% difference). Figure 3.8 shows an angle bolted to a wall. In practice the wall was uneven, and the angle was packed out with shims thus adding a significant bending moment to the supporting bolt, which failed. Even the analysis of a simple cantilever can be wrong if the model is incorrect (see the Ramsgate walkway failure in Chapter 8). In contrast to these examples, Chapter 4 describes one case where positive benefit to safety was achieved by eliminating variability of load position on a cantilever. In the box girder bridge collapses of the 1970's (Chapter 8) the eccentricity of the bridge pier support reaction to the diaphragm overhead within the box was a feature of the failures. It should always be considered that tolerances exist and create a departure from the idealised structure.

In recognition of vertical tolerances (the fact that a building will always be erected out of plumb), the Eurocode has derived the notional horizontal forces that are used to test robustness. In older British codes, these were simply notional forces related to building mass at each level, whereas in the Eurocode they have a more tangible basis (described in Institution of Structural Engineers, 2010). The point is that inevitable deviations create both global and local force systems that might be overlooked. They may also be overlooked in complex modelling of variations of axial thrust along members if the model is too coarse to allow the inclusion of fine offsets (Figure 3.9).

3.4.3 Load paths

A key task in setting up any model is to have clarity of vertical and horizontal load paths from the point of load application to ground. These paths ought then to be reflected in the model and that might include some judgement. As an example, it would be normal to presume a building frame might be propped off a lift shaft such

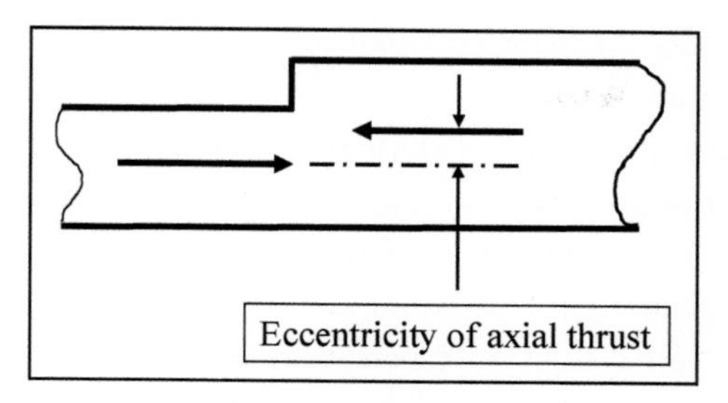

Figure 3.9 Eccentricity of axial thrust

that the frame itself only took vertical loads. Strictly, the validity of that assumption depends on the sway stiffness of the shaft. Looking at load paths and what provides restraint, and defining that, is very important not least for the construction phase when, for example, a floor slab intended to act as a horizontal beam or a diaphragm in the final condition may be absent. A classic case of error over load paths is that of the failure of the Hartford Civic Center roof (Chapter 8) where purlins supposed to be restraining the truss compression boom were in the wrong plane to perform that function. Another classic case of failure consequent on not fully understanding a load path is that of the collapse of the Hyatt Regency walkway (Chapter 8).

3.4.4 Modelling

The accuracy of elastic analyses is linked to prescribed boundary conditions. In reality, there are no fixed feet to columns; there is differential settlement; no connections are 'fixed' but neither are they 'pinned'. Column shortening may or may not be accounted for. These idealisations affect the output and thence the stress predicted and the deflections, but whether they are important or not depends on circumstance. An example of catastrophe linked to inadequate finite element modelling is that of the failure of the Sleipner Platform (Chapter 8). Other cases are known in which deflections have significantly exceeded computer predictions, usually because of false assumptions about steelwork connections. Such connections may be far from having perfect rigidity due to internal component deformations or due to bolt slip in clearance holes. Problems may occur when connections are assumed to be 'fixed' when their true flexibility results in real deflections being higher. Conversely, it is not necessarily safe to assume connections are 'pinned' since real connections (certainly at low load) have a significant degree of stiffness. This stiffness can, for example, create unanticipated column deflections at mid height via moment transfer from attached beams. A distinction must be drawn between assumptions that result in frames having adequate strength (at ultimate load) and assumptions that produce reasonable estimates of deformed shape.

The stiffness of concrete members, although essential to be included in any model, is of course not accurately predictable and will vary with time and will vary due to stress states in as much as these influence the amount of cracking. An example of a failure where excess deflections were not accurately foreseen is that of the Palau Bridge collapse (Chapter 8).

3.4.5 Component design rules

There is another general uncertainty; the one attached to the rules we use for component design. For some structural effects, designer confidence in prediction can be extremely high. We know, for example, that stress prediction in steel beams via stress = M/Z is quite accurate (at least if shear lag is low). Yet for other predictions, the 'research industry' is constantly trying to evolve new design rules which better fit the facts. Examples might be design rules for steel connections or design rules for assessing the effect of shear in reinforced concrete beams. In many situations, there is a scatter of test results and opportunities to evolve differing design equations to provide an acceptable fit. That fit might be a 'best fit' or a 'lower bound fit'. Either way, it is not generally possible to derive a theory which precisely predicts structural performance across the whole range of potential parameter variation. Some codes, notably American codes, cope with this by adding another factor to the theory equation to ensure predictions are conservative.

3.4.6 Summary

Although apparently a precise science, especially when impressive computer output is generated, a short reflection suggests that there is uncertainty and variation within the process of structural modelling and analysis just as there are uncertainties in the parameters governing input, such as the material properties and loading. Examples exist of catastrophic failure linked to inadequate analysis. Unfortunately, there is a modern-day temptation to presume that impressive model output must be accurate and a firm indicator of safety. Whether that is true or not depends on the significance of stress level and the type of stress, and that is discussed in the next section.

3.5 Role of stress

3.5.1 Stress

In day-to-day work, structural engineers design component strength and stiffness against a set of applied loads and the basic parameter they use as a pass /fail criterion is mostly stress. But focus on stress as an absolute criterion of safety masks some underlying principles of sound building and tends to obscure the need to look at the whole. Attributes that might additionally be considered are ones such as global stability and global stiffness; or insensitivity to settlement, moisture or thermal movements and minor alterations; or insensitivity to the inevitable errors that accompany routine construction – all these attributes might be grouped under the quality of 'robustness' or 'stability'. Moreover, although using stress looks simple, there are, as always, complications.

3.5.2 Elastic design

First of all it pays to examine some concepts of strength more deeply. We have seen above that early concepts were to analyse a structure under applied loads, assess its stress state and then apply a safety factor to make sure that working stresses were well below what the structure was considered capable of taking. Originally engineers would apply the factor to what they thought was the ultimate material strength. Later on, it became the practice to apply the margin to yield strength, certainly for steel. Although all this sounds perfectly straightforward and logical, over time engineers have realised that considerations of strength safety are more complicated.

The variables that safety factors / load factors need to cover are variations in material quality and material sizes, and variations in loading and uncertainty within the structural model. But we might add another variation and that is the ability of engineers to assess stress states accurately. Intuitively, if the stress state is highly predictable and the load is known exactly, and the material quality is known exactly, then a lower safety factor will be appropriate. The converse is obviously true for variables having a credible range of values. Having some background understanding of that allows us to make a rational judgement over the level of safety required (Chapter 13).

3.5.3 Analysis and the role of stress

We should distinguish between analysis methodologies for a whole structure, which will seek to establish axial forces and bending distributions around the frame, from the second task of analysing elements of the frame which must carry those forces. One of the very first concepts causing professional concern was the concept of elastic analysis itself. It is deceptively simple to suggest that a rational design method is to compute stress states in a structure based on elastic analysis and use these as a limit. In the early days this was all that could be attempted, yet even then, for redundant structures, the labour was intense and so various ingenious approximations were devised. Some of these were self-evidently 'nonsensical'. For example, in steelwork design, it became customary to design all beams as simply supported and then, if they were in moment frames, to assume the opposite and design the connections as fixed and capable of carrying wind-induced moments. Patently this meant that stress predictions in the beam and the connections were both seriously in 'error'. However, and overall, it was known that this design procedure resulted in safe structures at least insofar as there was no evidence of failures. Early concrete structures were designed based on frame elastic analysis followed by member designs assuming an elastic distribution which shared stress between rebar and concrete. But it was rapidly realised that elastic design of concrete members was quite unrealistic since all sorts of phenomena such as shrinkage, creep and cracking affected real stress states.

To tackle this irritating issue in steel frames, a Steel Structures Research Committee was set up around 1930; the committee procured an early form of strain gauge with which they could measure stresses in real buildings and measurements were taken on a London hotel under construction. Somewhat alarmingly, the results revealed actual stresses quite different to theoretical predictions, yet huge numbers of buildings had been constructed on approximate design methods and were performing quite happily. It was soon realised that one of the causes of discrepancy was the fact that the connections were nowhere near 'pinned' but neither were they 'fixed'. In fact, for the riveted cleat type connections in use at the time, behaviour could be characterised as semi-rigid (springs). Various attempts were made to measure typical connection rigidities in anticipation that this would lead to more accurate design methods and perhaps greater economy. However, there is considerable stiffness variation between even apparently identical connections and more generally, designers do not have access to relevant spring constants (and certainly not at a time when beams are being sized). Nonetheless, the code of the time (BS 449, first published in 1948) incorporated three methodologies for analysis: connections fully fixed, connections fully pinned and connections semi-rigid. There have been occasional attempts to trade on connection semi-rigidity since, but it has never been popular. Out of this process came the understanding that in truth the elastic stress existing in structures at working load was not that important. It was realised was that although engineers could not predict elastic stresses accurately, what they could do with reasonable success was predict ultimate capacity, always provided frame detailing was appropriate. Thus, if a steel beam is designed as pin ended, at 'ultimate load' its connections will have yielded enough for the beam to behave as a pin-ended member. The stresses en route to collapse may be uncertain, but the final ultimate load is reasonably certain (i.e. always provided the connections have the ductility to rotate yet still carry the presumed shear forces). As a further step, the philosophy of plastic design or ultimate load design was then born. Plastic design methodologies then became the design basis for a vast number of pitched portal frames where no attempt is made to predict elastic stress distributions under working loads; instead, it was only required that the frame's ultimate capacity be computed and matched to a set of applied factored loads. It is implicit in this raw approach that part of the frame may be yielded under working load. In terms of safety, the key was to switch to understanding that safety does not necessarily depend on keeping stress below certain limits or even keeping stress below the yield point, rather it depends on ensuring that the frame's ultimate capacity has an adequate over-strength demand. This thinking has evolved over time to encompass both steel and concrete structures in that instead of assessing stress at working load and looking to ensure a margin against the material's failure stress, the reverse process is undertaken whereby the loads deemed applicable are factored up to represent the ultimate state and then the ultimate strength is compared with that value. This approach offers advantages:

1. It becomes rational to factor up the various kinds of loads by different factors (partial load factors) and it becomes possible to include other global factors to represent uncertainties in material and other variables.
2. It allows use of a member's ultimate capacity rather than its capacity at working loads, e.g. a steel member plastic section modulus may be different to its elastic section modulus. Similarly, in reinforced concrete, a section's ultimate capacity is assessed presuming a stress distribution different to that which exists at low stress.
3. It allows use of a frame's ultimate capacity, which may take advantage, say, of successive plastic hinges developing and redistributing load as in plastic portal frame design, though that approach has limitations in reinforced concrete structures.

Globally, this process is more rational and permits uniform margins of security of different sections under different load conditions. It is implicit in this approach that there is no single numerical value relating the safety factor with the load factor.

3.5.4 Plastic design / ultimate load design

Although offering powerful insights into structural behaviour, adopting plastic design / ultimate load methodology is not a panacea. First, to gain advantage from continuity, the connections or joints within frames must carry the ultimate forces. Secondly, they have to be capable of rotating as plastic hinges to allow enough distortion to take place locally compatible with moment redistribution and finally, because of high stresses, local stability can become more of an issue. On top of that, design of any member constrained by buckling still requires an elastic approach: no design philosophy is without its caveats. When first introduced for portal frames, plastic design was considered elegantly simple. However, theoretical objections soon began to be raised. It was known that mild steel had adequate intrinsic ductility, but questions were raised about the capability of higher-grade steels. Although such concerns proved unfounded, higher-grade steels do strain harden. In terms of member strength this is beneficial – there is more strength. But there is a drawback. If a plastic hinge develops near a connection and is required to rotate, then it will increase in strength, but that increase must be carried by the connection. How much will the increase be? That is related to the amount of hinge rotation, which is not straightforward to calculate. Then it was observed that the geometrical changes in shape of portal frames, as they approached collapse, had a detrimental effect on the plastic collapse loads, so compensations had to be made. Finally, although plastic design of beams is simple, it is not so with columns, and dealing with the interaction between axial loads and plastic hinges is complex. Mathematically, a frame's capacity

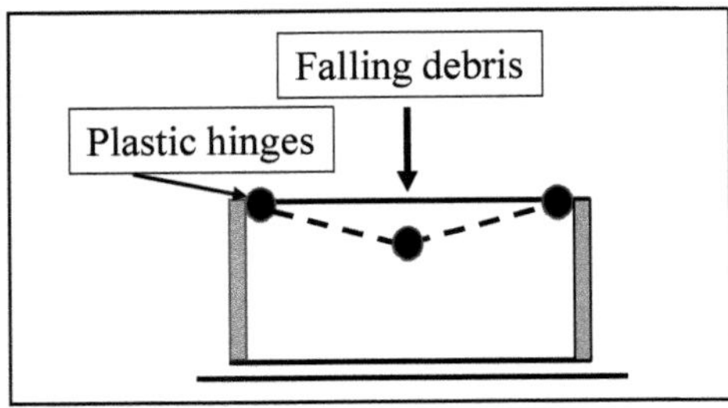

Figure 3.10 Morrison shelter. (Imperial War Museum D002058)

when governed by bending and based on permissible stress is a lower bound to its actual strength, whereas its capacity based on full plastic redistribution is an upper bound to true strength.

One early use of plastic design concepts taking advantage of steel's innate ability to deform and absorb energy was in the design of the wartime Morrison Shelter (Figure 3.10) which was a steel frame covered in mesh into which a person could crawl during a bombing raid. If the building collapsed on top of it, the loads were absorbed by a succession of plastic hinges in the frame at (hopefully) finite displacement. No attempt was made to check or constrain stress limits. What is easily missed is that the whole premise of the design methodology is that there must be ductility in the system; without that, the design assumptions are invalid and that applies just as much to larger-scale frames.

Ultimate load design is also used in reinforced concrete, but in a different way. Sections are designed on their ultimate capacity justified by test results that demonstrate redistribution of stress between the rebar and surrounding concrete. But because concrete sections lack the ductility of steel sections, only limited redistribution of bending moment is permitted within overall frames, save in special cases such as under-reinforced slabs (which can be designed by yield line theory). In slabs, the proportions of reinforcement are generally low such that enough ductility exists to justify a plastic-type analysis.

At its birth after the Second World War, exponents of plastic design had no guidance on what an appropriate load factor should be. The number adopted was a uniform 1.7, evaluated from the then safe stress on simply supported beams and the fact that the code (BS 449) mentioned a value of 1.7 for its derivation of safe stresses for columns.

3.5.5 The nature of stress

We know from classic theory that non-uniform states of plane stress exist in all structures. There are stress concentrations at all re-entrant corners, there are stress concentrations around holes and so on. In routine design, we usually ignore them and just work on the average stress in a section, for example: stress = M/Z. Even via that equation, we ignore the fact that there is some shear lag in beams. The stress across any flange is actually non-uniform: but we ignore that except in special circumstances such as wide box girders. We know there is residual stress in steel members from welding and uneven thermal cooling. We know there are shrinkage stresses and creep effects and thermal stress effects in concrete members and that the distribution of load between rebar and its surrounding concrete changes over time (to the extent that early theories of elastic design of reinforced concrete members were soon discarded). We know that so-called secondary stresses develop in steel connections even though the full structural member may be modelled with ends pinned. From this background it is clear that not all stresses are treated equally and not all matter to the same degree. But it is quite a difficult question to assess which stresses matter and which do not. Modern analytical tools such as FEA (finite element analysis) thrust this issue to the fore since that tool can give us output as complex stress patterns (which to some extent are dependent on how the structure is modelled). Engineers then face the task of reviewing the output to decide which stresses can be ignored and which need accounting for. Codes do not presume allowable stresses or capacities are based on FEA (though Eurocodes now give guidance). An example of confusion linked to FEA is given in Chapter 8 (Section 8.8.7).

The test of acceptability is really one of understanding displacement and ductility. Under static conditions, local stress concentrations can usually be ignored if the structure is ductile enough to accommodate the strain (to relieve the stress) without gross global distortion (though that ductility needs to be ensured by adequate detailing in both steel and concrete). The ability to verify this is by and large covered by codes which set out formalised rules or procedures for the design of members. Perhaps the easiest way to understand this is by considering plastic design methods defined in terms of energy whereby the balance is between the work applied in deforming a frame, set against the energy implicit in the structure available for resistance (Figure 3.11). Using this nomenclature, 'stress' does not appear and concerns about local stress concentrations do not appear either as they do not affect global displacement. That does not always mean that localised stresses can be completely ignored. In bolted connections, limits are set on bearing stresses under bolts to prevent

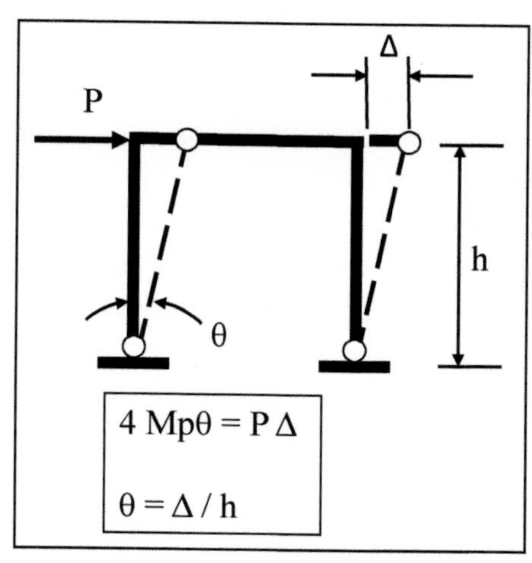

Figure 3.11 Plastic collapse of a portal frame

excessive deformation in the plies. In concrete, the whole question of understanding how shear stresses are computed is complex and indefinable by a straightforward assessment of stress state. There is also a distinction between the stress state which will cause unacceptable cracking and the stress states which will cause failure in terms of strength.

All this sounds very confusing and it can be. What should be appreciated is that in effect codes offer standardised procedures for evaluating structural capacity as a tool for design. Long successful experience of designing structures by such rules is the test by which we know the procedures are adequate. It is important to take code provisions as a whole, not mixing and matching; and it is important to be aware of the limitations of use which are not always explicit. This applies especially when rules are extrapolated to design members beyond the norm. Scale effects matter: see SCOSS Alert, 2018. It is also important to be aware of certain fundamental situations when absolute values of stress are more important than others. The key is to understand what the governing limit states are and thence grasp how critical the absolute stress state is to that performance. In turn that has a link to the method of analysis, noting that certain analytical methods will be perfectly adequate for routine structural design but may not necessarily allow prediction of precise elastic stress states.

Certain obvious states for which absolute stress levels are important (in steelwork) relate to assessments against fatigue and fracture. In both those conditions, the absolute value of stress, as influenced by stress concentration, has a significant effect on performance. This understanding then, in turn, influences the form of modelling adopted. For most steel beams in routine structures an adequate strategy is to presume that members are pin ended, However, that is not acceptable if fatigue life is an issue since then any secondary moments developed at connections are of true significance because they create stresses that will influence life. Fatigue and fracture are not issues for concrete structures, but appreciation of a realistic stress distribution is important if concrete structures are to be detailed in a manner that minimises cracking.

Section 3.3.9 gave an outline of strain-controlled loads and stresses, suggesting that in most cases they could be ignored. This is amplified below with reference to thermal, residual and secondary stresses:

Thermal stress. In a constrained member, thermal stress is simply evaluated as:

$\sigma_T = E\,\alpha T$ where E is Young's Modulus, α is the coefficient of thermal expansion and T is the imposed temperature range.

If we take steel as an example:

$E = 21 \times 10^4$ N/mm^2

$\alpha = 12 \times 10^{-6}$ / °C

Yield strength = 250 N/mm^2.

Thus, the temperature range to cause a constrained steel member to reach yield is:

$$T = 250 / (12 \times 10^{-6})\ (21 \times 10^{4}) = 250 / (1.2 \times 2.1) = 99\ °C.$$

From the same equation, any typical thermal range of say 30 °C on a structure is going to generate a substantial percentage of yield strength in a constrained member. Note: there is no length involved in this calculation, which seems rather unfair as thermal problems are typically associated with long lengths.

The strain to cause yield is $250/21 \times 10^{4}$.

So, if a steel member is 1 metre long, its displacement to cause yield is:

ε = extension / original length

so extension = $250 \times 10^{3}/21 \times 10^{4} = 1.2$ mm.

In other words, if a 1 metre long heated-up member expands by just 1.2 mm but is constrained, it will yield. Nonetheless, the high yield stress can only exist if the member is fully constrained and if there is any flexibility in the end constraints to accommodate part of the 1.2 mm, the developed stress will be much reduced.

Overall, it is unlikely that conditions of full restraint will exist in most structures and any slack (e.g. in bolt holes) will negate generated thermal stresses. One design strategy is to eliminate constraint by adding movement joints. Even if that is not done, practical experience suggests that imposed thermal ranges within buildings are not extreme, so generally thermal stresses are ignored in routine designs. As always, designers need caution when they depart from the norm. Long structures and external structures exposed to large thermal ranges should consider the consequence of imposed dimensional change (in the extreme heat of summer 2022, UK rail companies considered the possibility of railway track buckling).

Residual stress. This is a special case of thermal stress and a special case of strain – controlled stress. When steel members heat and cool (as they will do during manufacture) they usually do so at different rates such that internal restraints are set up. These create a self-equilibrating set of internal (residual) stresses. These stresses exist in all steel members after manufacture and may thereafter be modified by hot fabrication processes such as welding. Peak residual stress may equal or exceed yield strength (if strain hardening is accounted for). Yet there is no effect on a section's strength in bending provided the section is ductile because the gains and losses on the tension and compression sides just balance out. If the residual stress is compressive and an external compressive stress is added to the member, the region with residual compressive stress reaches yield rather earlier than it would have done otherwise and thus loses stiffness earlier than it would have done otherwise and starts to buckle prematurely. This is why the strength of rolled and welded struts of the same cross section differ; struts fabricated by welding are weaker. Residual stresses also affect a steel's performance in fracture and fatigue. Hence in some circumstances residual stresses matter, whilst in others they do not. The effect of shrinkage stresses in concrete is analogous. Shrinkage stresses might affect cracking but not usually strength. They do affect strength in the sense that an assessment of low tensile stress in unreinforced concrete can be totally misleading if the section happens to be pre-cracked from shrinkage or thermal effects.

Secondary stresses. Any steel member which is designed as pin ended but has restrained end joints (and all joints have some restraint) will develop moments at the joints, perhaps to a high value, all linked to the relative stiffness of the connected members and the stiffness of the connection components. In terms of strength, such stresses from the restraint are normally ignored because under overload, a modest amount of yielding permits the structure to revert to its modelling assumption of being pinned, plus the upper bound theorem applies. There is some art and judgement in assessing whether the degree of imposed yielding is modest enough to be ignored, and for a full assessment of safety, it would have to be ensured that the connection components could sustain the imposed distortions without shearing or cracking in any way. This approach is not usually used in concrete because the degree of cracking associated with large inelastic rotations is judged unsightly and because reinforced concrete is not truly a plastic material. Nonetheless, concrete codes do allow a limited amount of rotation to recognise the capability of the member to redistribute strength demands, as in a fixed-ended beam between supports and centre.

Localised stresses. Very large local stresses develop in all materials under point contacts, for example, under supports on concrete members and where bolts bear on steel. If an elastic analysis of a raft foundation is carried out with a fine mesh in the model, very high peaks of bearing pressure will be found. It can be hard to judge whether any of these are important or not. Partly that depends on understanding of consequence and local arrangements. In the extreme, materials under uniform compressive triaxial stresses can sustain very high values without failure simply because the material surrounding the stressed zone confines it and limits its ability to strain far enough to fail. Conversely, high triaxial tension stresses are potentially dangerous in all materials because they transform the failure mode from ductile to brittle. High localised bearing stresses under foundations, say in raft centres, are unlikely to be important because the criterion of raft failure is not really pressure but settlement: calculations are just couched in terms of permissible bearing stress for convenience. Localised high pressures under raft centres will have no effect on global settlement so can be ignored. High localised stresses in concrete can cause spalling or cracking but might not cause a global failure and a skill in all concrete detailing is in the placing of rebar to constrain crack widths under a variety of stress-inducing conditions. Very high stresses can be tolerated at points in steelwork if the steelwork is constrained as it is by the plies within bolted joints. The consequence of such high stress is just some minor distortion of finite magnitude.

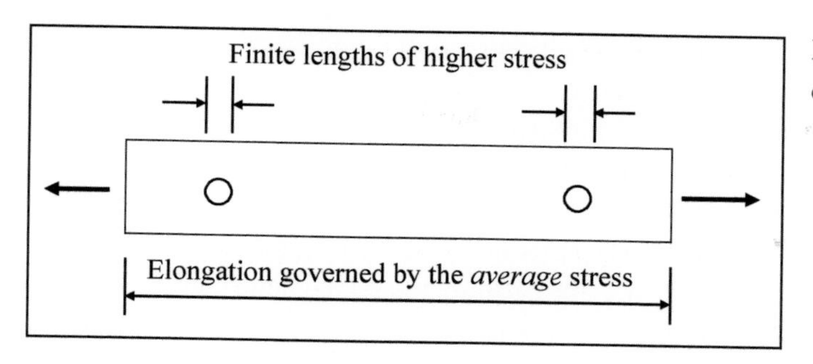

Figure 3.12 Elongation of a tie

Dynamic stress. This can be very confusing, as explained in Section 3.3.5.

Displacement. Displacement is a feature of acceptability in all the cases above. In the standard case of a steel tie with holes in it, the test of acceptability under tension is whether there is yield on the tie gross section before ultimate stress is reached on the net section (Figure 3.12). The rationale for this is that under test, failure is defined as the point at which the tie, as a whole, starts to extend excessively. The amount of stretch contributed by yield of the small length of tie next to the holes is trivial as compared to the stretch computed from lower stress in the tie as a whole. Thus, even if the net section yields, no excessive change in tie length will be observed provided that the net section does not fracture (and codified rules ensure this mode of behaviour). Hence in some components local stresses even above yield can exist with the component remaining safe. It can also be seen from this that a special case exists for structures supported by very long cables. For such items, the extension to failure can be so great as to render global ultimate load design meaningless: structural distortions of supported structure overall may be easily large enough to constitute failure well before the cables have even yielded (Mann, 2002).

Stress concentrations. The discussion on the tie above is a special case of how steel copes with stress concentrations. On true elastic analysis these exist around every bolt hole and at every re-entrant corner. In ductile materials, as far as strength is concerned, they can be ignored since under overload, the material's plastic ability will allow redistribution until the whole section is at a uniform yield stress. But when failure modes involve cracking, stress concentrations do matter since the highest actual stress drives crack propagation. In concrete structures, stress concentrations also exist and promote cracking, again influenced by true stress, be it applied externally or generated internally by shrinkage. The uncertain nature of such stresses and their effect on concrete capacity is a major reason why no advantage is taken from concrete tensile strength.

3.5.6 Summary

It sounds simple just to assess stress magnitudes and keep stresses low enough to ensure safety. Alas structural performance is not as simple as that (Heyman, 2005). Stresses arise under several loading conditions and although numerical values might be equal, that does not mean the stresses are all of the same significance. In some cases very high stresses, even above yield, are perfectly tolerable; in others they are dangerous. One key explanation lies within an understanding of the role of displacement and whether this is enough to be excessive and therefore count as 'failure'. Another lies in understanding the role of stress in particularly undesirable failure modes such as fatigue or fracture. Certain codes recognise these effects by distinguishing between primary and secondary stresses. It is normally possible to ignore a number of types of stress but safety must be achieved by making sure that implicit assumptions such as force redistribution are justified by commensurate detailing and ultimately via ductility.

All this is too complicated for explicit inclusion in codes of design practice that target routine design. Codes for building structures and bridges differ as real stress distributions tend to matter more in bridge structures than in building structures, so bridge design codes generally require more complex stress consideration (Bourne, 2019). Putting that to one side, engineers come across all sorts of unusual problems, and safety demands that they have a proper training so as to identify occasions when the application of routine code rules may be insufficient or unsafe.

3.6 Safety factors / limit states

3.6.1 General

Self-evidently the prediction of structural capacity carries uncertainty with it. Hence if we want structures to survive reliably, there is a need to compute capacity and then include a safety margin. In the main, that margin needs to cover uncertainties in:

- dimensional properties
- material quality
- applied loads
- method of analysis
- quality of construction.

Loads can of course be variable, but to different degrees depending on their type. Acknowledging that, safety margins applicable to different loading types can be adjusted accordingly; these load factors can be further adjusted when the loads are applied in combinations. As a further refinement, in some codes an importance factor is introduced. For example, in seismic engineering, the importance factor (> 1.0) applied to the earthquake load computed for a dam will be higher than that adopted for a routine structure. This has the effect of raising the design forces to ensure greater reliability, recognising that the consequences of dam failure are less tolerable than they would be for more routine structures. In effect some codes use the same strategy, though in reverse, by lowering safety margins required for some agricultural structures.

The introduction of stress nomenclature was a breakthrough in our tools for structural analysis and in our ability to describe structural behaviour. To students today it must seem obvious, but the discovery that a structure's state could be described by one parameter and compared to an absolute acceptance criterion was indeed a major step forward. Combined with the development of structural analysis methodologies and stress prediction, it allowed engineers to introduce a margin between what they thought was the capacity of the structure and what they thought were its likely stress states in service. They could design rather than guess.

The adopted margin was first formalised as a simple parameter, the safety factor, the meaning of this being that a safe working stress was presumed to be the failure stress divided by that factor. However, what the safety factor ought to be was originally just based on intuition. Chapter 1 describes historical debates about the factor for cast iron structures. Not everyone had the same idea and part of that variance stemmed from knowledge that the material quality itself was considerably uncertain. Not only that, but the mode of cast iron failure was brittle, and collapses were sudden. Engineers became much more comfortable once ductile wrought iron was introduced. Not only was wrought iron a more reliable material, it also had better ductility and ductile failures are much more benign than brittle failures. Intuitively, ductility is a good property to have for safe structures but neither its presence nor its absence is reflected in normal calculation. However, as we shall see in Chapter 4, some measure of ductility is presumed in our modern design methods and structures are less safe when ductility is diminished.

The terms safety factor (SF) and load factor (LF) are both in use but are not synonymous. The definition of safety factor is:

allowable working stress = peak elastic stress / SF.

The definition of load factor is:

allowable member capacity = actual capacity / LF.

If there were a linear relationship between ultimate member capacity and capacity at working stress the relationships would effectively be the same, but that relationship is not linear. The differences might apply to sections, to members or to frames. It is well known that the elastic stress distribution in a reinforced concrete member is a poor representation of realistic capacity because the distribution of load between rebar and concrete changes with time and because at ultimate load there is a redistribution of load towards the rebar and away from the concrete. Likewise, in steel sections, the plastic section modulus of a cross section differs from the elastic section modulus to a degree linked to cross section shape.

There are also differences when considering members. The redistribution of beam bending moments as loading increases to ultimate is well known. In a fixed-ended beam with a uniformly distributed load (udl), there is eventual equalisation between end and centre moments. In a fixed-ended beam with a central point load, the end and centre moments are the same at all loading stages. For the case of a beam loaded with a udl, the ratio between the safety factor (linked to end moment W L/24) and the load factor (linked to end moment W L/16) differs by a ratio of 1.5, i.e. the same beam is 1.5 times 'stronger' if computed using an ultimate load methodology permitting plastic redistribution. In this example, the amount of hinge rotation required of the end hinges is the demanded 'rotational capacity'. If the rotational capacity cannot be realised in practice, then the presumption of plastic redistribution capability is invalid.

3.6.2 Safety margin magnitudes

Although Section 3.6.1 gives important background understanding, taking account of all the described effects is too complex for normal commercial design. For such structures, codes just define appropriate load factors geared towards the limit states being considered. Overall, these are not 'scientific' values. Widely adopted values have their origins in safety margins that have been customary for a century or more but have been refined over time (Beal, 2011). By experience over that long period, it has been found that the margins are adequate, i.e. if they are adhered to and structures designed properly to them, then experience shows that by and large structures perform safely. It is useful to observe that the values are 'round numbers'– 1.6, 1.4 and so on – and they therefore serve as arbitrary definitions of the margin required. It would not be correct to say that a structure with a load factor of 1.59 was unsafe simply because the code requirement was 1.6. It is necessary to have a standardised value for convenience of design and to ensure a level commercial playing field. That was one imperative of the Eurocode introduction. It is also hard to create the balance between the materials of steel and concrete so that each is not competing against the other in terms of lower safety margins.

There have been studies attempting to define rational values for partial load factors (CIRIA, 1972) but these have never been fully adopted. Mostly it seems that using such approaches is too complicated. That has not, however, eliminated discussion (Structural Codes Advisory Committee, 1975, 1979; Taylor, 1994; Beeby, 1994; Beal, 2001). With the Eurocodes, the hope was that, as more was learned about materials and performance, then the various partial factors could be adjusted to reflect improved knowledge. When required, it is possible to calculate a required load factor and a methodology for doing this is given in Eurocode 0. (See Chapter 4, Section 4.5.9 for an example.) The basis is to use judgement on the failure consequences (which might suggest a higher/lower reliability is required) and knowledge of potential variability in loads and materials and so on. Clearly, if there is high confidence in the accuracy and applicability of the loads and in the accuracy of the analysis, then a lower safety margin can be appropriate.

3.6.3 Role of testing

In appraising existing structures, we might make use of similar Eurocode approaches. Alternatively, we might opt to judge existing structures' adequacy solely based on safety principles. Although in design it is vital that codes are looked at as a whole and that clauses are not used selectively, it is also legitimate to realise there is an inbuilt expectation of a certain material quality and an inbuilt expectation that a standard of quality will be achieved via appropriate QA schemes and post-construction QC. However, once a structure is built it is possible to reflect on the various uncertainties, perhaps eliminating some of them via test records of material strength or by load testing, and then to consider whether codified load factors are truly applicable. They may not be. There is a possibility for arguing for lower standards or indeed higher ones (especially if there has been degradation). Conversely, if material strength is significantly higher than presumed in the design (a not infrequent occurrence) then there is no intrinsic reason why advantage should not be taken of that.

3.6.4 Summary

The load factors we use in codes are semi-arbitrary values proven adequate by long experience. They are essential for standardisation of design and for commercial reasons. But they are not totally scientific and not perfect arbiters of a division between a structure being safe or unsafe. Competent design might suggest increasing the standard factors when there is great uncertainty or when the consequences of failure might be excessive. Conversely, in appraising existing structures for safety, it would be legitimate to be aware of the uncertainties that the load factors are intended to cover and appraise whether any deviations are justified. Load testing can play a part in this.

3.7 Chapter summary

Some reflection will reveal that what constitutes structural failure is not always obvious. Overload under gravity loading can certainly cause physical failure. But in other cases, 'failure' may be better defined by excess/unacceptable displacement. Some modes of failure are less preferable than others. The margins we seek between service conditions and whatever is failure are also subject to judgement based on uncertainties.

Materials vary in strength, so safety demands that precautions are taken to verify design assumptions. Such assumptions may only be true at a point in time, often at the time of initial construction and many weakening degradation processes are possible thereafter.

Loading is a key variable affecting the capacity of structures overall and hence their safety. However, many different sources of loading exist and uncertainty as to magnitude will vary in a manner related to type. As a complication, some loads can produce high stress yet be of little consequence.

All structural design requires some sort of analysis to predict how forces are carried through the structure. Examples of failure consequent on inadequate analysis exist. There is a modern-day temptation to presume that impressive model output must be accurate and a firm indicator of safety. Whether that is true or not depends on the significance of stress level.

Assessing stress is a task underpinning many of our judgements on structural safety. The term 'overstressed' implies failure. However, when looked at in detail, the role of stress is more complex than it first appears: some stresses matter more than others even if they have the same magnitude.

To cover inherent uncertainties, engineers have always applied a safety margin above stress levels predicted to exist under defined loading conditions. What that margin should be varies according to circumstances, with circumstances encompassing both uncertainty and knowledge of how a structure might respond under excess loading and the consequences of failure. A different design approach is required when a structure is brittle as opposed to when it is ductile and a conservative approach is required when the consequences of failure are excessive.

References

Alexander, S. J. (2002), 'Imposed floor loading for offices: a re-appraisal', *The Structural Engineer*, **80**(23/24).

Beal, A. N. (2001), 'Factors of ignorance?', *The Structural Engineer*, **79**(20).

Beal, A. N. (2011), 'A history of the safety factors', *The Structural Engineer*, **89**(20).

Beeby, A. (1994), 'γ factors: a second look. Partial safety factors: an overview', *The Structural Engineer*, **72**(2).

Bourne, S. (2019), 'An introduction to bridges for structural engineers' (part 1 and part 2), *The Structural Engineer*, **97**(1 and 3).

Carpenter, J. (2008), 'Safety risk and failure: the management of uncertainty', *The Structural Engineer*, **86**(21).

CEN (2002), EN 1990 Eurocode 0: 'Basis of structural design'; establishes principles and requirements for the safety, serviceability and durability of structures.

CIRIA (1972), 'Rationalisation of safety and serviceability factors in structural codes', CIRIA Report 63, London.

CROSS database: www.cross-safety.org/uk.

Hewlett, B. (2021), 'A checklist of water damage', *Proceedings of the Institution of Civil Engineers – Forensic Engineering*, **174**(2).

Heyman, J. (2005), 'Theoretical analysis and real-world design', *The Structural Engineer*, **83**(8).

Institution of Structural Engineers (2010), *Practical Guide to Structural Robustness and Disproportionate Collapse in Buildings*, IStructE Ltd.

Institution of Structural Engineers (2013), *Manual for the Systematic Risk Assessment of High-Risk Structures against Disproportionate Collapse*, IStructE Ltd.

MacLeod, I. A. (2005), *Modern Structural Analysis: Modelling Process and Guidance*, ICE Publishing.

Mann, A. P. (2019), 'Safety of hanging systems: lessons from CROSS reports', *The Structural Engineer*, **97**(9).

Rail Accident Investigation Branch (RAIB) (2013), 'Report 13/2013: Partial failure of a structure inside Balcombe Tunnel'.

SCOSS Alert (2014), 'Tension systems and post-drilled resin fixings', www.cross-safety.org/sites/default/files/2014-03/tension-systems-post-drilled-fixings.pdf.

SCOSS Alert (2018), 'Effects of scale', www.cross-safety.org/sites/default/files/2018-11/effects-scale.pdf.

Structural Codes Advisory Committee (1975), 'Partial safety factors in structural engineering codes of practice', *The Structural Engineer*, **53**(8).

Structural Codes Advisory Committee (1979), 'Code Servicing Panel on Rationalisation of y-Factors – Second Report', *The Structural Engineer*, **57**(9).

Taylor, J. C. (1994), 'A new approach', *The Structural Engineer*, **72**(2).

Taylor, R. (2010), 'On the rack – structural failure of pallet racking systems', *The Structural Engineer*, **88**(22).

Temporary Works Toolkit (2017), 'Part 13: The importance of understanding construction methodology', The *Structural Engineer*, **95**(7).

Temporary Works Toolkit (2019), 'Part 6: Back propping of flat slabs – design issues and worked examples', *The Structural Engineer*, **95**(1).

4 Other attributes relating to safety

4.1 Introduction

Chapters 2 and 3 discussed uncertainty in structural behaviour and the limitations of ensuring safety by relying solely on a numerical factor applied to computed stress. It was argued that safe structures additionally require to be stable and to incorporate a measure of ductility and robustness if they are to be reliable. This chapter now looks at those additional attributes in more detail.

4.2 Stability

4.2.1 Introduction

It is occasionally remarked that more failures result from errors in stability than errors in strength. Hence a grasp of stability concepts is key in any test of structural safety. Checking stability is a separate activity to checking strength. Alas, *stability* is one of those words like *robustness* that engineers can recognise but not always fully define, albeit the word *unstable* has decidedly negative connotations. We can define three general forms of stability:

- local stability
- member stability
- global stability.

These first two states can be mathematically defined and linked to a member's proportions and stress states. Recalling the uncertainty of stress state prediction, for safety a structural section should not flip suddenly from a stable to an unstable state, which is one potential problem with very slender members under compression. Likewise, it would be undesirable to use sections which were highly efficient in carrying load but whose sub-elements (say stiffening lips) could be so easily damaged that the consequence would be a rapid change in global member state or functional capability (hence such members are only normally used as secondary components). Chapter 8 relates some example of roof failures from member instability.

4.2.2 Global stability

Generalities

The concept of global stability is easy to grasp; essentially it means something not toppling or moving bodily sideways; it embodies the concept of lack of gross movement. Thus, a routine check on all structures is to ensure they are 'globally stable'. This is quite a separate check from ensuring structures are strong enough. As always, there are subtleties which might cause instability to be overlooked. The collapse at Ronan Point might, for example, be considered a stability failure in that once a key component was lost, a chain reaction of consequential failures was set up (Chapter 8, Box 8.9). A check on stability must necessarily consider some horizontal force for disturbance and this will be counteracted by a gravity force, as shown in Figure 4.1.

- The only true check on global stability is one verifying that gross correcting moments exceed gross overturning moments by a suitable margin and that global resistance to sliding exceeds applied lateral force.
- If the forces are exceeded, the consequences might be lateral sliding, or they might be overturning/toppling. In some circumstances, sliding might be thought of as less critical than overturning.

For global stability, common sense warns of the dangers of structures simply overturning. There have been cases of unsecured items just falling over and killing people. There have been cases of stacked items collapsing in an uncontrolled manner. There have been many cases of walls simply toppling (low masonry garden walls are a routine risk). There have been cases of reinforcement cages swaying sideways to trap and injure operatives working inside the cages: the danger is most acute when the top mat is heavy and supported solely on vertical links (Soane, 2013). In several reported accidents, loose packs or shims supporting elements under construction have been disturbed and precipitated major failure. Hence, as a general principle, a potential hazard is to consider the movement of anything not

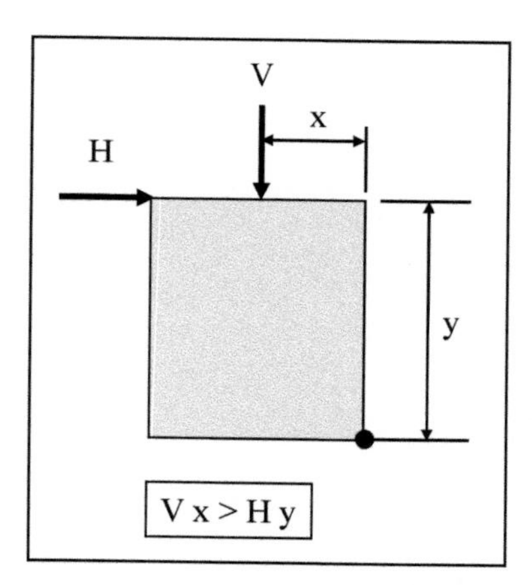

Figure 4.1 Overall stability check

fixed down firmly. As another principle, everything should be considered to have a horizontal load on it for the vertical inclination of any structure consequent on inevitable tolerances will always create an overturning tendency; there may also be just an accidental knock. Verification should be made to ensure objects cannot overturn or inadvertently displace under such horizontal loads. Good building practice should make sure no beam is supported on a narrow ledge since a minor shift in position transforms into a large shift in state. An example of what can happen if this rule is not observed is shown in the failure of the roof of the assembly hall in Camden School for Girls (Chapter 8).

The global stability problem is most serious when the toppling of any one object sets off a chain reaction like the toppling of dominos. Such failures have occurred in high-bay warehouses (McConnel, 1983; Taylor, 2010) and in a series of tall plate girders positioned vertically ready for bridge assembly (Short, 1962). There have been several cases of whole truss roofs collapsing sideways in a progressive manner.

In certain structural applications, horizontal sliding is not catastrophic especially if the resulting displacement is finite. This may occur for example under earthquake loading where sideways bodily movement can not only be acceptable but often entirely desirable as it isolates the structure from the ground-imposed forces. However, that is not always true if the structure is rigidly connected to services, whose rupture may create a separate disaster.

In many structures, prudence also suggests redundancy. Figure 4.2 shows that removal of a single restraint to the row of columns (or they might be trusses) has the potential to initiate collapse of the entire row. Robustness demands care that such an event cannot happen. A comparable example is illustrated by the retaining wall shown

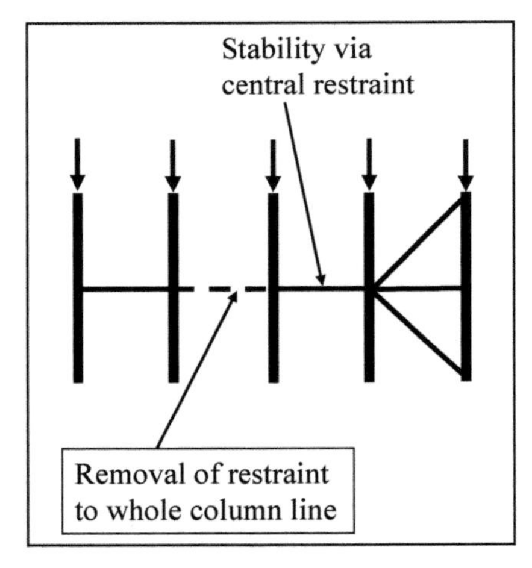

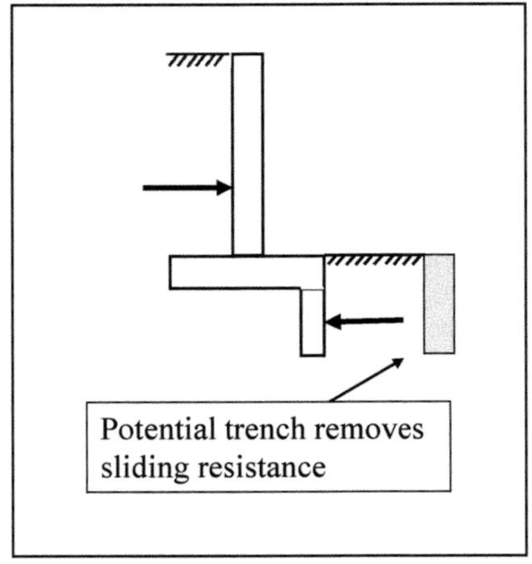

[Left] Figure 4.2 Restraint to a line of columns

[Right] Figure 4.3 Restraint to a retaining wall

in Figure 4.3. It might be unwise to ensure sliding stability by a front toe if it is possible for others to excavate a trench (say for services) along the front line (Chapter 8, Section 8.9.2).

The Institution of Structural Engineers has produced three reports giving overall guidance on maintaining stability in frames (Institution of Structural Engineers, 2015).

Checking stability

The danger of side sway promoted by top-heavy loading manifested itself widely in the 1960s and 1970s via a series of falsework collapses. In response, the government appointed Professor Bragg to prepare a report and recommendations. Bragg's report, published in 1976, (Bragg, 1974; and SCOSS Topic Paper, 2010) emphasised the need for designers to consider lateral stability (in all directions), taking account of realistic build imperfections. A major recommendation was for designers to consider 'a horizontal load of 1% of vertical loads plus calculated horizontal loads'; or '3% of vertical loads', 'whichever is greater'; however, BS 5975 uses 2.5% in place of 3%. In other words, even where no defined horizontal loading existed, designers were obliged to calculate a notional value as a tool for a stability test. Soane (2017) provides a valuable introduction to the stability of temporary works.

Normal checks on stability are just arithmetic and essentially presume displacements are small. However, some accidental eccentricity or some elastic displacement potentially worsens the problem. Figure 4.4 shows how a frame may be subject to sway during construction when vertical loading is unbalanced. Excessive sway in tall structures adds even more to base overturning by introducing P-Δ effects. This is an issue which must be addressed in seismic design or perhaps under wind loading because large lateral displacements may well be the norm. In the water tower illustrated in Figure 4.5, where there is a heavy mass at the top, there is a need to consider the global overturning moment as $H \times ht + P \times \Delta$. By analogy, a more common case arises in relation to the reinforcement cage referred to earlier (and illustrated in Figure 4.6) where the heavy top mat can sway sideways. Any sway is, in itself, a progressively destabilising process and designers should always be cautious when anything is 'top heavy'.

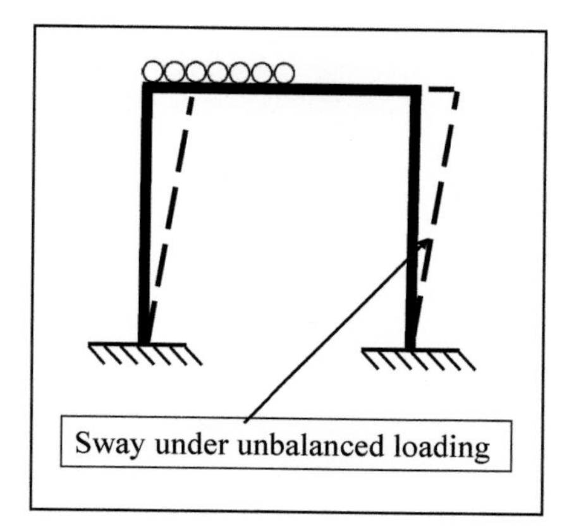

Figure 4.4 Frame sway

The test of overturning is *not* one of controlling uplift on any component although that might be relevant for other reasons. In a foundation subject to axial load and overturning (Figure 4.7) there is a requirement to ensure overall stability but that does not equate to ensuring that there is no 'tension' on the underside of the base.

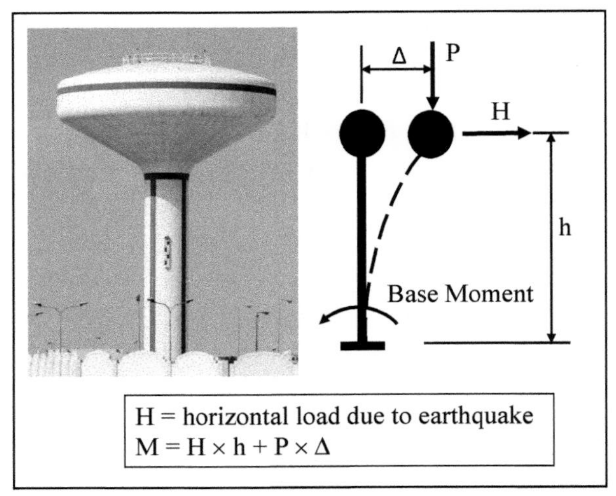

[Left] Figure 4.5 Effects of excessive sway (in the photograph there is a heavy mass right at the top). (Dreamstime 64043914)

[Below] Figure 4.6 Rebar cage stability (courtesy of Temporary Works Forum)

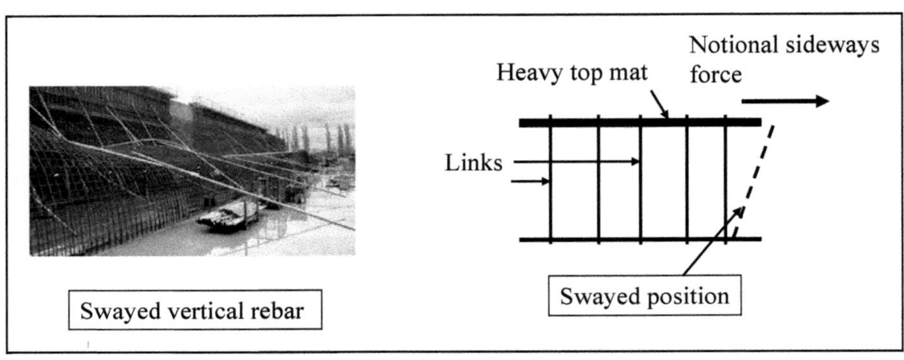

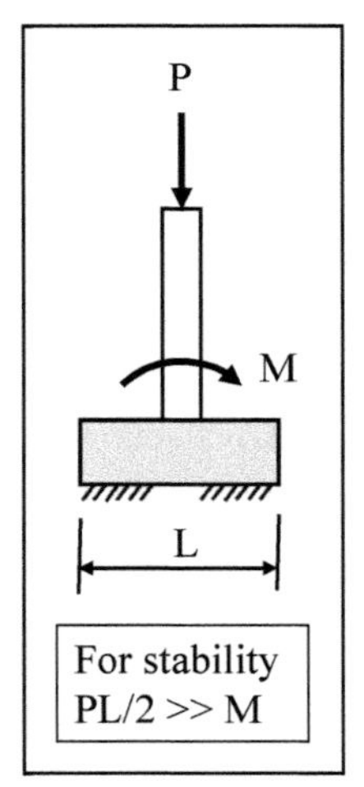

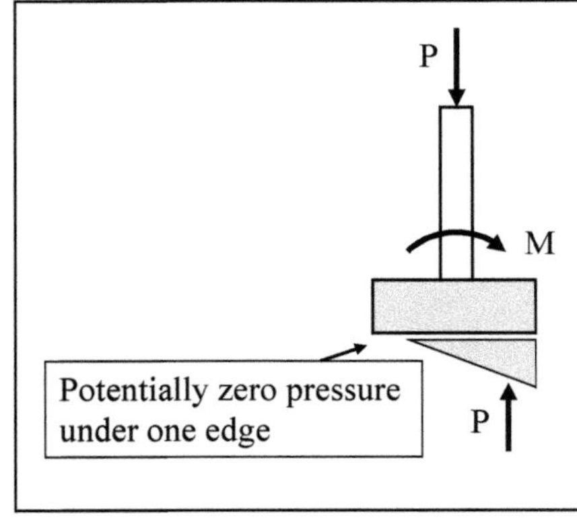

[Left] Figure 4.7 Foundation stability

[Right] Figure 4.8 Significance of 'zero pressure'

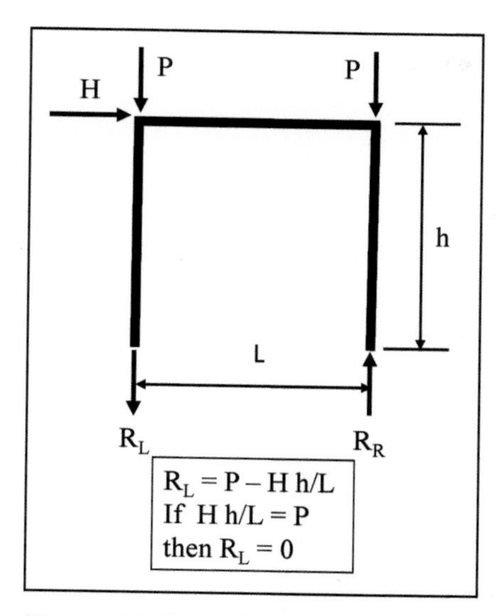

Figure 4.9 Criterion for stability in frame foundation

Indeed, it is quite common, as in Figure 4.8, for there to be zero contact pressure under parts of a spread foundation. Whether this matters depends on circumstance. In chimneys, where the wind base moments may frequently fluctuate, it can be detrimental for there to be zero pressure if this leads to ground pumping below the base, but in general that is not an issue. Even so, care is required in any arithmetic test. Figure 4.9 shows a portal frame where the force in one leg is zero under the applied loading. If, for example, the test of stability had been defined as requiring a foundation to take (say) twice the maximum uplift force (which looks eminently sensible), the test would have been wholly inadequate (the foundation weight would be zero) since the structure is actually unstable in the sense that any marginal increase in the horizontal force would have permitted the structure to overturn.

Nonetheless, a check on overall stability considering only global overturning resisted by global gravity load is only truly sound if the entity can act as a single unit like a solid block. It is easy to get confused about this and to overlook the role of foundations simply as a large mass to anchor the structure down locally. Figure 4.10 shows a lean-to portal founded on rock (with very high bearing pressure capability). Looking solely at bearing pressures, the demand on the foundation is small. However, a wind suction case would only be resisted by substantial anchorage for which a heavy foundation is required. (In recent years, child deaths have occurred when inflatable rides have blown away or overturned.) Similarly, single column foundation design is frequently governed by overturning stability. In the temporary construction case, foundations must be big enough to resist the overturning of columns directly after they have been erected (as vertical cantilevers) and before the frame is complete (neglecting this de-

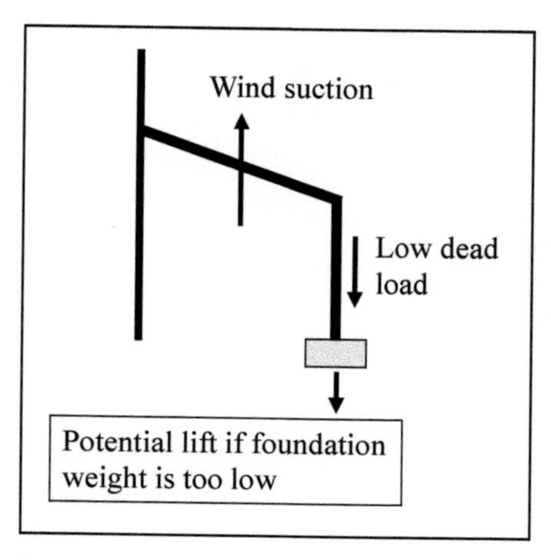

Figure 4.10 Role of foundation mass in providing stability

sign case has been a feature in lamp post overturning). A relatively modern trend has been to supply double height columns to speed erection and save a single storey splice, but the added dangers of temporary instability and implications for foundation sizing are obvious.

In checking stability, a critical case might occur if the gravity load is less than assumed and the horizontal load more than assumed. Codes give advice, but it is important for designers to be conversant with the nature of the structure and its use. A not infrequent failure has been flotation of basements as ground water rises during construction, i.e. before the basement is weighted down. Flotation is a form of instability failure. Common cases arise in civil engineering when large pipes are laid in the ground. A potential vertical stability failure exists if the excavation floods before the (empty) pipe is weighted down by backfill/surcharge.

Sensitivity

The objective of assessing global stability is to avoid toppling under defined horizontal forces. A sensible additional check is to ensure that overturning (a gross change in state) cannot occur under a *marginal* increase in the applied load. Self-evidently this is important for tall structures where it would be extremely unfortunate if a wind force slightly larger than the presumed design value caused a gross change in state.

Great care must thus be taken with destabilising loads that might be in serious error due to a small change. Wind and water pressure are two examples. Wind forces are proportional to the square of the wind speed, so variation in wind speed prediction can be critical. This applies especially for temporary load cases. Say the wind is presumed to be 12 metres/sec and there is actually a gust of 15 metres/sec. Subsequently the forces then rise by $(15/12)^2 = 1.56$, a magnitude which can easily exceed any load factor. There have been cases of temporary structures being blown away or roofs ripped off whilst under construction because this effect has been overlooked. In a similar manner, the water pressure on the back of a retaining wall is height × density (γh). The force is $\gamma h^2/2$ and the wall root BM = $\gamma h^3/6$. If a wall is 1 metre high with freeboard of 100 mm then the maximum pressure ratio may increase by 1.1, the maximum force ratio may increase by $(1.1)^2$ and the maximum BM ratio by $(1.1)^3 = 1.33$. However, if the freeboard is 200 mm and is realised in practice, the BM ratio error can be $(1.2)^3 = 1.73$ – a serious change. Conversely, if a wall is 10 metres high with freeboard of 200 mm, the worst error ratio is only $(10.2/10)^3 = 1.06$. The point is to understand the *sensitivity* of potential instability from small errors in design assumptions and to design prudently. It is a legitimate test to question: what is the worst that could happen? In one sense, providing for water overtopping is the equivalent of providing a safety valve.In hazardous industries, where exceptional reliability is required of structures subjected to extreme loading, formal studies known as *cliff edge studies* or *margins' assessments* are carried out to inform about sensitivity. The objective is to assess the structure looking for a gross change in state consequent on some slight change, say a wind load marginally higher than assumed. This ensures that the structure is not at some 'cliff edge' point when a slight change in design assumptions would precipitate calamitous failure.

4.2.3 Load path

A requirement in any stability check is to define the horizontal load path to ground. There is always a need for a competent path for transmission of horizontal loads from their application (at height) through the structure down to foundation level. Thereafter, it needs to be ensured that each structural component in the path chain can carry that load. In some structures the path is obvious. For example, in Figure 4.11 (standard portal shed) the load path for gable wind is via the eaves horizontal bracing to the eaves members then down the wall bracing to ground. However, in the same figure there is a wind load on the roof central chimney and no obvious structural route for that load to pass back to the side bracing system. In modern buildings, the route to ground for horizontal loading may be quite complex, involving transmission through floors as horizontal diaphragms and so on. A danger exists in that all these disparate elements may be in different work packages and there may not be unity of understanding of the roles. That danger might be most acute during construction when the load path may be incomplete or during alteration/demolition when a key part of the load path might be inadvertently removed. Figure 4.12 illustrates a pair of semi-detached houses whose owners desired the back walls to be altered to insert patio doors. The architect had proposed removing the rear wall and inserting a beam to carry the vertical loads back to the side walls. But in this proposal, what stops the whole combined structure swaying sideways? Figure 4.13 shows a cellar conversion. The contractor removed the basement floor and excavated, whereupon the bottom of the basement wall kicked in sideways: the builder had not appreciated the floor was propping the basement wall and the whole structure above collapsed. As always, it must be quite clear what holds what up and it must always be presumed there is a horizontal force and a need for a competent load path transmitting that force back to ground. Some of the examples in Chapter 8 illustrate what can happen when load paths are not adequately thought through.

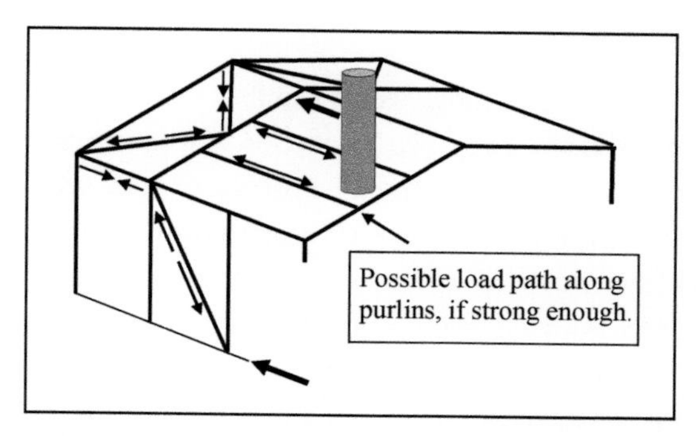

Figure 4.11 Load path to ground via bracing

Defining the load path for the permanent condition is a demand on all designers. That path needs to be recorded so that future alterations do not inadvertently destroy the assumptions. A case is illustrated in Figure 4.14. The lift

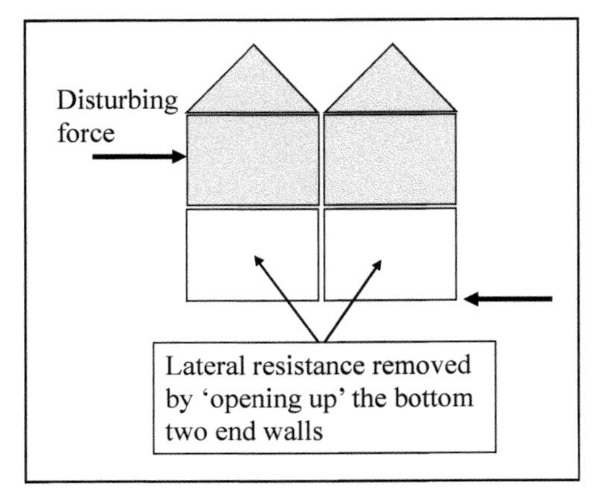

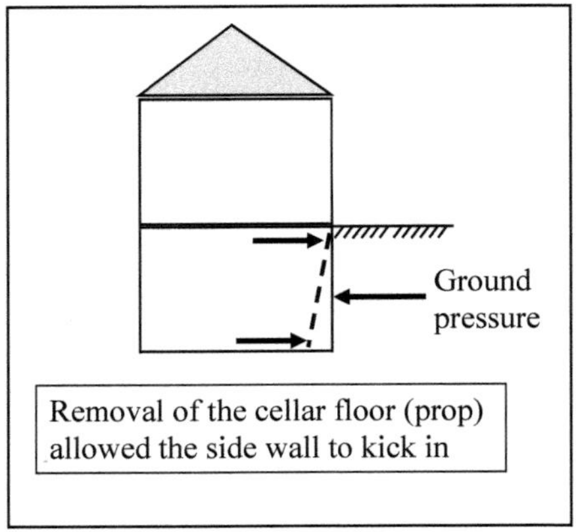

[Left] Figure 4.12 House stability

[Right] Figure 4.13 Cellar stability

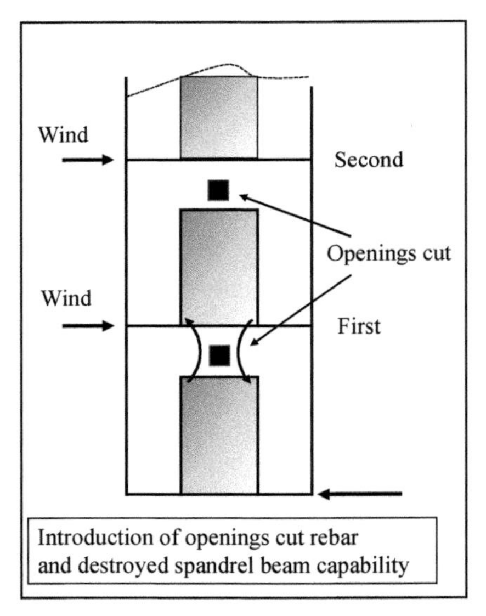

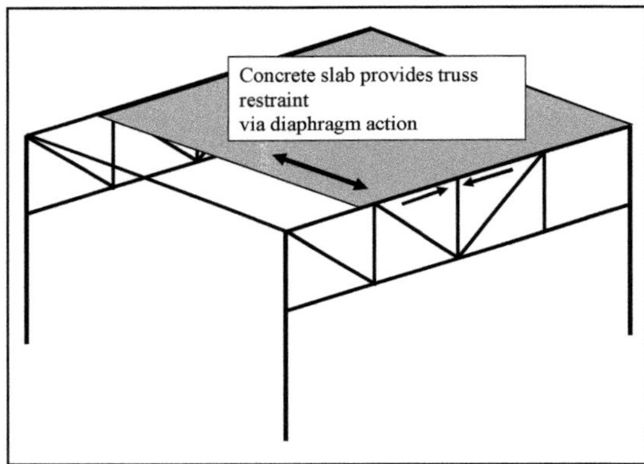

[Left] Figure 4.14 Destroyed lift shaft resistance path

[Right] Figure 4.15 Stability during construction

shaft was the main stability system for the entire office block. A lift supplier was engaged and required an indicator box to be cut out over the doors. To insert this, the contractor cut through the lower rebar (thinking it was a trivial beam member) at every floor level in the building. In doing so he destroyed the ability of the spandrel beam to transmit moment and thereby undermined the whole structural integrity of the building. A difficult case also arises in construction. Figure 4.15 shows a tall frame where lateral stability of the deep roof trusses was to be ensured by in-plane action of the cast concrete diaphragm. The deck was to be cast on a slope. During construction there was a lateral destabilising load associated with the wet concrete, yet at that stage, the trusses had no provisions for stability. The required construction check is one for an unsupported truss with destabilising load: fine after the concrete has hardened, but what about the intermediate stage? Figure 4.4 shows that when casting concrete on a deck it is not even essential that the deck is sloping to create horizontal loading. Any portal frame with an applied load just on one side (consistent with phased concrete pouring) will tend to sway sideways under the elastic load conditions created. The moral is there is always a horizontal destabilising load from some source, so all structures need to be checked for lateral stability (using some horizontal load) during the construction phase, i.e. when they may well be incomplete.

4.2.4 Role of stiffness

Stability is normally ensured by checks on strength or overturning. But as mentioned previously, it is evident that stiffness is also a factor. A very flexible structure can sway sideways and add to instability by introducing P-Δ effects. In the standard check on lateral stability of compression members, the value of the restraint is 2.5% of axial load as in Figure 4.16. Although this is couched in load terms, the real demand should be one of stiffness. If the restraint were provided by a super-strong long piece of wire, that wire would do nothing to ensure that the strut was maintained in a straight line (i.e. to contain P-δ instability) simply because the wire would stretch so far under the 2.5% load that δ would be uncontrolled. In contrast, a short stubby piece of material (stiff but weaker) would be a far more effective restraint. A practical case of this exists when frames are stabilised by crossed bracing wires as shown in Figure 4.17 and here the *length* of the load path is a feature: too long, and the stability system just becomes too flexible to act effectively.

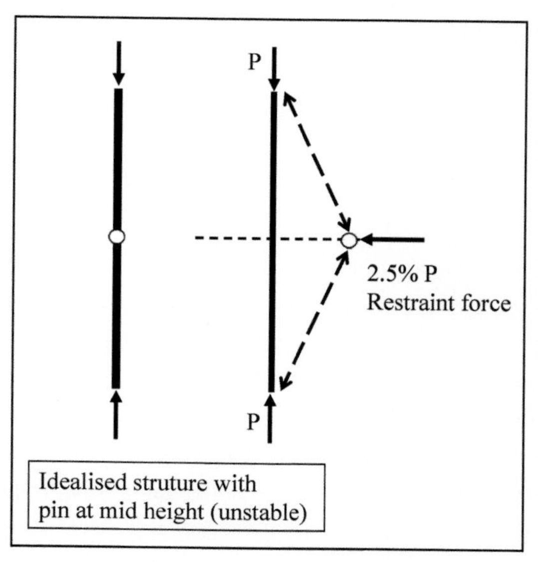

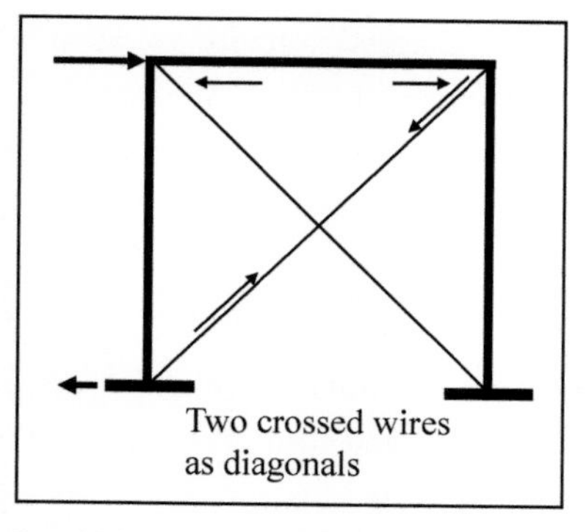

[Left] Figure 4.16 Restraints against axial buckling

[Right] Figure 4.17. Length of load path

Consideration of P-δ and stiffness are relevant to the design of individual members. The idealised member of Figure 4.16 (strut with pure pin in centre) is unstable as any slight displacement of the alignment would create an unsustainable bending at the mid height pin and the strut would collapse. For this reason, any splices at strut centres need to be able to carry not only an axial load but also the presumed mid-height moment consistent with strut action, and they need to be stiff.

4.2.5 Summary

Failures by gross instability occur and have the potential to cause total devastation. Perhaps the time of greatest danger occurs during construction (or demolition) when items that will be stable when incorporated in the final works are necessarily unstable during the assembly process. Individual steel columns and precast units may be vulnerable. Cases are reported of stored items, or walls, just toppling over and these have caused death. Anything 'top heavy' represents danger and lateral sway of wet concrete on shuttering (as a heavy mass high up) has occurred much more than once. Safety checks require the mental process of presuming there is always a lateral load on anything in both orthogonal horizontal directions and then making sure there is an adequately strong, and stiff, load path back to ground. During construction, if anything can move, it will, so great care is required.

4.3 Ductility

4.3.1 Introduction

The summary of Chapter 3 argued that for structures to be insensitive to strain-controlled stress, and for them to be able to redistribute peak gravity-induced stress, they require ductility. Significant safety benefits accrue where elements are ductile rather than brittle, irrespective of stress levels. Accordingly, it is worth examining *ductility* in more detail.

4.3.2 Elastic stress and redistribution

The fundamental dilemma of stress and displacement was aptly illustrated by Hambly in his three-legged stool paradox (Heyman, 1996). Hambly contrasted two seating stools, one with three legs and one with four, questioning what load was in each leg (Figure 4.18). For the three-legged stool, the answer can be W/3 (i.e. assuming the applied load is centroidal to the three legs). In contrast, a four-legged stool is statically indeterminate, so the distribution of load depends on the seat stiffness. More than that, the load in any of its legs depends on the support assumptions. Hambly postulated that one leg might not quite touch the ground. In which case, the force in opposite leg could be computed as zero with the force in the remaining two legs as W/2. An elastic analysis in full contact

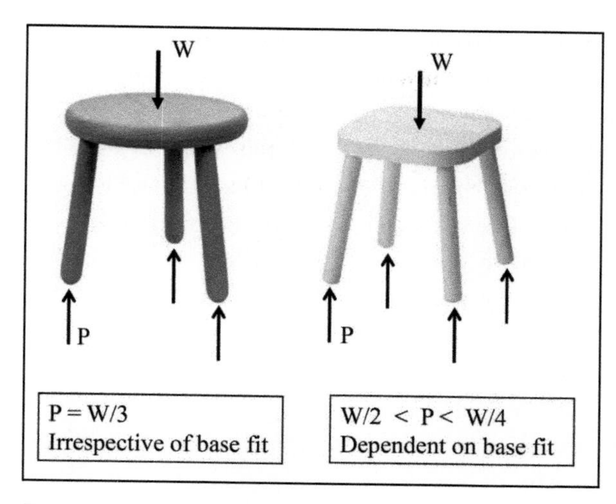

Figure 4.18 Hambly's paradox

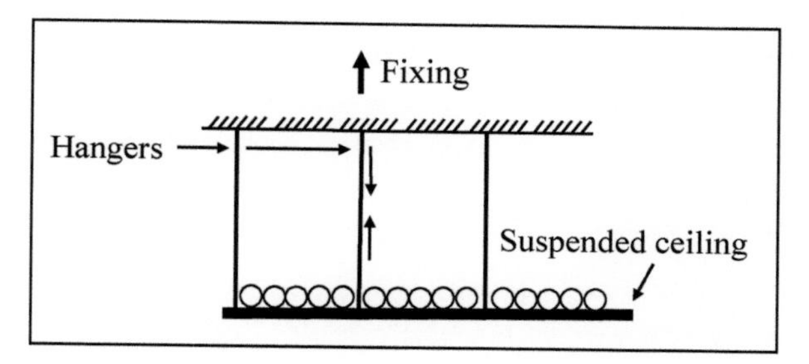

Figure 4.19 Tension structures

would give W/4. The point is that the load in a four-legged stool can be computed as worse than a three-legged stool and in a four-legged stool, the actual load can be anywhere between W/2 and W/4 depending on the boundary conditions and on the seat stiffness. In practice, if the seat (or legs) are ductile, eventually leg loads will reduce to W/4 which most designers would presume. Overall, it is not difficult to devise mathematical models which predict unrealistic elastic forces simply because the boundary conditions are inaccurate. All this reinforces the case for looking at structural performance at ultimate conditions and for realising the enormous benefits, indeed the actual reliance, on ductility and the capability of structures for realistic load redistributions. It is that property of ductility which largely governs their safety.

When a structure is in tension (Figure 4.19) the same arguments might apply that ductility will allow a redistribution of loads if any one hanger is overloaded. This is one possibility for overcoming the reality that loads in a series of hangers can differ if the source of the loading is stiff when spanning over several supports as described in Chapter 3, Section 3.3.4. But a hanging system fundamentally differs from a compression system in that key reliance is placed on hanger terminations. These must be stronger than the hangers if the benefits of ductility are to be realised. Where this is not the case, and a hanger fails, a cascade-type failure process can be set up. Examples of these are described in Mann (2002).

4.3.3 Available ductility

A fundamental weakness in the argument about relying on ductility is that the amount of ductility presumed must really be available (and this is rarely calculated). That ductility might be expressed as an extension or as a rotation in a beam subject to a sequential load redistribution process. A second demand is that the required amount of ductility must be tolerable as a design criterion. With regard to the first demand, the quality of available capacity in plastic hinges is termed *rotational capacity*. In a pure piece of steel there is considerable capacity, for under the right circumstances a piece of steel (such as rebar) can easily be bent through angles greater than 90 degrees. Nonetheless, that capability can be restricted by bending through too tight an angle or by, say, welding the steel and thereafter cooling it too quickly, leaving it in a hardened, brittle, condition. In many practical steel connections, there is a limit to the amount of rotational capacity available, linked to local buckling of outstands, excessive strain applied to connection bolts or via premature fracture of welds. There are detailing rules available for steel connections which ensure adequate rotational capacity in structures of normal dimensions (Steel Construction Institute and British Constructional Steelwork Association, 2013, 2014).

Likewise, in concrete structures, the amount of hinge rotation available across a reinforced section is constrained by the detailing and certainly the section overall must be under-reinforced (an example of controlling failure mode) if any significant ductility is to be achieved (Beeby, 1997). In general, there is a significant link between the manner of detailing concrete structures and the ability of those structures to behave as desired in overall analysis. If the detailing is inadequate, unsafe presumptions will be made about behaviour (Taylor and Clarke, 1976; Somerville and Taylor, 1972). A typical example of the relationship between detailing and capacity is that of the performance of opening corners (Jackson, 1995) and known weaknesses with bridge deck half-joints (see failures in Chapter 8).

The demand for hinge rotation in terms of θ is also complicated; an example of calculation is given in Figure 4.20. Any redundant structure being pushed towards

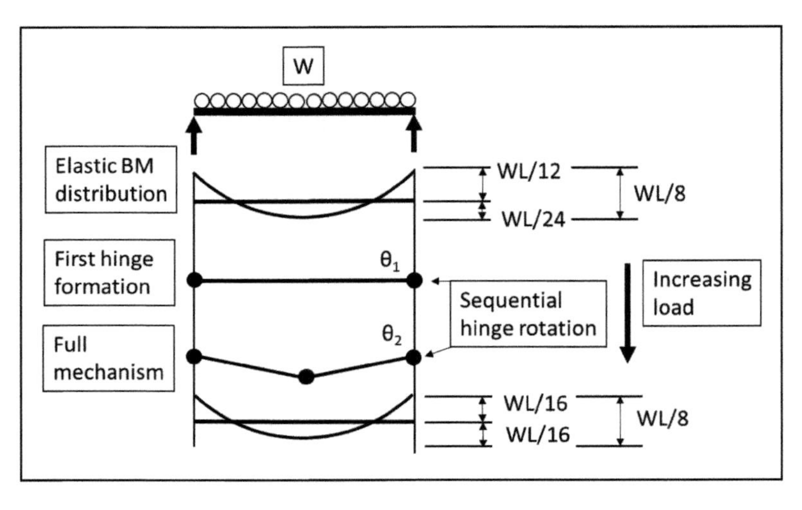

Figure 4.20 Successive hinge formation in plastic collapse.

its ultimate load will experience a successive generation of hinge formations with the highest rotational demand normally imposed on the first hinge to form. Clearly it is not at all easy to evaluate the rotational demand in complex structures. The essential point is that in safety terms, it cannot be automatically presumed that sufficient ductility or rotational capacity exists within the hinge zones to justify full attainment of the applied load linked to generation of a complete mechanism (the so-called *push over* load).

There is also a limit to the amount of *useful* ductility that ought to be claimed within design which is the second demand mentioned earlier in this section. Concrete slabs can be designed by the presumption of a yield line mechanism and this method works well for slabs of normal proportions. But on a very wide slab, the amount of sag required to generate hinges may well be such that the slab has really exhausted all its useful strength in terms of being a usable structure well before the initiation of yielding in the last hinge to form. Thus, although the yield line mechanism might be a backstop against total failure, it is hardly a useful or reasonable design strategy in larger slabs. Generally, there must always be a check to ensure that any engineering theory being used lies within the limitations of its validity and that there must be no violation of fundamental assumptions. Caution is required in dealing with any structure of larger scale than the norm (SCOSS Alert, 2018).

A property linked to rotational capacity (as in Section 4.3.3) is the amount of energy absorption capability embodied within the system. As shown in Figure 4.21, this is a function of the area below the load displacement curve and is governed by the length of the yield plateau. Energy absorption is the property desirable for adequate performance of the crash barrier post described in Chapter 3 and for structures which must sustain blast or earthquake loads. Indeed, it is that property which provides overall 'robustness'. If a structure has energy absorbing capability (high ductility), it will normally deform a finite amount (and remain standing) under any overload, rather than 'collapse'.

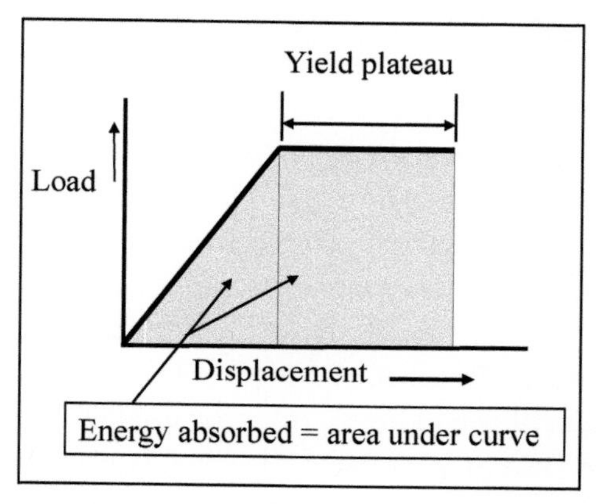

Figure 4.21 Energy absorption during plastic deformation

4.3.4 Limitations

Not all structural problems directly benefit from having ductility. In buckling problems, the real stress distribution matters (as does the residual stress distribution). In fatigue and fracture problems, the real stress distribution also governs performance. Nonetheless, and overall, having a ductile system is always a plus.

It may seem that one design solution to avoid these complications is always to specify a really strong material and rely on that. But this too has drawbacks. First, high strength may be gained only at the expense of ductility loss and that is a poor bargain. Second, in some applications, it is highly desirable that certain modes of failure are ensured and kept below a certain value. Situations crop up in car crash barrier design, in blast-resistant design and in earthquake engineering. In those circumstances, having over-strength steel or concrete can be a positive disadvantage and may be dangerous because in such cases, the safety argument often relies on confidence of a preferred form of structural failure under overload. This can be seen quite easily in the example of the crash barrier post (Chapter 3, Figure 3.1). On impact with a post, the objective is for the post to bend at the bottom and absorb the applied kinetic energy by a plastic hinge rotation. For this strategy to work, the base connection must be stronger than the post in bending otherwise the post will simply pull out of its anchorage. Hence, unwittingly putting in stronger post material to raise Mp (the plastic hinge value) could be detrimental since that would change the mode of failure from energy absorbing to brittle and the car may not stop, which is not the desired outcome. What is required is a controlled stop. In this mode of performance, it is also clear that the elastic stress endured under the design condition is really of no interest. All that is of interest is that members can bend, can develop a plastic hinge and have sufficient rotational capacity to bend considerably at semi-constant moment capacity. The connection must be able to withstand the forces generated by the bending of the post.

Of course there are also limits. In steel tension members, code rules effectively ensure that premature physical failure cannot occur across net sections before yield spreads across the whole member (Figure 3.12). Thus arithmetically, there is a maximum ratio of ultimate tensile strength (UTS) to yield stress that can be exploited. Although rarely explicit, there is a safety need to maintain a margin between yield strength and UTS so that a zone of useful plasticity can be preserved (the same principle applies in concrete). This is perhaps most obvious in bolts where grade 8.8 is common (yield/proof 80% of UTS). Grade 10.9 can be useful (yield/proof 90% of UTS). But use of stronger, higher-grade, bolts is frowned upon as they may be potentially too brittle for structural applications. Similarly, very short bolts, which lack global elongation capability, need to be treated cautiously and codes impose minimum grip lengths on pre-tensioned bolts to avoid the danger of them snapping during the tightening-up process. At the opposite end of the spectrum, members (such as cables) may be so long that yielding merely produces significant amounts of stretch and the members can hardly snap physically because the extensions involved to fracture can never be realised. In reinforced concrete design, rules on detailing ensure that, under overload, failure preferentially takes place in the rebar rather than the concrete so that a measure of ductility is ensured. The rebar itself will be highly ductile: it has to be, being bent beyond yield to form shapes even before placing within the concrete. Caution is required in trends to produce stronger and stronger concretes if these do not possess the strain capability to allow stress redistribution.

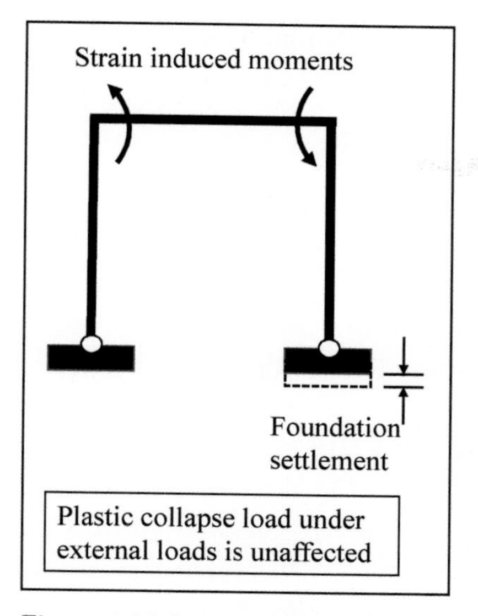

Figure 4.22 Strain-induced effects

As with pre-bent elements, the general presumption is that stresses due to settlement and thermal effects and lack of fit are 'strain controlled' and as such cannot cause global collapse and so can be ignored. Academically this is true. Given a form such as that shown in Figure 4.22, by the bound theories (Chapter 3), any set of pre-existing self-balanced stresses in the structure will not modify the plastic collapse load, though they will alter the load deformation curve en route to collapse. So, for this situation, settlement stresses and thermal stresses are justifiably neglected. But there are exceptions. The logic only applies to frames in flexure; the same argument does not apply to struts and bracing sets. For, as we have seen previously (Chapter 3), residual stresses do affect column behaviour and it might therefore be argued that compressive strain-controlled stresses will have the same effect. For practical designs in ambient conditions, any reduction in strut capacity may be enveloped by the standard safety margins particularly given the low probability of maximum thermal stress coincident with other loading. But this does not follow for the high expansions linked to heating from fire; and in fires, reductions in member capacity require consideration.

4.3.5 Summary

The concept of ductility does not usually feature within the standard numerical design process so an appreciation of its contribution to safety can be hidden. The reality is that the protection we enjoy from the many uncertainties governing stress levels is dependent on having some ductility within the structural system. Safety is prejudiced by brittleness. It is therefore important to ensure adequate ductility by appropriate detailing and construction methods. Safety is promoted by making sure that, under overload, the consequence is excessive component distortion, not fracture. In routine design work, no explicit assessment of ductility is carried out, it being assumed that compliance with code or industry standard detailing rules will suffice. In design for hazard conditions such as impact or earthquake, more stringent detailing rules exist in relevant codes and these are defined to ensure appropriate levels of ductility are built in. In complex structures relying for integrity on a sequential set of plastic hinge development (steel or concrete), it might be necessary to compute the theoretical joint rotational capacity demands and ensure these can be accommodated.

When appraising existing structures (perhaps where elastic stress levels are high) an appreciation of the role of ductility can be useful in assessing a structure's potential safety.

4.4 Robustness

Chapter 2 introduced the concepts of robustness, citing unease about highly optimised structures which have no capacity reserves. Such structures might meet the letter of a code but are not necessarily good designs. Section 2.1.1 of that chapter noted a description of much Victorian engineering as being built to last and capable of taking a good deal of punishment. The paragraph also noted that whilst experienced designers can intuitively make structures 'robust', it is a quality not easily defined and thus hard to implement formally, certainly via equations, even if building regulations demand it.

A good example of a structure with deliberate in-built robustness is the Second World War Wellington bomber frame designed by Barnes Wallis (Mann, 2015). The bomber's structure was designed as a highly redundant geodesic frame to give it both strength and high survivability during attack. The strategy certainly worked (Figure 4.23). Another example is the configuration of the Ribblehead Viaduct as shown in Chapter 1, Figure 1.2.

Figure 4.23 The value of 'robustness'. (Imperial War Museum CH 009867)

In May 1968 came the infamous collapse of the Ronan Point block of flats in London (Chapter 8) which thrust concepts of robustness to the fore. After the Second World War there had been a great need to increase the housing stock and one way of doing this was by building high-rise blocks of flats using industrialised methods. Many construction companies produced their own systems which were essentially a kit of parts – precast elements ready for site assembly. That morning, there was a gas explosion in a kitchen on the 18th floor of Ronan Point. The pressure generated blew out a wall panel and in doing so, support for the floor above was removed; there followed a spectacular progression of failures to all the storeys above. Following this incident, the professions grasped more urgently than ever the need to consider hazards and their consequences and they grasped the need to impose concepts of 'robustness' into design practice as a complement to considering strength and stability. Building regulations were changed to reflect this demand. The regulations state that: 'the building shall be constructed so that in the event of an accident the building will not suffer collapse to an extent disproportionate to the cause'. In the investigations into Ronan Point (Box 8.9), much was discovered about shoddy workmanship and that reinforced the message yet again that safety as a concept cannot be achieved by provision of numerical design margins alone.

Approved Document A (HM Government, 2004) provides one means of meeting compliance with this key regulation in the UK. This document categorises buildings into classes 1, 2 and 3 and provides guidance on what types of buildings are to be included within each class. For example, Class 1 structures are low-rise domestic structures whilst Class 2 covers higher-rise structures. Class 3 structures are unusual ones, where, say, many people may be at risk if the structure fails. The guide then suggests notional rules for tying structural components together as an empirical contribution to 'robustness' plus it offers alternative design strategies for satisfying the robustness demand. In doing this, it essentially recognises a link between risk and safety. More attention is paid to complicated structures where more people are at risk if the structure fails and less attention is paid to low-rise domestic structures where fewer occupants are at risk. There is also an essential link between cost and risk, more money being assigned to limiting the risk of failure for structures where the consequences are highest.

Although the intent of the regulations cited above is clear, it is not immediately obvious how that intent should be implemented. The words state a general principle which is perfectly understandable. But the wording is non-prescriptive, leaving designers uncertain as to precisely what they are to do, especially, and perhaps surprisingly, for higher-risk structures. What sort of accident needs to be considered? Is it a typical occurrence like a gas explosion? If so, what pressures should be considered? Or is the accident something much more severe like an aircraft impact? And then, what is the interpretation of *disproportionate*? Was the collapse of the World Trade Center (Chapter 11) disproportionate to the cause of impact from a hijacked jetliner? Of course, the responses must all be subjective but even so have a profound effect on safety and a significant effect on the work designers must do in satisfying safety demands. To assist design engineers who have run into circumstances where interpretation has proven hard, the Institution of Structural Engineers has produced two reports providing authoritative advice (Institution of Structural Engineers, 2010, 2013). The Eurocodes also include regulations on robustness and highlight that achieving robustness is a key structural aim.

Overall, designers need to apply some common sense. It is foolish to skin certain elements down if their failure, from causes that cannot be fully forecast, results in injury or significant commercial loss. The nursery rhyme reproduced in Chapter 1 (Section 1.6) says it all.

4.5 Reliability

4.5.1 Reliability and uncertainty

Ultimately all engineers are highly interested in having certainty that their structures will perform as intended: we want them to be *reliable*. In civil/structural engineering we want to be sure that loads applied to a component will be carried without failure and we want to be sure that structures will not fail in service. Since it is impossible to load test everything (certainly to overload conditions), the design/construct process must be configured to give high confidence in outcome prediction. Basic uncertainties have been discussed in preceding chapters. Academically, since there are uncertainties, whether a structure is reliable enough can only be couched in terms of probability. However, for all practical purposes, supported by long experience of the process, we can be fully confident in the capacity of most building components provided standardised rules and good practice are followed. Nonetheless, intuitively, most engineers would consider that there ought to be more confidence in the reliability of nuclear reactors than domestic lintels. Techniques are available to ensure this (Chapter 13).

In some branches of engineering, the concepts of reliability have a more obvious role than in structural engineering. We know, for example, that most of the time cars start when the ignition key is turned, but not always. We know that most of the time when we switch on a domestic lighting circuit, the light will illuminate, but sometimes not; either because the bulb has failed, the switch has failed, the fuse has failed, or the main lighting circuit has failed or there might just be a loss of power. Intuitively we might expect the likeliest cause of failure to be the bulb or the fuse. A lighting circuit might be thought of as a *system* in which each component has a probability of failure and if the reliability of the various components is known, the probability of an entire circuit failure can be assessed. The raw data of component failure can be assessed from information known as the *mean time to failure*. Intuitively we can reduce the odds of failure by simple means using the concepts of *redundancy* and *diversity*. For example, modern cars have dual braking circuits in case the pipes of one corrode and the odds of in-service failure are reduced even more by imposed inspections which ought to detect corrosion. We could reduce the risk of a light system failing to illuminate 'on demand' by having two bulbs, or even better, two bulbs each fed by separate circuits.

4.5.2 Redundancy and diversity

Such concepts exist for structures partly by the *redundancy* of ensuring different load paths. The concept of *diversity* is hardly used in structural engineering, but the idea might be useful. In mechanical engineering, there can be two identical items performing the same function and the chances are that both will not fail at the same time. However, as they are identical items, they might fail by some *common cause* failure mechanism or might both have the same manufacturing fault. If instead of having two identical mechanical items there is, say, one mechanical one and one electrical one, the systems are then both *redundant* and *diverse*. Hence the odds of failure are much reduced; probably one of the two will work. We might use this concept in checking work where the probability of 'not detecting a mistake' is reduced if checks are carried out completely independently by separate persons in different ways. It is one level of checking to verify a set of calculations. It is a better check to do them independently and see if the conclusions match.

4.5.3 Material reliability

In any discussion on *reliability* a fundamental topic is our reliance on materials. We almost take it for granted nowadays that materials will come with set of defined properties. A large part of the history of engineering relates to both standardisation and the development of test types and testing regimes that have given designers confidence in material properties for the design phase. There has been a steady improvement over the years, yet we still have a bewildering variety of material standards on offer. Some known failures are linked to the use of inadequate materials, to faulty procurement or faulty application (Chapter 7). Other failures have been consequent on human error (simple mis-ordering) or even falsification of records (see Chapter 6).

4.5.4 Ensuring material quality

In buildings, we have to live with the fact that certain key procedures require human control with all the risk that entails: laying brickwork, mixing and placing concrete and welding steel are all basic activities critically affecting the strength of structures. The risks of inadequate workmanship are reduced in three ways: first, by requiring competence in the personnel; second, by executing the activities under a procedure that is known to work; and third, by sample testing. But whatever we do, we cannot test everything and certainly not test to destruction. Even the raw materials we buy are only sample tested.

Nonetheless, provided the materials are certified to a standard by reputable companies, the risks are miniscule (but see also risks described in Chapter 7). The 'invention' of QA as a concept was intended to set in a place a set of procedures which, if followed, would ensure the quality of the output.[1] Quality control merely checks that the product meets its defined guidelines when tested after production. Whilst QC is essential, it is in one sense 'too late'; the objective is not to have a problem in the first place (Chapter 13).

Confidence in material properties clearly affects the selected value of a load factor. Material testing is bound to show a scatter of results and this is managed by using concepts such as the 95% confidence level, which often defines a material's characteristic strength.

4.5.5 Component reliability

Reliability in structural engineering (under static loading) in terms of components can mostly be thought of as the probability of coincidence of a severe overload combined with a material at the tail end of its strength probability (Figure 3.2). Even if that should happen, we anticipate that the safety factor or load factor will still ensure avoidance of failure and by common experience that is so. Component reliability is also ensured by batch testing of manufactured components or, say, concrete cube strength testing. Raw products will routinely be supplied with a manufacturer's certification of factory testing. When faced with the difficult task of assessing existing structures, it pays to assess the likelihood or uncertainty in the many variables that govern safety. Risk is reduced if the state of existing materials is thoroughly examined and tested. In all structures there are degradation mechanisms (see Chapter 7). For example, there have been problems with concrete losing strength over time, with rebar corroding, with structural steel corroding or any manner of degradation causes linked to water. Components may also fail over time via fatigue damage which results from multiple applications of stress. The causes of fatigue damage are not always obvious (see Chapters 7 and 8). Consequently, long-term reliability is linked to regular inspections. Any structure supporting an item that is moving should be periodically inspected. Ensuring 'inspectability' is part of the design process (an inability to detect early degradation signs has been a cause of failure: see Chapter 7). Items such as bridges are routinely inspected, as are cranes and crane support structures, fairground rides, and any lifting device structure must by law have periodic inspections and certification (cf. Lifting Operations and Lifting Equipment at Work Regulations 1998 (LOLER); HSE, 2013). Some of our worst train crashes have come about through inadequate track inspection. Part of the design process for high reliability is to make sure structures are insensitive to minor flaws and ensure that there is a good chance that any initial defect or degradation (say a fatigue crack, or some

1 Modern concepts of quality assurance were adopted from Japan, which showed during the rise of their economy that it was possible to consistently produce better quality products than the West. Thus, Japanese cars became a byword for quality and reliability, performing consistently better than foreign competitors.

corrosion or material degradation) is detectable via inspection before it has propagated far enough to pose real danger.

4.5.6 Product reliability

Many of our building components are not just materials but products. Examples are bricks or steel beams, bolts, rebar and so on. Specific products might be items such as proprietary purlins. Although the strength of all such products can be calculated, load capacities are often more reliably justified by programmes of product testing, always provided any assumptions in testing the prototype are replicated in the production model.

It is extremely unusual to consider the probability of products being unreliable or how they might fail, at least in mathematical terms. But one example is the technique available to look at how products might perform under very uncertain loading conditions. In earthquake engineering, attempts have been made using *fragility curves*. A fragility curve is a graph which expresses the probability of product failure as a function of the loading. The idea is that rather than calculating, say, stress in a seismic event, a database is built up of how certain members have actually performed under different earthquake motions and that raw data is then used to predict likely component performance in future events.

Mathematical predictions are not always successful. In Chapter 6 there is a discussion on the new Dreamliner aircraft, where predictions were made over likely failure of its novel lithium ion batteries in service as one failure in so many years. Alas they don't seem to have been correct and there were problems.

4.5.7 How do we know?

A key question any designer should ask about any part of his design under execution is 'How do I know I have got what was intended?' As an example, a set of failures that formed a pattern reported to CROSS were hanger failures pulling out of concrete and allowing suspended ceilings to collapse. The whole safety of the system depended on the adequacy of the hanger fixing installation. It appears that no one proof-checked that the presumed pull-out loads could be achieved under site conditions and in the event, some could not. In any system where there is reliance within a load path on certain elements, the design/construction team need to be sure that failure risks have been minimised by imposing adequate testing and checking. Some of that is obviously judgemental and might be linked to the sensitivity of the system to potential variations in design assumptions. There is often a weak link, so it pays to identify it. A serious case of a construction differing from design intent was exemplified in failures of brick wall panels at some Edinburgh schools in 2016 (SCOSS Alert, 2017). In one failure, bricks weighing about 9 tonnes in total fell onto a path that was frequented by pupils. Fortunately, no one was hurt. The investigation revealed instances of the construction (wall ties) not being as the design intent and no one had inspected during construction to verify that what was being built was what was intended to be built. Post building completion, there was no way of verifying the internal status of the construction. All in all, there can be no high confidence in product reliability unless there is confidence in the build quality.

For some products, designers rely on industry standard QA schemes. Examples are the CARES scheme for rebar and Quality schemes for ready-mixed concrete. But thereafter, where workmanship is involved, procedural checks are required on the activity (as in *welder* qualification, ensuring personal competence) and on recognised procedures to ensure quality (as in *weld* qualification procedures; i.e. the process is known to work if executed properly) and supplemented by QC checks (such as non-destructive examination (NDE) of finished welds). In some circumstances, we already make codified allowances for execution class. Where there is more control, higher stresses can be justified. Masonry is an example. But overall, for any engineering product, from an aircraft to a bridge, there needs to be a thought-out plan that will give the designers overall confidence that the assumptions made during design have been realised.[2] Where components have significant uncertainty, and are critical, it is customary to test just for assurance. Pile testing is an example; designs predict that with certain lengths, in certain grounds, certain capacities will be realised. But true confidence only comes from sample installation and on-site test.

4.5.8 Assignment of load factor

In some codes, partial load factors are varied explicitly to take account of known factors affecting the probabilities. As an example, in masonry design codes, the partial factor for materials depends upon the workmanship (execution class) and quality control of blocks and mortar used. This approach recognises the overall lower probability of failure if workmanship is controlled. Partial factors are also varied in all codes for different load combinations (e.g. dead + live + wind) to reflect the lower overall probability of that combination of circumstances arising. It is important to be aware that some uncertainties do not have an identified partial safety factor, but are included within others, such as the accuracy of our models of structural behaviour.

Another reason codes might vary load factors is to offer a means of increasing reliability when the consequences of failures are high. A simple way of doing this is just to increase the load factor or to add in another partial factor with a value greater than one. Generally, this can be designated as an *importance factor*. Such factors are common in codes governing earthquake design (since earthquake design implicitly embraces a high probability of failure) (see Chapter 3, Section 3.6.1).

2 In some of the box girder collapses of the 1960s, the designers presumed the box girder diaphragms would align with supports. Because of tolerances, they did not.

4.5.9 Eurocode methodology

Use of an importance factor is a crude but simple device to use in practice. Eurocode 0 (CEN, 2002) discusses structural reliability generally (e.g. Annex B is 'Management of Structural Reliability for Construction Works'), plus it provides background on the selection of partial load factors and it provides a useful way of *calculating* a required factor based on the reliability required.

This approach was used in the London Eye spindle design (Berenbak *et al.*, 2001). The spindle is a single cantilever on which the whole wheel and passengers depend. It is thus potentially a significant single point failure item. Intuitively it needs to be reliable and on first consideration a large safety margin might be justified. However, there is a problem. The spindle is a very thick piece of steel and thus has two clear modes of failure; fracture and yield. Of these two, fracture is unacceptable, being potentially, and literally, disastrous whereas yield, whilst being commercially unacceptable, is unlikely to cause injury. A dilemma is that if steel were thickened to increase the safety factor (against yield), the risk of brittle fracture would be increased because thicker steel is more at risk (see Chapter 7; *Berenbak et al.*, 2001; Burdekin, 1999). To minimise the fracture risk, the steel needed to be thinner and to minimise the risk of yield it needed to be thicker. For safety and functionality, the spindle also had to stay elastic under working conditions. The reason is simply that the spindle supports mechanical bearings and for those to work adequately, the supporting structure strains had to be low.

The design objective overall was to provide very high reliability against fracture and this was linked to reducing the loading safety factor in a rational manner so as to minimise the steel thickness, which was part of the package for ensuring fracture toughness.

In terms of loading, the applied loads are dead, live and wind. The live load is insignificant, and the dead load was known with high certainty (not least since the rim components were check weighed); wind is not a significant component of the spindle stress so any potential variation on it would not cause a significant change in the predicted stresses. The spindle is a cantilever with loads applied through bearings on machined locations, so the stress prediction is simple and known with high confidence. The steel for the spindle was cast and heat treated so its properties were tailor-made and known precisely since samples were taken of the actual steel used rather than from a batch production run. The spindle was load tested to above its working load before use so ensuring confidence in capacity. These ingredients can be utilised within the Eurocode to calculate a safety factor. The code's methodology takes account of degree of redundancy, coefficient of variation and a reliability index within the calculation. It was this technique that was used to define a suitable safety factor for the Eye's spindle.

Carpenter (2011) discusses more generally the reliability aspects that underpin the Eurocodes by reference to BS EN 1990 (an opinion on this is given in Beale, 2010).

4.5.10 Statistical basis

Some engineers might be uncomfortable using statistical methods, although over the years they have gained more acceptability. We are all now used to concepts such as designing for the 1 in 50-year wind event and so on. In recent years, there have been so many flooding events that even the public has become familiar with concepts such as 'a 1 in 200-year event' or 'the worst storm in 100 years'. Those numbers graphically convey the rarity of events against popular understanding. There is nothing particularly significant about a 1 in 50-year event rather than say a 1 in 55-year event, it is just that a 1 in 50-year event has become accepted as a reasonable design basis offering a balance between cost and failure probability. Nevertheless, some rationality might be applied by noting that for wind, the 1 in 100-year event is not that much worse than the 1 in 50-year event. This does not automatically apply to all loading types. Earthquake loading is so unusual that it can only be considered on a statistical basis. Moreover, because the effects vary so much between 'every day' events and 'severe events', structural design is graded accordingly. It is customary to design for minimal damage for an 'everyday event' but to have varying degrees of acceptance criteria against possible, but rare, events. Thus, for high-intensity earthquakes that might be linked to a 1 in 1,000-year event, the structural acceptance criterion might be one of requiring the structure to just survive (albeit with permanent distortion) short of catastrophic collapse. In the UK we seem to have a flooding problem as more houses are prone to damage by occasional floods (Chapter 10). We can neither provide nor fund instant flood defences nor do we have enough land to relocate homes away from flood plains. The only rational way to deal with the problem is prioritising in terms of likely flood scenario (linked to storm probability) versus consequences. That prioritisation is used both in deciding what to protect and in deciding which scheme to implement first, since it is not possible to construct all schemes in any one year.

4.5.11 Statistical techniques

Some loadings are so uncertain that they can only be looked upon in statistical terms. UK tornado loading is one; it is very unlikely that any single building will be hit by a UK tornado (although these do occur, see Chapters 2 and 10), hence this is not a standard design case. A way to rationalise this ignoring of tornado loading, is to observe that within the stock of UK buildings, it is quite likely that one of them will be hit at some time but if it is hit, the damage cost is best covered by insurance. Likewise, it is most unlikely that any single building will be hit by an aircraft following mechanical failure, though it is not unknown for aircraft to crash into buildings (as at Lockerbie, Pan Am flight 103 in 1988, or the London helicopter crash of January 2013, in which a tall crane was hit). Despite the low probability of the event, because the consequences of aircraft impact are potentially so serious, the hazard is considered for certain structures where the consequences would be intolerable. We can also note that the nature of the hazard changes with time. Our skies are destined to become ever more crowded and anyone of us can witness the queues of aircraft lining up to land at busy airports with only minutes separating them. Although aircraft are very reliable, statistically one will fail and even Heathrow has had examples such as a bad crash in 1972 or a 2008 flight from China which landed short (the 1972 crash was a flight from Brussels and resulted in 118 deaths). Exactly where the balance is of failure probability accounting for increasing flight numbers versus increasing aircraft reliability is hard to say. Linked to that question is one of terrorism. Before September 2001, few would have considered even the remote possibility of deliberately crashing a large fuel-laden aircraft into a building, but after the New York Twin Towers incident (Chapter 11), we must now consider that for some structures that is a possibility. When dealing with difficult questions of loadings (say ship impact) that *might* happen, statistical techniques offer significant insights into what might or might not be a tolerable risk. Such approaches are used in probabilistic studies of risks to offshore installations or in assessing railways.

4.5.12 Low-probability events

Dealing with low-probability events raises intellectual difficulties. The event might happen, so what should we do? Where the consequences are trivial, we might just ignore the risk since unfortunately, as a society, we only have limited resources: we just have to accept that structures might fail, even with loss of life. We might, for example, ignore the risk of an asteroid hitting, even though it has happened.[3] Exactly what events we consider and what we ignore have arisen by custom and practice but is a source of real difficulty when dealing with failures which have high consequences, and in practical terms, nuclear safety is one of them. The Health and Safety Executive has published their thinking on this in *Tolerability of Risk from Nuclear Power Stations* (HSE, 1992).

4.5.13 Safety-related structures requiring extra reliability

In the nuclear industry, safety-related structures are designed to have a very high reliability against failure. The target is couched as a failure probability of 1 in 10^{-7} p.a. against a range of extreme loads to ensure there will be no radiation leak. Relevant loadings can be consequent on plant failure (such as blast or overheating) or they may be extreme environmental loads. Since the presumed normal failure risk of structures

3 In 2013, an asteroid described as equivalent to a 50 metre-long lump of rock approached close to Earth but was predicted to miss, and it did. Scientists calculated that asteroids that big might strike the Earth maybe once in 1,200 years. However, in 1908, a large asteroid did hit Siberia, where it levelled 80 million trees over an area of 830 square miles. About 66 million years ago, a six-mile-wide object smashed into the Yucatán peninsula in Mexico, wiping out most of life on Earth. The same NASA source reported that 'Basketball-sized objects come in daily', whilst 'Volkswagen-sized objects come in every couple of weeks'. A large meteorite, the Chelyabinsk meteor, struck in Russia in 2014 and caused quite a lot of structural damage. The event can be seen on YouTube. Hailstone damage is relatively common.

is 10^{-3} p.a., the numerical target of reliability of 10^{-7} p.a. implies the application of loads which have a probability of occurrence of 1 in 10^{-4} p.a., rather than the more normal 1 in 10^{-2} ('1 in 100' or '1 in 50' return periods) we use for everyday structures. The arithmetic for extreme events is $10^{-3} \times 10^{-4} = 10^{-7}$ where 10^{-3} is taken as the structural reliability; 10^{-4} is the probability of the event and 10^{-7} is the desired overall failure probability. It is thus customary to consider 10^{-4} extremes of wind, temperature and so on (often referred to as 1 in 10,000-year events) and techniques exist via extreme value statistics (Gumbel distributions) to predict magnitudes. In the UK, adopting a target return period of 1 in 10^{-4} p.a. means that very rare earthquake loads become a significant design case.

4.5.14 Techniques to add reliability

The strategy in nuclear engineering to achieve reliable performance design under low-probability events also includes an intent to provide simple robust structures relying on known technology. This eliminates some uncertainties. Structures should also be ductile (Section 4.3) and the design process includes checks to ensure that the designs are insensitive to variations in assumptions (so called 'margins' or 'cliff edge studies'; see also Chapter 13). The acceptance criterion for structures under the applied load conditions is adjusted to suit the safety consequences of the anticipated failure. If the structure is a containment one, it may be that the acceptance criterion is 'no leak' and that implies a nominal elastic response under load. On the other hand, structures which just 'have to not fall down' may be tested against a ductile response with a defined finite post-yield displacement. Considerable attention is given to long-term durability by design, to ensuring that structures go into service in their intended state and that thereafter in-service inspections and monitoring are an integral part of the safety management.

Part of the strategy of coping with low-probability / high-consequence events is by adjusting the mode of failure of the structure, and engineers often have real choice over this. We can design structures to have greater amounts of energy absorption capability if we wish. This does not stop the structure 'failing' but might control the consequences to something tolerable (Chapter 5, Section 5.2). That philosophy can only be followed if there is understanding of the structure's function. For example, no great harm is done if a standard concrete structure cracks severely but does not collapse on top of people. However, if that structure were a dam, or other containment structure, severe distortion may not be allowable since copious water could still be released.

4.5.15 Human error

In the assessment of any system, it is essential to factor in the performance of humans. There is a relatively high probability that they will make a mistake and that probability is likely greater than the probability that the inert materials themselves will fail. Mistakes can be made at any stage of the process: in concept; during detailed design; in assessment of the computer programs; in communication; in implementation; or in operation. This is discussed in Chapters 6 and 13. In a very real sense, humans are usually the weakest link.

4.5.16 Summary

All stakeholders want engineering products to be reliable – to work as intended. If a bridge or tall structure turns out to be incapable of performing its function, the safety and commercial consequences are disastrous. During design, we take it for granted that the mathematical tools we adopt will suffice. Yet many of the techniques we routinely deploy have only been discovered over a current engineer's lifetime, so there is risk. Industry is producing new products and new materials constantly. Hence it is always a legitimate question to look at the whole complicated task of building a structure and ask, 'Is it reliably safe?' Engineers need strategies for answering that question positively. For most of the time, use of tried and tested design rules, material control, and site inspection and testing will suffice as a package. But we must also design certain structures to cope with events that are scarcely definable and for those, more thought is required. Some structures need to be more reliable than others simply because the failure consequences are less tolerable. Techniques available to assist are the use of statistics and concepts of probability, which can provide valuable insights (Chapter 13).

4.6 Chapter summary

At first sight, a safe structure can be thought of as one having a strength greater than the loading demands imposed upon it. Indeed, for many minor structures a simple check of strength is all that is required. However, for many other standard structures (and certainly for high-integrity structures) that simple check is insufficient. The concept of 'safety' demands rather more thought and assessment, and that assessment is not readily definable by mathematics alone. Judging 'safety' then involves recourse to concepts such as 'stability', 'ductility', 'robustness' and 'reliability'. Understanding these terms involves some numerical expression but also demands a more fundamental grasp of other basic concepts. Ignoring these wider demands can have serious consequences and many examples will be given in later chapters.

References

Beal, A. (2010), 'Reliability: the practical application of Eurocode BS EN 1990', *The Structural Engineer*, **90**(1).

Beeby, A. (1997), 'Ductility in reinforced concrete: why is it needed and how is it achieved?', *The Structural Engineer*, **75**(18).

Berenbak J., A. Lancer and A. P. Mann (2001), 'The British Airways London Eye. Part 2, Structure', *The Structural Engineer*, **79**(2).

Bragg, S. L. (1974), 'Interim report of the advisory committee on falsework', HMSO.

BS 5975:2008+A1:2011, Code of practice for temporary works procedures and the permissible stress design of falsework.

Burdekin, F. M. (1999), 'Gold medal address. Size matters for structural engineers', *The Structural Engineer*, **77**(22).

Carpenter, J. (2011), 'Reliability: the practical application of Eurocode BS EN 1990', *The Structural Engineer*, **89**(15/16).

CEN (2002), EN 1990 Eurocode 0: 'Basis of structural design'.

Heyman, J. (1996), 'Hambly's paradox; why design calculations do not reflect real behaviour', *Proceedings of the Institution of Civil Engineers*, **114**(4), 161–166.

HM Government (2004), *The Building Regulations: Approved Document A*.

HSE (1992), *Tolerability of Risk from Nuclear Power Stations*, HSE Books.

HSE (2013), *Lifting Equipment at Work. A Brief Guide*, INDG290 (rev1), HSE Books.

Institution of Structural Engineers (2010), *Practical Guide to Structural Robustness and Disproportionate Collapse in Buildings*, IStructE Ltd.

Institution of Structural Engineers (2013), *Manual for the Systematic Risk Assessment of High-Risk Structures against Disproportionate Collapse*, IStructE Ltd.

Institution of Structural Engineers (2015), *Stability of Buildings, Parts 1 and 2: General Philosophy and Framed Bracing; Stability of Buildings, Part 3: Shear Walls; Stability of Buildings, Part 4: Moment Frames*, IStructE Ltd.

Jackson, N. (1995), 'Design of Reinforced Concrete Opening Corners', *The Structural Engineer*, **73**(13).

Mann, A. P. (2002), 'Safety of hanging systems: lessons from CROSS reports', *The Structural Engineer*, **97**(9).

Mann, A. P. (2015), 'Engineering victory: structural advances during World Wars I and II', *The Structural Engineer*, **93**(11).

McConnel, R. E. (1983), 'Structural aspects of the progressive collapse of warehouse racking', *The Structural Engineer*, **61A**(1).

SCOSS Alert (2017), 'Inquiry into the construction of Edinburgh Schools', www.structural-safety.org/media/397456/scoss-alert-inquiry-into-the-construction-of-edinburgh-schools-final-20-february-.pdf.

SCOSS Alert (2018), 'Effects of scale', www.cross-safety.org/sites/default/files/2018-11/effects-scale.pdf.

SCOSS Topic Paper (2010), 'Falsework: full circle?', www.cross-safety.org/sites/default/files/2010-08/falsework-full-circle.pdf.

Short, W. D. (1962), 'Accidents on construction work with special reference to failures during erection or demolition', *The Structural Engineer*, **40**(2).

Soane, A. (2013), 'Collapse of tall reinforcement cages', *The Structural Engineer*, **91**(7).

Soane, A. (2017), 'Temporary Works Forum: Part 7: Can we learn?', *The Structural Engineer*, **95**(2).

Somerville, G. and H. P. J. Taylor (1972), 'The influence of reinforcement detailing on the strength of concrete structures', *The Structural Engineer*, **50**(1).

Steel Construction Institute and British Constructional Steelwork Association (2013), *Joints in Steel Construction: Moment-Resisting Joints to Eurocode 3*, P398, SCI and BCSA.

Steel Construction Institute and British Constructional Steelwork Association (2014), *Joints in Steel Construction: Simple Joints to Eurocode 3*, P358, SCI and BCSA.

Taylor, H. P. J. and J. L. Clarke (1976), 'Some detailing problems in concrete frame structures', *The Structural Engineer*, **54**(1).

Taylor, R. (2010), 'On the rack – structural failure of pallet racking systems', *The Structural Engineer*, **88**(22).

5 Concepts for assessing structural safety

5.1 Introduction

Chapters 1 to 4 introduced a range of issues expanding on what 'safety' means in the context of designing safe structures. This chapter collates several attributes that ought to be exhibited by such structures and having them should help engineers verify that their designs are in fact safe.

Over the last century or so, engineers have moved into designing structures by scientific means. At first this was relatively crude process; sizing being guided by experience, assisted by simple calculations and rudimentary understanding of structural behaviour. Significant research over the last century has brought us to a position today whereby structural behaviour is very well understood. Over that same period, the engineering community has also tried to develop its thinking about the more abstract qualities which make structures safe. The community recognises that despite our best endeavours, the forces structures must resist can be uncertain, the performance of materials can be uncertain and the way we design and construct can lead to error and poor performance. We can also probably say that our modern structures are very much more sophisticated than they used to be, and that this evolution alone requires that we address safety more proactively than hitherto. Given that failure is intolerable or at least very expensive, engineers have developed various techniques for minimising the probability. Generally, it is insufficient (or uneconomic) simply to apply a large numerical safety margin and hope for the best.

The overall objective must be that no structure will 'fail' (however that is defined) due to error in any of the factors that affect performance. At a high level, there must be no gross change in state based on a small error. To achieve those aims we need to look at structures overall plus the performance criteria set for individual components.

In his 1966 book on 'The Safety of Structures', Pugsley suggested that key attributes should be:

- that the structure shall retain throughout its life, the characteristics essential for fulfilling adequately the purpose for which it was constructed, without abnormal maintenance cost;
- that the structure shall retain throughout its life an appearance not disquieting to the user and general public and shall neither have nor develop characteristics leading to concern as to structural safety;
- that the structure shall be so designed that adequate warning of danger is given by visible signs and that none of these signs shall be evident under design working load (Pugsley, 1966).

The first two attributes might be thought of as overall objectives whilst the third is a direct call for controlling the mode of component failure under overload. Generally, the structure should deform and warn of distress before it collapses. Despite bullet 1 being a key statement of desirability, there are many records of failures due either to poor maintenance or poor durability or to designers not adequately considering maintenance needs.

Putting Pugsley's observations into a modern context:

Bullet 1 harks back to Chapters 2 and 4, and the need to understand structural function, particularly when this is part of another system. It also highlights in the comment about maintenance that a safety attribute only exists at a point in time; most structures degrade over a time span that can be short. Lack of maintenance on highway bridges and on railways has been the cause of severe failures. We can read into Pugsley's statement that unless a structure is *inspectable* in areas where degradation may put it at risk, then the design is flawed in terms of its safety (as discussed in Chapter 4).

Bullet 2 could be construed as an exhortation to restrict deflections which might cause disquiet but not be unsafe. If we widen this understanding to, say, excess displacement of structures during crowd response where dynamic effects matter, then we can see that several safety concerns on the dynamic response of floors, grandstands and bridges would also be covered.

Bullet 3 is really a key observation in that we should design structures so they cannot exhibit a gross change of state under marginal changes in the design assumptions such as material properties or loading. It can also be read that the design codes and designers need to consider ensuring some forms of failure in preference to others. It is an exhortation to ductile design in preference to brittle design.

Whilst all these attributes are high-level desirables for safe structures, they don't necessarily inform designers how to go about achieving them. Designers may be helped by looking at some additional terminology which is common when considering safety.

5.2 Preferred modes of failure

In reinforced concrete design, the rules for beams are configured such that under overload, beams will fail by rebar extension and cracking before concrete crushing. That mode of failure is preferred for ductility and gives warning of collapse. There has to be a balance between concrete strength and rebar amounts and that includes a minimum amount of rebar and a balanced amount of rebar to ensure that neglecting concrete tensile capacity is not in itself dangerous. Thus in very large concrete members, it is necessary to provide rebar above the notional section capacity based on concrete tensile strength. If this is not done, any cracking of the concrete suddenly transfers load to a rebar area that is unable to cope, and that rebar will just fail instantly: an undesirable form of failure.

In steelwork, failure by bending is always preferable to failure by fracture and, ideally, members should deflect and yield before supporting connections shear.

The values of detailing on member performance and creating a preferred mode of failure can be shown vividly in the effect that links have on column failures. Under severe loads (as from earthquakes), links need to be increased and anchored better to ensure containment of longitudinal bar under axial load. If this is not done, then under cyclic loading, first the concrete cover spalls off, which then allows the links to open out, which in turn permits longitudinal bar buckling and thence central core disintegration; thereafter column capacity is totally lost (see Chapter 10 on natural disasters). Additional anchoring of links shifts the whole performance of columns under severe overload to one of enduring damage but maintaining axial capacity.

Under extreme loading conditions, designs are ideally configured so that structures fail in bending before they fail in shear. In seismic engineering, it is frequently a requirement that connection capacities exceed member bending capacity (as also in all plastic design) so as to ensure a ductile failure in the member rather than a brittle failure in the connection. Even in standard applications such as concrete flat slabs, it is clearly preferable that under overload, the failure mode should be slab bending, not punch through at column supports. And in seismic engineering the axiom 'weak beam – strong column' is cast as indicating that under overload it is preferable to retain vertical carrying capacity and avoid instability in columns whilst accepting excessive beam deformation under overload. We have seen in Chapter 3 the everyday example of a crash barrier post where the preferred mode of failure is post bending and other failure modes need to be excluded. Taken to its logical conclusion it can sometimes be argued that a structure is safer if certain components are weaker since weakness can define the failure mode. To further that objective, some codes put upper limits on material strength as well as lower limits. In normal design under gravity load alone, no harm is done if components are stronger than assumed. But that presumption breaks down when considering performance under extreme conditions. Under blast loading, blast relief panels can be put in to release pressure rather than impose excess load on the structure and this is a standard aircraft solution to promote survival in case of blast in the cargo hold. Modern cars are designed to incorporate crumple zones, partly to absorb crash energy and partly to minimise the risk of forcing the steering wheel column back into the driver.

5.3 Single point failures

Chapter 4 promoted the benefits of robustness and redundancy. Care is required in any engineering system when significant amounts of structure are dependent on the capacity of a single point of resistance; this is the concept of single point failure. Failure risks are reduced if the possibility of load sharing exists. The idea is frequently used in assessing complicated mechanical and electrical systems which ought to incorporate redundancy and diversity, but if there is a single component anywhere within the system, then reliability of the whole reduces to the reliability of that single component. A significant requirement of any safety review is to identify any such parts. In buildings, a failure example might be 'loss of power' which might, if it occurred, knock out all electrically operated doors, fire alarms, lifts, ventilation systems and so. To minimise this risk, it is common to have stand-by generator power as an alternative to the grid and UPS (uninterruptable power supplies: batteries) for emergency lighting. It is also common to make sure modes of escape are not dependent on electrical power or on a single exit route (in the 2017 Grenfell Tower fire disaster there was only one exit route from upper floors, but this rapidly filled with smoke and became unusable). In purely structural terms, single point failures are not always avoidable: in some senses every cantilever might be a single point failure. Where this is critical, it pays to think of how reliability can be ensured (Chapter 4, Section 4.5.9). Much can still be done; if the cantilever mode of failure is ductile bending (rather than connection snap), then the consequences of overload might be benign. Moreover, if several cantilevers are joined together at their tips, there are possibilities for load sharing in the event of any one being faulty. Configuring the structure to accommodate isolated failures from whatever cause is one task embodied within the idea of creating a robust structure.

5.4 Common cause failures

A single point failure is one where a whole system fails if just one part fails. A common cause failure is one where a single event (such as hazard) creates multiple failures in what might otherwise be thought of as separate systems (or defences). The nuclear power plant at Fukushima (Chapter 11) was prevented from overheating by cooling circuits dependent on electrical power. To ensure the power supply, there were grid supplies, back-up diesel generators and facilities to hook up emergency generators if the two main systems failed. However, the tsunami that knocked out grid supplies also flooded the location of the back-up generators and hook-up facilities. Subsequently all power and all three supposedly independent power systems were lost. The tsunami acted as source of common cause failure.

The idea of common cause failures comes from safety studies on mechanical and electrical systems. Redundancy can exist via two equal switched electrical circuits. But what happens if both switches are from the same batch and both have the same manufacturing fault? Diversity is one solution to this potential problem. In civil/structural engineering, we might have a faulty batch of bolts or a faulty batch of steel with the nightmare that if one component is faulty, all are faulty. Guarding against that means in practice having adequate QA on material control and it means on-site installation having enough procedures to ensure that safety is being confirmed. In any design that requires significant repetition, it is prudent to take some measures to ensure that whole batches are not faulty and that no design fault is repeated. One example of this is the failure of connection bolts from hydrogen embrittlement on London's Leadenhall Building ('Cheese Grater Tower') (2015). A costly bolt replacement programme had to be carried out after building completion.

Some of the worst failures to have taken place are those where the error has been repeated many times over. This might be because of over-reliance on a computer program or a standardised design technique or by just overlooking some key aspect of design and repeating it lots of times (the failure at Paris's Charles De Gaulle Airport was one; see Chapter 8). In considering the safety of any structure, it is prudent to look at all parts of the process which represent considerable repetition and to ensure correctness by alternative means. This is especially wise if any design feature is novel.

5.5 Chains of events

It is bad enough one member failing; it is a nightmare if a single event sets up a progressive chain. It is not hard to find clips on YouTube of ships breaking loose and then careering into others to create significant collateral damage. In 2018, a typhoon hitting Japan's Kansai airport caused a large ship to break loose, which then collided with the main bridge connecting the airport to the mainland. Closer to home, boats breaking loose in Holyhead (North Wales) marina in 2018 during Storm Emma destroyed much of the marina infrastructure. The gas explosion at Ronan Point set off a chain event (Chapters 3 and 4 on progressive collapse). In any safety study, a question should always be to consider what happens after the postulated failure? There should always be attempts to contain the consequences. After all, that is the function of any safety barrier. An errant vehicle cannot be ruled out but crashing into a bridge pier and taking out the bridge as well can be avoided. A vehicle going off the road is not generally as bad as one going off a road and onto a railway track, where the consequences, i.e. a train collision, are likely to be more severe.

5.6 Fail-safe

Fail-safe embodies the concept of a mode of failure that, if it occurs, causes the system to default to a safe condition. An obvious example is that of the dead man's handle on a train, whereby the train only moves if the handle is depressed and the default position is 'off'. The notion is clearly linked to the idea of a driver becoming incapacitated, so the train would then default to stop. In passenger cars, foot brakes are only 'on' if the driver depresses them. But in many moving structure safety systems, the brakes are configured in reverse, i.e. always 'on' unless there is power to release them. This guards against lack of control in the event of power failure as the brakes spring 'on' automatically on loss of power and the system defaults to a safe (stationary) state. The use of lifts in tall buildings became a lot safer with the invention of the safety brake. If the hoisting cable fails, side brakes spring on and prevent the lift from falling catastrophically. In many systems of which structures are a part, assessing safety means looking at all modes of failure (such as cable failure) and assessing the consequences and applying mitigating features.

At a simple domestic level, fail-safe requires the use of overflow devices on domestic sinks and cisterns, or at a larger scale, on any tank. Safety requires that consideration be given to potential overfilling and of what happens in that event. Dams will always incorporate spillways to limit the head of water that can build up. Proper consideration to potential failure of the primary system has not always been given. At a domestic level there have been cases of incompetent plumbing when hot water cylinders have not been provided with proper expansion facilities with the consequence of explosion when the immersion heater failed to switch off (i.e. the system was not fail-safe). At Buncefield, UK (2005), (Chapter 11) one fuel storage tank was overfilled but the electronic device supposed to detect overfilling had failed and fuel spilled over the sides, causing a massive fire; this system again was not 'fail-safe'.

5.7 Leak before break

Other techniques are also used. The phrase leak before break expresses the same idea as Pugsley's Bullet 3. It conveys the idea that a failure might occur but if it does there will be warning. Thus, it is desirable that a water-retaining structure leaks and warns before it breaks and floods. In high-integrity liquid-retaining systems (such as used in nuclear plants) water-storage structures are placed on porous beds which channel any leaks to the perimeter for the purpose of making leaks observable and thence giving time for response (see Section 5.9). This leak before break analogy is often extended to welding and fatigue design. For safety, it is highly desirable that fatigue cracks are detectable and visible on initiation before becoming long enough to be unstable. This can be achieved by designs adopting details whereby any crack will start on a weld's outside surface and work inwards rather than on the (unobservable) inside surface and work outwards.

If that latter circumstance is permitted, cracks may only be discovered too late. It is also desirable that cracks can propagate some distance before representing real danger, i.e. the tolerable crack length should be long. Codes recognise these possibilities by permitting design options of acceptable damage and repair.

5.8 Defence in depth

Defence in depth is a concept used in hazardous industries. It is an idea as old as military engineering when two rings of castle defences were considered better than one. We might say that ideas such as using bunds around oil tanks embody both defence in depth and fail-safe concepts. Clearly the idea is that if the main tank fails, any spillage will be contained. Ships are designed with watertight bulkheads such that if one compartment is holed, the ship should still float (this concept was violated on early roll-on roll-off ferries with consequent disaster (Chapter 6). The idea can be extended into structural engineering in a number of ways, one of them being redundancy and having alternative load paths such that if one path fails for any reason, the structure may still survive, avoiding catastrophic failure. A criticism of the Ramsgate disaster (Chapman, 1998 and Chapter 8) was that a catastrophic failure of the passenger boarding ramp was inevitable on failure of the support components that did fail. A recommendation of the subsequent inquiry was that all such future ramps should have end safety chains fitted such that future structures would be caught in the event of similar failures (namely two barriers before catastrophic failure). Two barriers are better than one, but they might both fail from the same common cause. If the two barriers differ (have diversity) so much the better.

The benefits of Defence in Depth are often depicted by the 'Swiss cheese model' conceived by James Reason. The model is drawn as parallel slices of Emmental cheese where each individual slice includes a random set of holes, of various positions and sizes. A unique pathway through all the slices (the failure analogy) only exists when separate slice holes align. The more slices there are, the less likely this is: each slice adds to Defence in Depth. A single slice of cheese can be said to represent a Single Point Failure. In very rare cases, for extremely hazardous circumstances, it is possible to make an 'incredibility of failure' case.

5.9 Reaction time / rate of failure

Pugsley's third point (Section 5.1) was to require that 'adequate warning of danger is given by visible signs'. That objective is twofold: (a) to allow those in danger to escape and (b) to give those who might have to manage the structure time to save it. Brittle failures clearly violate this safety attribute and must therefore be avoided. A further good example of (a) is to reflect on the way fire precautions are designed to promote both rapid escape and rapid response by fire brigades, and to give firefighters a chance to contain the blaze before it gets out of hand. Alas, in many fires, even a rapid response time has been insufficient (Chapter 9).With knowledge that all structures degrade, an objective of in-service inspection is to detect the onset of degradation before it progresses far enough to pose a danger: to give time for repair. An objective of emergency preparedness against man-made and natural disaster is to have contingency planning available to maximise the time available for search and rescue to save lives.

A question to be asked of any identified hazards is: if it happens, what is the 'rate of failure'. The failure of Genoa's Morandi Bridge in Chapter 7 clearly violated Pugsley's principle of having adequate reaction time: the bridge's failure was instantaneous. In the Grenfell fire of 2017 (Chapter 9), the rate of spread of flame (the hazard) across cladding surfaces was very rapid: so rapid that even though fire brigades arrived on the scene swiftly, there was little they could do. The 2018, wild fires in California spread so rapidly they could not be contained (Chapter 10).

5.10 Hazard and risk

A major evolution in all engineering work has been to embrace the concepts of hazard, risk and consequence (Carpenter, 2008) as a basis for producing appropriate designs. Hazard is defined as anything which might cause harm: it might be a loading condition for environmental effects or plant failure or just an error that operatives can make. Risk is the chance that someone or something is adversely affected by the hazard. Under this definition, risk = hazard × probability, i.e. this differs from common usage where the words risk and probability might be synonymous. Whether a risk is tolerable or not is obviously linked to the consequences (considered below).

Experience tells us that all enterprises are risky. Now it is possible to look at hazards and risks in terms of probability as discussed earlier: the chances of a New Orleans-type disaster (Chapters 10 and 13) consequent on a low probability natural event and so on. But at a more general level, we know that many building projects go wrong for any number of reasons: a loading condition no one envisaged, a fault in manufacture, poor installation, design assumptions not realised or someone just making a mistake. The consequences can be awful, as at Abbeystead in 1984 (see Chapter 10). This was a pumping station where there was an explosion of methane in an underground room which killed several visitors on the day the station opened. The gas hazard had not been identified. However, this was not a risk in the sense of something that could never have been foreseen. With hindsight, it was a fairly obvious factor to consider as a possibility.

The objective in any risk analysis is first to identify all possible hazards and then consider what the implications might be and what mitigating factors might be deployed (presumably it would not have been difficult at Abbeystead to include methane detection equipment). Identification of course begs the question of whether any procedure will spot all the hazards and consequences that flow from them (see Chapter 11 on the Fukushima disaster in Japan, where predictions were not as thorough as they could have been). A full risk assessment can only be completed on the finished design, but the hazard and risk assessment really needs to be completed in advance since retrofitting a design to minimise risk is not at all desirable or may even be impossible. In 2004, a very severe explosion occurred in the Stockline Plastics factory in Glasgow (Chapter 11). The building was destroyed, with nine people killed and several injured. The cause was a gas leak from a corroded pipe under the factory floor. If a formal hazard analysis had been carried out, the possibility of pipe corrosion might have been spotted and it can be assumed that the severe consequences would have been thought of. If the question had been asked as to how anyone would know the pipe was in sound condition, the response would surely have been there was no way of knowing as it was just buried. Thus, a process was capable of identifying the hazard; of identifying a risk and severe consequences; and of identifying that there was no control over that risk since inspection was impossible. This process of hazard and risk assessment is a very common technique to apply to complete electro-mechanical systems, often incorporating control systems as well (Chapters 8 and 13). Structures frequently act as one component in such systems and it is vital to understand what safety functions are demanded of them or what the consequences might be attendant on failure in any part of the system (including the structure).

A part of any risk analysis is to consider the consequences. These might be benign or disastrous. They might involve injury to persons or they might just entail commercial loss. If the latter, parties may elect to accept a much higher probability of failure (Chapter 2 includes the example of dealing with a power station's faulty cooling tower where the possibility of failure was just accepted). The whole question of what is a tolerable risk is a difficult one and the HSE have published their thoughts on this (HSE, 2001; HSE, undated). An essential point is that where injury to persons is predicted, steps are mandatory to reduce risks. However, engineers are not expected to eliminate all risk. Risks need only be 'ALARP' (as low as reasonably practical).

Any risk assessment should include the role of people and human error, since human failures are nearly always a root cause of any failure. Chapter 6 deals with human error with examples of several failures stemming from that cause.

Once a schedule of hazards is identified, a normal process is to review them and take some design management action to deal with them. There are six principles of inherent safety with regard to hazards:

1. Elimination: avoid the hazard completely.
2. Substitution: reduce the hazard severity by changing the nature of the hazard.
3. Minimisation: reduce the hazard severity by changing the scale of the hazard.
4. Moderation: reduce the hazard by minimising the impact of a release or hazardous event.

5. Segregation: limitation of the effects reducing potential for hazard to cause harm.
6. Simplification: reduce the hazard likelihood by inherent design features of the design.

There are many variations of this hierarchy. A simple one used frequently is ERIC: Eliminate / Reduce / Isolate / Control (Chapter 12).

Although six principles look abstract, they are no more complicated than saying 'Don't keep all your eggs in one basket.' The hazard is dropping the eggs and breaking them all. Elimination: avoid carrying the eggs. Substitution: if you must transport the eggs, don't use an open basket that is capable of being dropped. Moderation: wrap the eggs. Segregation: use more than one basket (so you don't break them all). Simplification: design a new basket that is incapable of being dropped.

A troublesome question is how can we know that any hazard and risk assessment is complete? (SCOSS Alert, 2017). How can we know that all the hazards have been spotted? It is clear from study of accidents and failures that one frequent cause is that the scenario causing the failure had just not been thought about (as in the Financial Crisis of 2007; and see Chapter 6). But the risk of omission itself can be reduced. If we look at the many failures cited in this book, it should be self-evident that there is commonality. We can learn. Moreover, these days, frequent Web-based news feeds provide a ready source of examples. Bearing in mind the Titanic, future captains will be risk averse when it comes to icebergs and naval architects will ensure there are enough lifeboats. One purpose of this book is to assist engineers with learning what has gone on before, so as to avoid problems in the future. In civil/structural engineering, the study of failure isn't voyeurism, it is a pragmatic task intended to preserve safety and careers.

5.11 Chapter summary

Pugsley defined some general principles of safe structures and these are reproduced in Section 5.1 above. Taking account of modern thinking and the discussion of Chapters 1 to 4, we may extend and summarise these principles by using certain key words / concepts.

Safe structures should:

- have strength capacities that are inherently conservative against defined demand;
- have identified load paths for vertical and horizontal loading;
- be stable and insensitive in strength or stability to minor variations in design assumptions;

- avoid being vulnerable to single point failures;
- be configured to ensure preferred modes of failure;
- appear sound;
- give warning of failure (so there is time to react);
- be robust;
- be durable through life (and so be inspectable and maintainable).

System attributes can be examined by using key concepts such as:

- 'fail-safe';
- 'leak before break';
- having 'defence in depth';
- 'controlling the consequences';
- ensuring that whatever might go wrong does go wrong in a way that permits adequate reaction time for evacuation and management response.

Safe structures are ones that avoid:

- the consequences of human error;
- being vulnerable to chain reactions;
- being vulnerable to repeated error;
- being sensitive to minor variations in design assumptions.

Thereafter a safe structure should:

- be inherently buildable;
- have the quality of their products verified;
- have the build quality verified to ensure that it meets 'design intent'.

These attributes are not achieved without conscious effort and Chapter 13 discusses various management techniques which ought to be deployed to an extent commensurate with the consequences of potential failure. It will be found that all the above principles are incorporated with the 'Safety Assessment Principles' used for nuclear plant, albeit the full range of those principles is considerably more complex (Office for Nuclear Regulation, 2014). In life, many failures still occur and many of them occur because designs have violated one or more of the above principles. This will be illustrated by examples given in the following chapters. Chapter 6 (on human error) has failure descriptions which illustrate the application of the key concepts set out in this chapter.

References

Carpenter, J. (2008), 'Safety, risk and failure', *The Structural Engineer*, **86**(14).

Chapman, J. C. (1998), 'Collapse of Ramsgate Walkway', *The Structural Engineer*, **76**(1).

HSE (2001), *Reducing Risks, Protecting People: HSE's Decision-Making Process*, HSE Books, www.hse.gov.uk/risk/theory/r2p2.pdf.

HSE (undated), 'ALARP "at a glance"', www.hse.gov.uk/risk/theory/alarpglance.htm.

Office for Nuclear Regulation (2014), *Safety Assessment Principles for Nuclear Facilities*, Rev 0, Office for Nuclear Regulation.

Pugsley, A. (1966), *The Safety of Structures*, Edward Arnold.

SCOSS Alert (2017), 'Hazard identification for structural design', www.cross-safety.org/sites/default/files/2017-09/hazard-identification-structural-design.pdf.

6 Human error

6.1 Introduction

Unfortunately, human beings make mistakes, often catastrophic ones. This may be at an individual level or may be via corporate failure. Sometimes the cause has been pure duplicity or dishonesty, occasionally personal arrogance or incompetence, but most commonly just fallibility. Sometimes the cause has been a failure within the overall management of the hugely complex process that is required to create a great structure. Sometimes the error has been linked to poor risk assessment of plant and the consequences of failure. Whatever the cause, history gives us many examples and for sure those are being repeated up to the present day. Many studies have concluded that, to some degree, the most frequent root engineering failure cause is one of human error.

A clear example for most people occurs at every filling station, when error might result in filling a petrol engine with diesel. Ideally, nozzle design should make this impossible, but filling cars with the wrong fuel has happened. An extreme example of this class of human error occurred at Camelford (UK) in 1988. In this incident, the town's drinking water was contaminated when a sediment-settling chemical was erroneously dumped in the wrong tank at the treatment works. Those who then drank the water were put at considerable health risk.

The Camelford incident occurred due to human error combined with poor management control. The site was unmanned and the delivery driver had been given a key with a simple instruction: 'Once inside the gate, the aluminium sulphate tank is on the left.' The driver was confused, but found that his key unlocked what he thought was the correct manhole cover and so he poured his load of 20 tonnes of aluminium sulphate into that. The consequence of putting his chemical in the wrong tank was the town's water supply then contained a chemical concentration some 3,000 times over the admissible limit.

Human error was obviously a factor. But that possibility was foreseeable, and it appears no attempts were made to account for it in the plant design or management. Since the early days of engineering, safety-conscious engineers have sought to prevent humans from making disastrous mistakes. Early railways used a unique token to permit access onto single tracks that would be used by multiple trains. This procedure ensured that only the engine with the token was allowed on that track at any one time, which eliminated collision. To be effective, the token had to be single and unique to the stretch of track being protected. Many mechanical devices contain interlocks to prevent danger following error. In domestic fuse boxes, opening the box is an action that by itself isolates the electric supply. At Camelford, the design should have made it impossible to open the wrong manhole cover with the key provided. Additionally, documentation should have been available to direct the driver to the right manhole, which ought to have had unique labelling. All such precautions would have eliminated the error. However, nobody appears to have looked at the system in advance to consider what might go wrong.

The world's financial markets went into turmoil after 2007 and in the frightful aftermath the Queen asked why nobody had noticed it coming. A professor is reported to have answered: 'At every stage, someone was relying on somebody else and everyone thought they were doing the right thing.' The credit crunch and financial crisis applied not just to individual banks, but across many banks and across governments, regulators and most of the world as well. Everyone had hitherto assumed that these financial high-flyers must have known what they were doing, yet it turned out they were making elementary mistakes and taking unwarranted risks, and eventually they were found out – disastrously. A parallel with our own profession is that after the investigation of any civil or structural engineering failure (like Camelford), the same observations are frequently made, i.e. that other parties assumed those responsible knew what they were doing or were too cowed to ask 'experts' the obvious questions. It frequently transpires that elementary mistakes had been made with undue risks being taken and frequently without adequate monitoring or supervision by the companies involved. Sometimes no one questions anything. In 2017, the Grenfell Tower fire occurred (Chapter 9). It might well be asked how nobody amongst the professional design/procurement teams spotted what appears to be an obvious mistake: enveloping an entire tower block in highly flammable material.

Human error was also exemplified by the events at Fukushima, Japan in 2011. On 11 March that year, one of the world's most frightening failures occurred at the nuclear plant. At the time it seemed as if the nuclear nightmare might really happen. Certainly, there was core overheating accompanied by reactor material meltdown and it wasn't until the following December that Japanese authorities considered the event might be under control. The disaster was precipitated by a natural occurrence, a large earthquake followed by an enormous tsunami (see Chapter 10). These twin events caused plant failures that initiated the subsequent catastrophe. Yet the official parliamentary report was unequivocal that this was not a natural disaster, but a man-made one. The report concluded that the crisis was 'a profoundly man-made disaster' and the disaster 'could and should have been foreseen and prevented' and its effects 'mitigated by a more effective human response'. The report catalogued serious deficiencies in both the government and plant operator, claiming that wilful negligence had left the plant unprepared. Additionally, the report blamed cultural conventions for a reluctance to question authority[1] and criticised the role of the government regulator. The deficiencies in the plant's safety had been known about and instructions issued for rectification, but they had not been acted on. Neither had the regulator made any attempt to enforce them. There was neither preparation nor proper planning for response in the event of failure. Admittedly, the tsunami was much larger than thought possible and the plant's defences had been overwhelmed, but the consequence of that could have been foreseen and the weakness rectified. The report also blamed plant staff and regulators for not learning from previous accidents and not taking benefit from worldwide experience, all factors that should be promoted in any commitment to safety in any branch of engineering.

Thus there are obvious parallels with the financial failures: an initiating event that got out of control; faults that in hindsight could all have been anticipated;[2] consequences that could have been foreseen; a failure to assess risk properly; and a failure of the regulatory system. Furthermore, in both situations, the consequences were global and after both events those in charge scrambled to evolve a suitable response.

A lesson from all these diverse failures (from a water treatment works to a nuclear plant) is to consider everything as a system and to consider not only the possibility of engineering error within that system, but the possibility of human error creating a catastrophe. Techniques have been evolved to consider the safety of systems as a whole and these are discussed in Chapter 13.

6.2 The reality of human error and its consequences

Early in 2018, the population of Hawaii were officially alerted to an imminent ballistic missile threat. There was said to be wholescale alarm and chaos, with residents seeking shelter. The alert turned out to have been false ('the wrong button was pressed') but it wasn't rescinded for around half an hour. An official investigation cited 'human factors' amongst the causes. It can only be speculated what would have happened if the military defence had responded.

Sometimes human action seems almost inexplicable, as in the case of the Herald of Free Enterprise disaster (Box 6.1) or the sinking of the liner *Costa Concordia* (see below).

1 An apt proverb says: 'It is not disgraceful to ask, it is disgraceful not to know.'

2 Perhaps that is unkind: another (Turkish) proverb says: 'Only after the earthen water jug gets broken on the road to water, do the many advisors appear for a better route.'

Figure 6.1 Herald of Free Enterprise capsize. (Alarmy G4HKD4)

Box 6.1 Herald of Free Enterprise, Zeebrugge, Belgium (1987)

The Herald of Free Enterprise was a ferry plying between Dover and Zeebrugge. Shortly after leaving Zeebrugge it dramatically capsized, killing 193 passengers.

Cars entered over a back ramp which was supposed to be pulled up to form the back (stern) door to seal the ship before the voyage. The disaster (Figure 6.1) was caused by the ferry sailing with these doors open, allowing seawater to enter the car deck and generating a severe instability that allowed more water to pour in and led to the ship capsizing (this is about shallow water on a wide surface and is called the 'free surface effect', amply demonstrated when trying to carry a domestic tray with some enclosed water). This whole disaster might be put down to the human error of leaving with the doors open, but the failure reveals issues of wider generic significance.

The whole ship's design could have been treated as an engineering system and subjected to a HAZOP study (Chapter 13). If this had been done, the headings of '*human error*' and '*departing with the door open*' ought to have been identified. An examination of the *consequences* should have identified capsize, not least as the stability of an open roll-on roll-off ship was *sensitive* to water ingress. A further examination of *consequences* might then have concluded: loss of life, loss of asset and bankrupting the company (all of which happened). Collectively, that would suggest great design attention needed deploying to minimise the risk. Mitigation at modest cost was perfectly possible either by interlocking the door to the sail instruction, or by deploying alarms/sensors/CCTV to feed information to the bridge that the door remained open (albeit those options had been ruled out by management). Ships are generally designed with several watertight compartments to provide *redundancy* and *robustness*. But these benefits were not available on this open car deck if water flooded in (an inquiry recommendation was to fit separation screens into car decks to be activated before sailing). The *mode of failure*, capsize, was undesirable as it prevented lifeboat launch and trapped passengers on board. Furthermore, in capsize, the *rate of failure* is rapid, again inhibiting rescue.

All the *italicised* words above are key ones defined in Chapters 4 and 5.

Finally, this failure exemplifies what happens when commercial considerations override safety. The management imperative was a rapid port turnaround to maximise revenue, so it had become customary to depart and raise the rear doors en route, this time with fatal consequences.

Similar failures of roll-on roll-off ferries have occurred, such as in the Estonia (1994, 852 lives lost). In this case the bow caved in during a storm, again allowing water to pour into cargo decks. The official inquiry found that door locks had failed, but the mode of failure had not been detected by sensors and the bridge was set too far back for anyone to see what had happened. Again 'free water effects' played a role in the mode of sinking. In September 2000, the Express Samina sank in the Aegean Sea (82 dead). In this case, rocks ripped through the hull. But the ship sank because several of the ship's watertight compartment doors had been erroneously left open (a common cause failure). Moreover, once water spread, power was lost and with that loss, the ability to close the doors was also lost (an example of how sequential consequences, chains of events, need to be considered: again, see headings in Chapter 5).

Although these ship incidents would not normally be considered in relation to 'building structures', they are useful indicators of safety. First, they are all systems: collections of multiple engineering disciplines all performing a function. Along with that function there are a range of hazards that the system might be subjected to; some physical, some consequent on human performance. In assessing consequences, the mode of failure needs to be understood and some modes are less desirable than others (that might be considered a measure of 'robustness'). All modes of failure need, if possible, to be detectable so as to give time for reaction to pre-empt disaster, perhaps simply by saving lives. Finally, reaction to severe effects plays a part in controlling (or not) the scale of the disaster. When applied to building structures, we might think of the issues of failure being detectable either via inspection (spotting the onset of degradation) or as in Pugsley's concept that 'the structure shall be so designed that adequate warning of danger is given by visible signs and that none of these signs shall be evident under design working load' (Chapter 5).

In early 2012, the modern cruise ship Costa Concordia sank spectacularly off an island close to Italy, and 31 passengers died. This was clearly a major disaster befalling a modern ship in calm seas. An obvious question is how on earth could such a catastrophic event occur? It would be thought that sinking such a major structure was virtually impossible, the general assumption being that the design teams must be really competent on such large vessels and the concept of a major failure is therefore bewildering. The ship sank when it was travelling too close to a shore line and hit a rock, a predictable hazard. Any risk register ought to have identified the hazard. There is no suggestion that the rock was unmapped, so the charted course should have avoided the shoreline and the course could have been monitored both automatically and by human control. It has been reported that the captain departed from the ship's pre-programmed course and steered close to the island as a favour to friends. He was indicted for manslaughter, found guilty and imprisoned. The captain admitted his departure from course but is reported to have blamed others for failing to inform him of impending disaster. Whatever the truth, as usual it seems there were multiple human errors and it also looks quite possible that an automatic satellite check on the ship's position must have been overridden. (If there were not multiple human errors in steering the course, then there must have been a management failing in not ensuring reliance on more than one person against such obvious danger). Twin generic safety issues are first: human failings (potentially mitigated by more than one person being a decision maker) and second (as with the Herald of Free Enterprise), the potential use of a backup control system to guard against foreseeable human fault. Additionally, and again as with the Herald of Free Enterprise, the failure mode and rate of failure (capsize) were undesirable and for the same reasons.

In the *Costa Concordia* case, any overriding of position alarms would be culpable. But in terms of 'safety', there have been corresponding airline incidents whereby crews have either misread instruments or not trusted them and subsequently reacted incorrectly. A good example is that of the crash of Air France flight 447, en route from Brazil to Paris, in 2009. The airbus crashed in the Atlantic, killing all passengers and crew and almost disappearing without trace. Eventually the flight recorder was recovered, and the investigation pieced together what had gone wrong. It appears that there were inconsistent airspeed sensor readings and so the autopilot was disengaged whilst the pilots reacted. But the pilots took the wrong action, putting the plane 'nose up', despite stall warnings. The plane then lost critical speed and made a deep dive into the water. The pilots in charge at the time were inadequately trained on how to respond in the circumstances they found themselves in. The inquiry predicted that if the pilots had been properly trained, they could have recovered without crashing. It is reported that the exact cause of faulty airspeed indicators is unknown but suspected as being due to ice formation in the pitot tubes; a problem that has contributed to other crashes.

Thus pilot error inadvertently caused the Air France crash, though in this case, it appears not to have been a culpable action. Nonetheless, there are generic safety lessons beyond the one relating to human error:

a. the initiating event (blocked pitot tubes) must have been a foreseeable form of failure as it had occurred before. The consequences of that (faulty speed indication) are predictable.

b. There is a generic safety issue in any 'system' of always considering the possibility of faulty instrumentation that needs to be mitigated. Behind that is the science of designing control systems that are safety related (see Chapter 8): this is relevant to any 'moving structure'. Most of us are familiar with sensors in the standard family car of, say, a red light coming on: is the oil really low? Or is the sensor just faulty? Or if the sensor is faulty and does not light up, what is to warn against low oil levels? It might be argued that the sinking of the Thetis, as described in Chapter 2, is another example of gross failure consequent on a faulty reading. In the case of the Thetis, the reading suggested a torpedo tube was dry when it was already flooded.

c. Within the process of control system design, there is the essential step of considering the consequences of sensor failure (either failing 'off' or 'on' or just giving an incorrect signal).

d. Finally, as in the financial crisis, the Fukushima crisis and the sinking of the *Costa Concordia*, there is the question of anticipating the worst and planning and training an appropriate response. The process for 'anticipating the worst' is via a form of risk assessment (see Chapter 13).

There are strong parallels between the schedule of errors in Air France flight 447 and the causes of the Boeing 737 MAX crashes in 2019. In its final report on these, the US House Committee was highly critical of the performance of Boeing in both design and management (US House Committee on Transportation and Infrastructure, 2020).

The sinking of the *Costa Concordia* and the crashing of flight 447 illustrate how generic issues of human error and automatic control can follow in the same pattern. If the possibility of human error is acknowledged, it can be mitigated by design or by the deployment of better validation controls. But that then leads on to a certain risk, either of over-reliance on such controls or of the controls themselves being misleading.

Control engineering is an integral part of the design of any moving structure. There are reported failures of cranes where safety systems have just been overridden. There have been many train failures due to signals passed at danger (SPADs). In some cases, this has been due to fatigue or carelessness (human failings), but in other circumstances it has been shown to be impossible for drivers to read light signals when obscured by low sun; errors certainly, albeit not culpable ones. Recognising these possibilities, coupled with their consequences, leads to appropriate mitigation measures being required. The lesson is that in any system, proper design requires consideration of the consequences based on either human reaction, faulty control or inadequate human response to control.

A classic failure consequent at least partly on failure to respond to instrument readings is that of the tunnel collapse under Heathrow Airport in 1994 (Chapter 8). The tunnel was being built under the main runway and collapsed catastrophically, bringing chaos to the airport. So much so that the HSE described it (HSE, 1994) as one of 'the worst civil engineering disasters in the UK in the last quarter century'. One feature of the inquiry singled out that 'Warnings of the approaching collapse were present from an early stage in construction but these were not recognised.' The tunnelling technique adopted (the new Austrian tunnelling method, NATM) relies in part on observations of ground movement and continuous monitoring and reaction to the readings in terms of the details of excavation and lining. Such monitoring was in place, but the warnings provided were not acted upon. Those in charge of monitoring were very heavily fined. Additional conclusions were of 'Poor design and planning, a lack of quality during construction and most importantly, a lack of safety management was present' (see also Temporary Works Forum, 1994).

The Heathrow incident provides another major lesson about safety. No one should assume that the worst can't happen in any project. This includes 'large' ones, since in those there is an added danger that overall safety objectives and risks get overlooked because parties perhaps assume someone else must know what they are doing (see Section 6.1 above). In the aftermath of the crisis the contractor stated: 'We will never try to change our plea of guilty but it is clear that the collapse was the result of organisational blindness' (see also Temporary Works Forum, 1994).

The errors in instruments and monitoring described in Box 6.2 below are one aspect of the problem. Car drivers pass traffic lights for a variety of reasons and train drivers pass track signals for a variety of reasons as well. There is some science behind the human aspects of how visible certain signs might be, and in the development of the UK motorway network, great thought was given to the size, style of lettering and colour of motorway signs to maximise their ability to be read properly at speed. Failures in the ability of users to read signals has resulted in disasters such as at Ladbroke Grove in 1999. In this crash, 31 people were killed and 520 injured. The Cullen Inquiry (Cullen, 1990) examined all the evidence. It is known that the primary cause of the crash was the train driver passing a yellow signal which warned of an

Box 6.2 Severn Tunnel Rail Crash, UK (1991)

In this incident, two trains collided in the very long tunnel under the River Severn (the tunnel is about 4.5 miles in length). The crash can be primarily linked to faulty control. One driver entering the tunnel was advised of signal failure but given permission to proceed with caution. Inside the tunnel, that train was hit by a following train which was proceeding as normal. Errors meant that two trains were on the same segment of track at the same time with the front one running slow.

One of the reasons may have been operatives resetting the system due to earlier errors thus allowing the rear train to enter, passing a clear signal as it did so. The alarm was routinely overridden because it frequently gave false signals: on this occasion, the alarm was for real. Other features of the incident were the delay in knowing there had been a crash and hence delay in rescue. Although there were signal problems, operator error guaranteed the accident would happen.

impending red but then passing the red as well. As the driver was killed, there has been no possibility of establishing for certain why the signals were ignored. It is known that the driver had only recently qualified, but the inquiry considered it more probable that his poor sighting of the signal and poor sighting in comparison with the other signals on the overhead gantry, coupled with bright sunlight at a low angle were factors that might have confused him.

As for the Ladbroke Grove incident, the Croydon tram crash of 2016 exhibited similar failings (Box 6.3). In December 2017, a US Amtrak train travelling from Seattle to Portland derailed on a bend on a brand new section of high-speed track. The carriages left the rails, killing six passengers. It appears that the train speed was far too fast, and it appears probable that the driver, who was under training, did not see a slow-down signal. The train speed was about 130 km/hr when there was a 50 km/hr speed limit. No automatic braking systems were operational.

BOX 6.3 Croydon tram crash, London, UK (2016)

In November 2016, a tram in Croydon derailed on a bend, overturned and killed seven passengers (Figure 6.2). The tram's speed was excessive being, 46 mph instead of the recommended 13 mph. It appears the driver lost concentration and missed the slow-down signal. This failure exhibits certain common features.

First, the possibility of drivers being fatigued or inattentive could have been appreciated as could the possibility of missing signals (exacerbated on the day by bad weather). Second, other faults emerged; it transpired that, given the signal's location any late driver reaction could never have slowed the train down sufficiently. In looking at the track configuration, the tram was travelling downhill. Hence the default in the event of a driver concentration lapse was for the train to go faster into the bend, not to slow down (the system was not *fail-safe*). The ratio of actual speed to required speed was 46/13 = 3.5. But radial forces are proportional to $(46/13)^2 = 12.5$, so the potential for overturning was much magnified by any speed error. In that sense, safety should have been assessed bearing in mind the attribute of *sensitivity*; i.e. stability safety in terms of excess radial force was very sensitive to variations in speed. Furthermore, the *mode of failure* was one of overturning, which was an undesirable mode in that it placed passengers in greater peril and hindered rescue. Finally, the *rate of failure* was rapid, which is again undesirable. These three safety attributes were all discussed in Chapters 4 and 5.

Another desirable safety attribute on any passenger carrying device is containment. On balance, in the event of a crash it is far better if passengers are retained inside. Hence in cars, passengers wear seat belts, and on roller coasters there are a variety of restraints. In the Croydon crash, some passengers were ejected. A parallel attribute is that after any crash (in which passengers are contained) they should be capable of being extracted. (See Rail Accident Investigation Branch, 2017).

Figure 6.2. Croydon tram overturning on a bend (courtesy of the Rail Accident Investigation Branch, 2017).

It is possible to fit trains with automatic braking systems that activate if a danger signal is passed. The train that crashed at Ladbroke Grove was not fitted with any such system and the inquiry raised the issues of the cost–benefit analysis of fitting (the expensive system) versus the safety gained. The basic safety issue is the impossibility of ever being able to guarantee that drivers will not make an error. A similar tragedy took place on the London Underground in 1975 when a tube driver drove his train right through Moorgate Station at speed into a dead-end tunnel (43 died, 73 were injured). Examination suggested the train was mechanically sound and the crash was caused by the driver, though no satisfactory explanation was ever forthcoming to explain his behaviour. Subsequent to the crash, London Underground fitted a mixture of controls to ensure that trains both slowed down and stopped automatically if signals were ignored.

On a roller coaster, no human decisions are normally taken at all apart from the decision to 'launch'. Even in that case, a variety of controls are installed to ensure that there is no other coaster on the track ahead of the new car. Once in motion, sensors automatically check the coaster's speed and location and apply trim brakes if speeds become excessive. On nearing stations, sensors detect a coaster's location, slow the coaster down and finally stop it automatically. It is technically possible to do all this, and the normal judgement is that a properly designed automatic system is safer than one controlled by a human (the London Docklands light railway is driverless). On all rides, the risk assessment that dictates the control system design will always have regard to what human activities are required and will consider the consequences of human error.

For example, on fierce rides, it ought to be impossible for the coaster to leave its station if passenger restraints are not in place (i.e. via interlocking); whereas for mild rides, it may be considered adequate for attendants to just visibly check that restraints are in place (albeit restraint design geometry should ensure this is highly visible to minimise the risk of attendant error).

Statistically, roller coasters are very safe, but they are complicated engineering systems and a key weakness within them applies when humans intervene. 'The Smiler' is a coaster in a UK theme park and it includes 14 inversions: a world maximum. The mechanical and control system design was configured carefully to ensure that no two cars could crash into one another. Yet in 2015, a loaded train collided with an empty train, causing life-changing injuries to some passengers. The empty train had been automatically halted by the safety system and this was recorded on the control system screens. Maintenance crews were called out to clear the track but in doing this, they overrode the safety system. A passenger-carrying train was then accidentally launched and it collided with the stationary train. Because of the track's complex configuration, the crew had apparently not spotted the stranded carriage. The accident was caused by their error.

6.3 Human error in design/construction

6.3.1 Introduction

All design is a complex business and it is not difficult to get it wrong as many incidents in this book testify. Such failures should not simply all be put down to culpability; rather the industry should recognise that design error is a standard hazard and endeavour to consider how errors can be minimized or to recognise that teething problems are almost inevitable. Even the best design teams face this issue.

6.3.2 Problems with aeroplanes

In 2011, Boeing launched their much-delayed Dreamliner (787), heralded as the new generation of planes offering advanced fuel efficiency and comfort. Unfortunately, a sequence of problems developed, cumulating in a worldwide grounding early in 2013. According to press reports, there was an electrical failure and an emergency landing during the test-flight programme. Then a Japan Airlines flight suffered a fire, traced to a lithium-ion battery used to start the aircraft's auxiliary power unit. Shortly afterwards on another Japanese flight, the pilot was alerted to smoke from an electrical compartment. Again, the problem seemed linked to battery performance. Meanwhile, another investigation by Japanese regulators was taking place into oil and fuel leaks. Press reports also suggested that sensors had detected problems with the plane's braking system and that there has been a cracked pilot's window. In themselves, these issues might turn out to be just teething problems. Undoubtedly, these prove two points: first, that despite huge efforts being devoted to design and development, it is to be expected that not everything will be perfect. The history of engineering is one of gradual improvement based on performance experience. Second, whatever the cause, loss of public confidence and fleet grounding on safety grounds is very expensive and Boeing had the task of arguing that their planes remained safe for passengers despite publicity over their faults.

No company is unique in facing development problems and it is as well to recall that the Dreamliner's main rival is the A380 Airbus. During early flights in 2012, cracks were discovered in the plane wings and safety checks were required on all operating planes. Reports suggest that the cracks were caused by stress generated between brackets and the wing skin, perhaps initiated during manufacturing (where a rivet interference fit was used), perhaps exacerbated by the flexing of the wings during flight. Whatever the cause, the costs of repair were significant, reportedly more than £200 million. Moreover, Airbus, like Boeing, had the task of arguing that their plane remained safe to operate on the premise that the cracks could be repaired progressively during routine maintenance. Like the Dreamliner, problems were not confined to just one issue and the plane has suffered certain engine malfunctions.

The overall lesson from these two giant engineering projects is that it is easily possible to get the design 'wrong' and at least one factor is the communication of intent amongst the many parties involved. Following on from that, it might be concluded that all novel designs (including civil engineering designs) require prudent in-service inspection to confirm that the design is functioning as intended. The contract form adopted may well need to consider that need and allocate risk funds accordingly. Another lesson, as discussed in Chapter 3, is the need for designers to be able to articulate a safety argument for continuing use (clearly only if appropriate), despite the evidence of problems.

6.3.3 Problems with Astute

Many big engineering projects seem beset by delays, budget and technical problems and many management studies have sought to investigate why this might be so. Certainly, it is not just aircraft that have setbacks. In 2013, newspaper headlines criticised the initial performance of Britain's latest nuclear power submarine, Astute: 'HMS Astute: nuclear submarine beset by design problems and construction failures. One of the Royal Navy's multi-billion pound nuclear-powered submarines has been beset with a catalogue of design problems and construction failures, it emerged last night.'

Whatever the truth, the MOD conceded that trials had uncovered teething problems. Somewhat embarrassingly, early in the trials the new submarine ran aground following a catalogue of (human) errors, and its commander was replaced. The substance behind later newspaper reports is unconfirmed but they range over issues from boat speed and hull corrosion to the safety adequacy of some systems. Being aware of this, the designers, BAE, are quoted as saying:

> Safety is of paramount importance to every stage of the design, build, test and trials of a submarine and is at the heart of everything we do. Before entering full service, every submarine is required to complete an exhaustive period of sea trials, which are designed to prove the vessel's capabilities.

In one confirmed problem, a cap on a pipe taking seawater from the back of the submarine to the reactor sprang a leak. As a result, a compartment began flooding, forcing a rapid surfacing of the submarine. The investigation apparently revealed that the cap was made from the wrong metal and had corroded. In safety terms, a number of standard generic issues are illustrated. First, a form of failure: corrosion; second, a consequence of failure: rapid flooding with perhaps disproportionate consequences; third, and perhaps of more concern, the response to the generic question, 'How does a designer know that what he thought was being built has actually been built?' (see Chapters 2 and 5). In this Astute case, it has been reported that the pipe cap metal used was at variance with the QA construction records. Some (human) appears to have erred.

6.3.4 Match between design and construction

The generic issue of what is built being at variance with what a designer thought was being built is a constant theme in reports made to the CROSS Reporting System. A typical example relates to the failures of external masonry walls in some Edinburgh schools in 2016 (SCOSS Alert, 2017); see Figure 6.3. Although no children were harmed, that was only by good luck. Falling masonry weighed some 9 tonnes in total. An inquiry found that the external wall skin was improperly tied back to inner supports even though the design drawings were correct. Although that was the basic fault, a more deep-seated problem was the lack of any contemporary site inspection to verify that what was being built accorded with design intent, and thereafter, the impossibility of inspection after construction: not technical errors but errors in execution by operatives.

Figure 6.3 Wall failure in Edinburgh schools. (Permission received from Blyth and Blyth)

Along with the sagas of the Dreamliner and the Airbus, the stories also show that for safety, it ought to be assumed that whatever the resources put into the design of an engineering product, validation that what is actually built meets design intent is an essential activity and supposedly one covered by a proper quality assurance system

(QA). It may be recalled that this point is also highlighted in the Hackitt Report (Hackitt, 2018; SCOSS Alert, 2018). For more complex products or structures, prototype production and prototype testing to prove desired performance may also be essential or highly desirable. Anything innovative brings risk with it. Moreover, it is to be expected that during that programme, problems will arise. Although those are due to 'human error', the only safe presumption is that they are inevitable. Furthermore, once in service, history suggests that most systems will not remain problem free so that monitoring and modification are the natural order (there have been several mass recalls of car makes due to design faults discovered in service). For civil engineering projects, this poses a real quandary. The Dreamliner, Airbus, Astute and mass car productions are massive repeat projects where design costs can be dispersed over the many units manufactured. Moreover, they are mechanical systems where prototyping, development and testing are the natural order. In many great civil engineering projects, these benefits are impossible. Every major bridge and building is a 'one-off' in terms of design and production and it has to work 'first time'. A major management task is therefore to try and minimise the consequences of human error in either the design or execution stages. Nonetheless, that does not exclude the ability, or need, to carry out in-service monitoring with one justification being to catch the consequences of error before it is too late. Drawing lessons from in-service performance, we might say that any package of measures that constitutes a safety case ought to include a period of in-service checking to ensure that the structure is working as predicted and intended.

6.4 Human error in civil and structural engineering design

Failures attributed to error in civil/structural engineering certainly occur. Sometimes they are simple errors of omission or confusion. Sometimes they arise because the design is sophisticated, and everyone assumes the computer analysis must be correct when actually it isn't. Sometimes they arise because design teams have failed to identify a critical design case. Sometimes they arise because of mismatch between what has been constructed and what was intended to be constructed.

One simple example is that of the complete collapse of a tower crane in Liverpool in 2009. The damage to nearby buildings was considerable whilst the crane driver suffered multiple injuries and paralysed legs. The toppling was initiated by failure of the tower base connection at the interface with its dedicated pile cap. The crane tower anchorage failed in tension and there were two primary causes:

a. an elementary error in the calculations had underestimated the anchor pull-out value by a factor greater than 2.
b. the original 32 mm anchor bars were in the wrong place and so were cut off and replaced by epoxied-in 20 mm bars.

> These were clearly weaker than the original bars but critically, base anchorage into the pile caps was inadequate with some embedment depths being around 400 mm, less than required to achieve full anchorage.

Error (a) arose because proprietary software was used and the wrong data entered. The calculation was in fact trivial and any manual check should have highlighted the error. Error (b) falls into the category of 'Does the designer know that what was assumed was actually built?' Error (b) also violates the principles of building in ductility and controlling preferred modes of failure.

The crane base error might be classed as 'gross'. But it is surprisingly easy to make an undetected gross error. This applies especially when there is blind faith in a complex analysis. In 1991, the Sleipner oil rig collapsed in Norway (MacLeod, 2005). This was a giant concrete platform some 210 metres high with the underwater part consisting of multiple chambers. Stress calculations were made using a finite element program and these examined shear stresses across the chamber walls generated by hydrostatic pressure. Unfortunately, the mathematical model was flawed so that real shear stresses were about double those predicted. Under applied water pressure, the walls cracked through catastrophically and this in turn led to the destruction of the entire facility when it sank to the bottom of the fjord. Direct and indirect costs were huge (Chapter 11).

Overlooking the obvious was a design error discovered on the Citicorp Building in 1978 (Box 6.4) as were the setting-out errors described in Box 6.5.

The issue of not being able to identify a fault or design case in advance, or of overlooking design cases, is well illustrated by the failures of the Tacoma Narrows Bridge (1940s, Chapter 8) or the Comet aircraft in the 1950s (see below) or even dynamic problems on the London Millennium Footbridge (2000) (Box 6.6).

Box 6.4 Citicorp Building, Manhattan, USA (1978)

This building was 280 metres high. Almost by accident it was discovered that the wind analysis was fundamentally flawed. In all towers there are two basic wind conditions to consider: (a) wind blowing face on which is resisted by a couple, normally provided for by (at least) two columns spaced apart by the building's width and (b) winds blowing diagonal on, usually resisted by one pair of corner columns but with a lever arm equal to the building's plan diagonal.

Rather unusually at Citicorp, base columns were not located in the corners but were centred on the four sides i.e. there were no corner columns. In this configuration, it turned out that wind blowing across the diagonal should have dominated the design, but this was overlooked. On re-checking it was found that the (completed) structure was far too weak and expensive remedial measures were required. The wind resistance had been miscalculated: it was a basic design error. Brady (2014) gives a full account.

Box 6.5 Case study: French trains (2014) and other measurement errors

In 2014, the French national railway company realised that its 2,000 new trains were too wide to fit through many regional stations. It had been overlooked that older platforms were matched to a narrower train gauge. Press reports suggest that about 1,000 platforms had to be adjusted at a cost of about 50 million euros.

Closer in time, the Laufenburg Bridge was being constructed between Germany and Switzerland. Instead of the two sides being the same height above 'sea level' they differed by 54 cm. It was apparently known that the German and Swiss definitions of 'sea level' differed by 27 cm, but compensation erroneously doubled rather than eliminated the difference.

The author once worked on major project where the site datum on two adjoining buildings was exactly 1 metre different: highly dangerous!

A lesson from all these incidents (and the Mars Orbiter (Section 6.5, below)) is to remind us how easy it is to have a 'gross error'. Never assume that the obvious has been checked.

6.4.1 The Comet Aircraft

The Comet was the first commercial jetliner (1952) and like the Airbus and Dreamliner, it was supposed to herald a new era of travel. Unfortunately, within a year of introduction, three jets crashed with great loss of life. The cause was metal fatigue, which was supposedly not well understood at the time, though certainly understood as a phenomenon. In safety study terms, a point of interest is that the cause of the first crash was incorrectly diagnosed. It was thought to be due to bad weather. Following later crashes, this cause was discounted and a very thorough programme of testing was conducted, pinpointing fatigue as the culprit; partly due to the stress variation driving the cracks, linked to changes in pressurising and depressurising the cabins and partly due to stress concentrations around the windows. Tests to failure indicated a fatigue life which was comparable with that endured by the in-service flights before they crashed. A further finding was that the punch rivet construction technique employed had probably been an exacerbating cause. Inevitable imperfections around the holes could have started fatigue cracks around the rivets (Brady, 2017).

A lesson we might draw from the Comet failure is that every failure needs to have a thorough examination and during that examination, investigators need to be careful not to jump to conclusions. True confidence can only be achieved when the form of failure is replicated. If the cause is known, then normally a solution can be found. In the case of the Dreamliner, although early suspicions focused on the battery (not least since this had been burnt out), investigations failed to identify a precise cause. In safety terms, intermittent faults are more dangerous than identified ones. The Boeing 737 MAX crashes of 2019 are another example. There were two crashes. The first was originally assumed to be pilot error and it was only when a second crash occurred that had strong

Box 6.6 Millennium Footbridge, London, UK (2000)

The Millennium Footbridge was an innovative design and the winner of a competition. On opening day, the deck became fully crowded and began to sway alarmingly, to the extent that the bridge had to be closed. From pictures, the cause of closure was a lateral bridge sway sufficient to persuade foot passengers to begin walking in synchronisation, which then had the effect of reinforcing and amplifying the sway motion. The solution was to add damping. The designers maintained the cause was a new phenomenon (Dallard et al., 2001).

similarities with the first that a thorough investigation was made (US House Committee on Transportation and Infrastructure, 2020).

6.4.2 London Millennium Footbridge

In everyday structural engineering, oscillation is a rare design condition. Many designers are unused to dealing with structural dynamics, so oscillations introduced by wind on slender structures or by footfall on long-span structures are (by error) not that uncommon. Occasionally even the most experienced of design teams are caught out as on the Millennium Footbridge over the River Thames in London (Box 6.6).

A second illustration of how even prestige structures can go wrong is given by the failure of the Charles de Gaulle Airport Terminal (Box 6.7). The cause was a detailing

Box 6.7 Charles de Gaulle Airport terminal failure, Paris, France (2004)

Charles de Gaulle Airport was the showcase airport for Paris. Early one morning, part of Terminal 2E collapsed completely, killing four people.

The airport's terminal building had a very architectural and novel form. In structural terms, the inner 300 mm precast concrete shell supported thrusts from metal posts supporting the outer glazing. It appears that failure was caused by these posts punching through the inner concrete, perhaps because they were embedded too deeply, perhaps because of some construction misalignments (the official report remains sub judice). There is also evidence that cut-outs in the concrete structure had weakened it and created cracking, which was steadily progressing under the influence of fluctuating thermal stress. Signs of trouble had been evident before the final failure, with some concrete spalling off. The temperature during the week before the collapse was around 25 °C, but on the morning of the failure it had dropped to around 4 °C and there are press suspicions that this shift was the final trigger.

A punching-type failure is not an unknown technology, so this failure can therefore be put down to human error. The costs were significant, as the authorities decided demolition was cheaper than repair, and reported costs were 100 million euros (Wood, 2005).

error so might be looked at under the same category as the Edinburgh schools' failures, i.e. a human error along with the need to match design intent with detailing and construction in practice.

An overall lesson we might draw from this is to be particularly aware of error when the element being checked is repeated many times, for then there is a chance of the same error having quite disproportionate effects. One very long bridge in the UK has an inspection gantry below it. To ease the running of the gantry wheels, the joints in the lower flange of the running beam were scarfed in plan. Unfortunately, this meant that as the gantry wheel pair crossed the scarf, there was a condition of load eccentricity on one flange side only. This load case was missed by the designer and by the checker and of course it applied for every single stretch of beam that made up the gantry support length. The error was not picked up until after installation and while the gantry was being trialled.

Most of the incidents referred to above were dramatic and costly failures; all were illustrative of the reality that gross error can occur and be overlooked very easily (just as in the Financial Crash described in Section 6.1).

6.5 Gross error and communication

In Chapter 2 references was made to an Institution of Civil Engineers report on 'Safety in Civil Engineering' (COMSAFCIVENG and Shirley Smith, 1969) highlighting ongoing problems and great cost. One of the cited possible causes of failure was:

> (c) lack of proper communication at all levels e.g. between Engineers and Contractor, office, site, supplier, and manufacturer.

A particular example of (c) showing the ease of communication errors is the story of the Mars Orbiter Climate Spacecraft in 1998. This is a near-legendary failure. The Orbiter went off course, apparently because the firm performing the calculations (Lockheed) were using imperial units whilst NASA was expecting metric units. Consequently, the spacecraft went into orbit around Mars too close to the planet's surface and disintegrated. Whilst this is a good example of human error and of how it is quite possible to make a 'gross mistake', an interesting supplementary question is to assess why the error was not picked up. Reports suggest the error was 'small' so the consequences did not stand out and deviations in the spacecraft's flight pattern crept over time. NASA's reported reaction was commendably not so much to assign blame but to look closely at its systems to figure out why the error in data transmission had not been spotted. For safety, management of the safety process is just as critical a task as is comprehension and confidence in the technology. Knowing data is correct is a vital ingredient in the overall safety process. On top of that, a lesson is that when the performance of any engineering system starts to look incorrect, it is as well to establish cause as early as possible, without assuming anything.

BOX 6.8 Pressure vessel, UK

A pressure vessel was designed in Switzerland and manufactured in Britain. The designer specified fillet welds of a particular size. Unfortunately, at that time, Swiss practice defined weld size by throat whereas British practice was to define by leg length (i.e. welds would be smaller). There is a 30 % difference in capacity and about 50 % difference in metal volume between the two nomenclatures. Each party simply misunderstood the other. It was fortunate indeed that the difference was spotted before the pressure vessel was tested.

Perhaps an even better example of human error / gross mistake can be seen in the failure of a Russian weather satellite in December 2017. The satellite contained 18 smaller satellites belonging to research and commercial companies from Russia, Norway, Sweden, the US, Canada, Germany and Japan. Unfortunately, contact was soon lost and it transpired that the on-board computer was programmed with co-ordinates from the wrong launch site. Losses have been put at $45 million (BBC News, 2017). (Another example of errors in dimensions is given in Box 6.8). Perhaps space vehicles are uniquely unlucky? In 2001, NASA's Genesis space probe was lost on its return to Earth because its parachute failed to open. Apparently, the switches designed to trigger release had been installed back to front. The lesson for civil engineering structures is obvious: be careful not to instal members upside down.

None of this is new; such mistakes have always been with us. In 1628, the brand new Swedish warship Vasa sailed from Stockholm but capsized after a short distance. It was said to be top heavy. But when the ship was raised in 1961, a bulge was found on one side; the hull was asymmetrical. It transpires that both sides had been set out in feet but the shipwrights on one side used feet in Amsterdam units whilst on the other side, the workers used feet calibrated in Swedish units. The difference was significant. As an example of lessons not being learned, the launch of Nelson's Victory in 1805 was nearly delayed when it was realised that the dry dock gates were too narrow (this was solved by some rapid hacking-off).

Currently, UK and US practice differs. Any project due to finish on 8/8/2020 will be acceptable to all parties. But one might be surprised to learn that a project apparently due to complete on 12/1/2020 (1 December in US usage) was intended to complete on 12 January (UK usage). CROSS has had reports of errors in textbooks, including one set of tables where co-ordinate properties were erroneously swapped over, and it's certainly not rare to have codes of practice issued with amendments or corrections.

Another, and earlier, NASA failure in which communication was cited as a contributory factor is that of the Challenger space shuttle. The seven-member crew was lost when the flight broke up during launch in January 1986. The failure was recorded on film and left the world stunned. President Reagan set up a commission of enquiry which eventually determined that the loss was due to failure of some O-rings whose function was to maintain a tight seal when rocket pressure distorted the structure of the solid fuel booster. Pictures from the moment of failure showed black colouration and smoke puffs suggesting that the rubber O-rings and joint seal were being both burned and eroded by propellant gases. The investigations revealed that the O-rings were unsuitable for use at low temperatures, losing their resilience to the extent of failing to provide a seal (at low temperatures they could not recover their shape fast enough). The temperature on the day of the disaster was much lower than it had been on previous launches.

Whilst this physical explanation was discovered with confidence, more disturbing was the background. The President's Commission (Rogers) (Presidential Commission on the Space Shuttle Challenger Accident, 1986) concluded:

> [T]he cause of the Challenger accident was the failure of the pressure seal in the aft field joint of the right Solid Rocket Motor. The failure was due to a faulty design unacceptably sensitive to a number of factors. These factors were the effects of temperature, physical dimensions, the character of materials, the effects of reusability, processing, and the reaction of the joint to dynamic loading.

It will be noted that the word sensitivity is used and it is a characteristic of safe design that the sensitivity of the whole to minor errors in the parts should be understood and addressed (see Chapter 5). The safety studies of the Challenger had shown the failure of the O-rings during the initial part of the flight was critical in the sense that no recovery was possible if anything went wrong. The Commission also concluded:

> The decision to launch the Challenger was flawed. Those who made that decision were unaware of the recent history of problems concerning the O-rings and the joint and were unaware of the initial written recommendation of the contractor advising against the launch at temperatures below 53 degrees Fahrenheit and the continuing opposition of the engineers at Thiokol after the management reversed its position. They did not have a clear understanding of Rockwell's concern that it was not safe to launch because of ice on the pad. If the decision makers had known all of the facts, it is highly unlikely that they would have decided to launch 51-L on January 28, 1986.

And 'That testimony reveals failures in communication that resulted in a decision to launch 51-L based on incomplete and sometimes misleading information, a conflict between engineering data and management judgments, and a NASA management structure that permitted internal flight safety problems to bypass key Shuttle managers.' There cannot be many clearer statements of the need for proper communication and clear comprehension of all the risks. After the accident, there were some who suggested it was the price mankind pays for progress. Equally there were many who argued it was a perfectly foreseeable and preventable failure. Overall, the Rogers Commission concluded that the Challenger disaster was 'an accident rooted in history'.

Other reports raised serious concerns about NASA management's perceptions of safety. One of the Commission members (Feynman) suggested their management perceptions of risk / shuttle reliability were significantly at variance with what their engineering teams thought. Feynman singled out the project's software development for praise because this was always thoroughly checked. But he observed that because the software tests had always passed, NASA management wanted to reduce the amount of testing to save money (a review of the structural technical press over the years will reveal those who argue that our material safety factors must be too high since nothing falls down!). Whilst there always needs to be a balance with resources deployed, it is absurd logic to stop testing just because no faults have been found. In another example, tests had shown the Challenger O-rings could partially burn through in operation. Rather than be disturbed that a crucial safety element was being subjected to damage with its safety margin being eroded, management had apparently taken the opposite view that partial burning through represented some element of robustness.

The collapse of the Hyatt Regency Hotel walkway in 1981 (Box 6.9) is the classic example of miscommunication in structural engineering and an example of failure consequent on change (Brady, 2015).

If the theme of ensuring design intent is pursued, it will be concluded that any design is only safe if the design demands are properly detailed and if those details are properly communicated to the constructors and if that execution is verified. All developing designs are subject to change. Therefore, communication is never more vital than when it relates to changes, as discussed below.

6.6 Management of change

The Hyatt failure (Box 6.9) came about at least partly because a change was made and that change inadvertently changed the load transfer system. Similarly, the Grenfell Tower tragedy (Chapter 9) came about when the original unclad building was changed by over-cladding with a flammable material. Hence, as a generality, all changes need to be managed in case the change inadvertently undermines the original concepts of safety. This certainly applies during project development until the project is completed. During this period, evolutionary change is continual; many different disciplines may be involved, and there is great danger that the 'wrong' design will get built. It also applies throughout a structure's life since many buildings and plant are in an almost continual state of upgrade and modification. There is nothing especially unusual about that save to say there are associated risks.

The Grenfell Tower disaster sparked a review of the Building Regulations. This was conducted by Dame Judith Hackitt and within her report (Hackitt, 2018) she concluded: 'Change control and quality assurance are poor throughout the process. What is initially designed is not what is being built, and quality assurance of materials and people is seriously lacking.' Alas, none of this is new.

A serious failure consequent on change occurred at the UK Flixborough Plant in 1974. The plant was destroyed, 28 people were killed and 36 were injured. Two months prior to the explosion a crack had been discovered in one reactor, so it was decided to install a temporary bypass to allow continuing plant operation. The official inquiry into the accident (Department of Employment, 1975) determined that the bypass pipe

Box 6.9 Hyatt Regency Hotel walkway, Kansas City, USA (1981)

This was a key collapse in the history of failures, which occurred in an elevated hotel walkway, killing 114 people and injuring 216 more. The cause was relatively simple. Originally there were two walkways above each other and both were suspended via a single tie rod passing through the supports (Figure 6.4). This proved difficult to erect and so the design was changed to that shown in the figure. A glance at the load path will show that this change has doubled the load on the original nut. Because of that extra load, the nut failed and the whole structure fell down. At the time it was considered the worst structural collapse in US history.

As with most failures, although there was a prime initiating cause, the notability of the failure was due to several effects. Both designs required load to be transferred through channels joined by welding. During failure, these welds split and the nut slipped through the gap. There was also considerable confusion in the communication and responsibility for the change. Sketches showing preliminary proposals were interpreted as final design. There was no review of the design and the change was made without any calculations being carried out.

Investigators concluded that one cause was lack of proper communication between the designers and the fabricators. The latter thought they had finalised drawings, but they did not.

Comment: This failure reflects several themes:

1. confusion in communication;
2. vulnerability to key details (see Chapter 5 on single point failures);
3. lack of robustness (Chapter 4);
4. change control (see Section 6.7);
5. interaction between design and construction (Chapter 8).

Although this tragedy is included under the 'change' heading, it is also a good example of problems due to communication, detailing, and interaction between design and construction, all of which are used as thematic headings elsewhere.

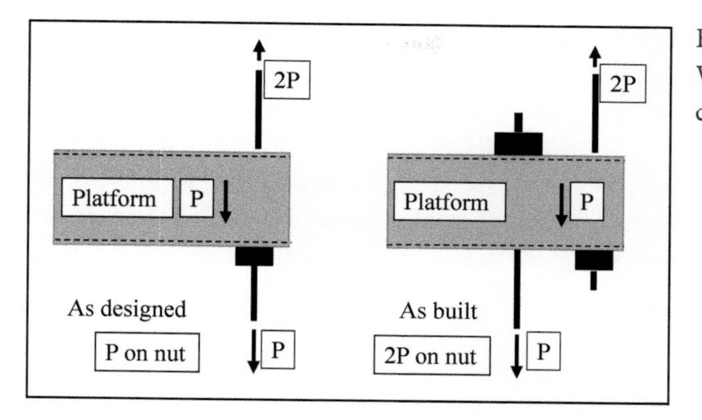

Figure 6.4. Hyatt Regency Walkway failure: original concept and 'as built'

had failed because of unforeseen stresses created by a pressure surge. The bypass had been designed by personnel who were not experienced in high-pressure pipe work, no plans or calculations had been produced. The pipe, which was not pressure tested, was only supported on temporary scaffolding poles, which allowed it to twist under pressure; and there had been no design review. We can see here the ingredient of commercial pressure to continue with operation combined with lack of control over changes being made.

An even worse incident occurred on 1988 when the Piper Alpha North Sea Oil Platform was destroyed (see Chapter 11 for a fuller report). The initiating event of the incident (a fire) was put down to failure of a pump that had been out of operation but was brought back in line. There were two of these pumps (A and B) and the A line had been taken out of service for maintenance, which included removal of its safety valve. This pump pipe outlet was temporarily sealed off with a disc. Documentation laid out that the pump was to be overhauled and supplementary transfer documentation laid out that it was not to be used under any circumstances (because the overhaul had not been completed). When the following staff shift came on, no one was verbally informed of this condition[3], though the non-use documentation was left in the control centre. During the shift, a pipe blockage caused pump B to stop and it could not be restarted. As the entire power supply of the offshore construction work depended on this pump, the manager had only a few minutes to bring the pump back online, otherwise the power supply would fail completely. A search was made through the documents to determine whether pump A could be re-started. Whilst the overhaul permit was found, the non-use permit was not found. The valve was in a different location from the pump and permits were stored in different boxes. Consequently, no one on the shift was aware that a vital part of the machine had been removed and the assumption was made that starting up pump A would be safe. No one noticed the missing valve, particularly as

3 A shift change was also a feature in the causes of the Chernobyl disaster which occurred during a test, The timing of the test was put back so the (experienced) shift was replaced by the less experienced night shift.

the metal disc replacing the safety valve was several metres above ground level and obscured by machinery. When pump A was switched on, gas flowed through it, bursting off the blanking disc. The exiting gas soon ignited, setting off the chain of events that cumulated in disaster.

As with most disasters, there was no single cause and many safety lessons can be learned from Piper Alpha. In this chapter, it serves to illustrate aspects of human error: failures of communication, failures in authorisation and documentation and the need to control all changes carefully. There is also the common theme of pressure to maintain production in parallel with making changes.

There have been several structural failures consequent on making changes when the consequences have not been considered. These range from domestic house collapse to large-scale structural disasters. There are many examples of basement excavations or house 'opening up', causing collapse. The cause is either no one knew what they were doing when making a change or they failed to grasp the significance. Chapter 8 will cite the examples of offshore rig collapses which came about by the change of welding on small attachments to underwater legs, thereby degrading their fatigue life significantly. The change consequences had not been understood or foreseen. The site changes on the West Gate Bridge provide another illustration (see Section 6.7.2).

Everyone working on documents and drawings will be familiar with changes being made in one area and not being carried through, thence leaving ambiguity or conflicting information. Any change risks switching inadvertently from a condition of acceptability to unacceptability. The principle should be that for any engineering enterprise, there should be no unauthorised changes, and no one should be making even authorised changes without fully understanding the implications. On properly run projects, there ought to be a design authority and a change order procedure. The role of that authority is to retain ownership of the overall safety logic and thence to verify and authorise any change just in case those wanting to make a change have not grasped all the implications.

6.7 Management overall

6.7.1 Design management

The need for management and control of data in this world of globalisation and many languages should be self-evident. Our inability to say what we want clearly and precisely leads to many misunderstandings. If the facts were otherwise, there would be nowhere near so many contractual disputes between parties who thought their conditions of contract said one thing whilst their opponents thought the opposite. Lack of data control was cited as an issue for the Mars Orbiter and a failure to use data, even when it was available, was cited as a contributor to the Heathrow Tunnel collapse. What should stand out is that in any engineering project, the needs of safety should be addressed positively as a special topic and the whole process managed so that decisions are taken against the background of what governs safety. That management process should include an overall strategy for ensuring safety and in turn, that needs to be based on clarity of what is safe and what is unsafe, what critical lapses might be and how those can be avoided. The safety attributes of sensitivity, robustness, etc., all as set out in Chapter 4, would all have been pertinent to the accidents discussed above. Overall, possibilities of human error abound.

6.7.2 Construction management

The principles of management and communication are relevant to all engineering projects, be they spacecraft or civil engineering. One example of tragedy consequent on human failing is the story of the West Gate Bridge (also known as the Yarra Bridge) collapse in Australia in October 1970. A fuller report on the collapse is given in Chapter 8 but for now it serves to illustrate that human failings were a major contributory factor and there was severe contemporary criticism of the main bridge designers over their role. As background, this was a complex bridge and at the time, the ability to understand how box girder bridges 'worked' was known only to a few, who were reluctant to share their knowledge for reasons of commercial advantage. The team in charge of assembling and erecting the bridge were relatively inexperienced and certainly lacked knowledge of box girder bridge design. Very little assistance was given to them by the designers. The plan to erect the boxes relied on them being fabricated in two sections (split down the middle of the bridge deck), being lifted up, and then seam joined down the centre splice by bolting. At the time of the incident, attempts were being made to join up two sections which did not exactly match in longitudinal profile. To try and achieve a match, bolts were removed and kentledge applied. The result was that the subsections were overloaded, buckled and collapsed. As with the Challenger disaster, the technical causes of the collapse were one thing, but the roots that led to it were another. The Royal Commission (Royal Commission of Inquiry into the Failure of West Gate Bridge, 1971) stated that:

> To attribute the failure of the bridge to this single action of removing bolts would be entirely misleading. In our opinion, the sources of the failure lie much further back; they arise from two main causes;
>
> Primarily the designers of this major bridge, FF&P failed altogether to give proper and careful regard to the process of structural design;
>
> A secondary cause leading to the disaster was the unusual method proposed by WSC for the erection of spans 10–11 and 14–15.[4]

4 FF&P: Freeman Fox and Partners; WSC: the original erection contractors, later replaced by JHC.

In addition, the Commission made other criticisms, including:

- the duality of responsibility between FF&P and Maunsell for the design;
- the poor personal relationships between the parties;
- the dismissal of WSC (for poor performance) and the subsequent appointment of the inexperienced JHC to complete the contract;
- the tardy way in which FF&P dealt with site queries and the fact that some queries were not responded to at all;
- the inadequate investigation into the use of kentledge before authorising its use;
- the use of K-bracing as part of the internal structure of the trapezoidal elements.

In short, relations between the designers and contractors were very poor indeed and that was a major contributory cause of the disaster.

The Royal Commission were scathing in their condemnation of certain parties (Royal Commission of Inquiry into the Failure of West Gate Bridge, 1971; Brady, 2016). Similar culpability could be assigned to the management of the Herald of Free Enterprise (Section 6.2 above) who had apparently been warned of the dangers of not closing the ship's bow doors before sailing, but ignored those warnings. The US committee of investigation into the Boeing 737 MAX incidents were highly critical of safety management within Boeing (US House Committee on Transportation and Infrastructure, 2020).

6.7.3 Maintenance management

All plant and all structures require maintenance and Chapter 7 will describe failures brought about by through life degradation. Systems are unsafe if maintenance is not carried out and systems are unsafe if maintenance and inspection are not carried out properly. Part of safety management is to make sure it is. One example of this was the train derailment at Grayrigg in Cumbria in 2007. In this accident, a passenger train came off the tracks and rolled down the embankment; amazingly only one person died. The investigation concluded that the derailment had been caused by a set of faulty points. The culprit was the bars which kept the twin running tracks in line (stretcher bars). Of the three bars present, one was not in position, another had nuts and bolts missing, and two were fractured. Underlying this accident was the fact that a scheduled inspection had not been carried out (the stretcher bar degradation had been taking place over about 11 days). The Rail Accident Investigation Branch (RAIB) (Rail Accident Investigation Branch, 2017) confirmed that a measurement train had surveyed the track just two days beforehand but the way this was done could not have picked up problems such as those causing the accident. That detection could only be done by slower, visual inspection. Network Rail was fined around £4 million.

The RAIB report includes the cause in more detail (referring to the faulty points numbered 2B) as:

> The bolts holding the third permanent way stretcher bar to the right-hand switch rail became loose, and subsequently completely undone. As a result of this, and the excessive residual switch opening, the left-hand switch rail was struck by the inner faces of passing train wheels, giving rise to large cyclic forces. As a consequence, rapid deterioration of the condition of the remaining stretcher bars and their fasteners occurred. This led to the left-hand switch rail becoming totally unrestrained.

The RAIB also stated:

> This situation arose at 2B points because of a combination of three factors. These were: the failure of the bolted joint connecting the third permanent way stretcher bar to the right-hand switch rail; incorrect set up of the points with excessive residual switch opening; and the omission of the scheduled weekly inspection on 18 February 2008. All three were necessary for the accident to occur.

Overall, the RAIB concluded with: 'There were a number of shortcomings in Network Rail's safety management arrangements which were underlying factors in this accident.'

Hence, as usual, a number of separate factors combined to initiate the failure, but the technical failings were not picked up because the inspection/maintenance regime was inadequate (human failings). The accident at Grayrigg in 2007 was similar in cause to that at Potters Bar in 2002 (seven dead) (HSE, 2003). The HSE established that the Potters Bar cause had been faults in the points, again due to bolts holding stretcher bars in place becoming loose, so allowing rails to move apart. As at Grayrigg, criticisms were made of maintenance and inspection regimes. A generic safety issue is how to spot that safety critical items are deteriorating before that degradation gets to a point that is dangerous. Nuts working loose under vibration are always a potential hazard, so ought to be a hazard considered in the design/maintenance of any structure supporting moving loads. In circumstances where this might be important, nuts are often coated with a paint stripe to make it obvious to an inspector that some rotation has occurred. Modern lorries have indicators fitted onto their wheel nuts to make it obvious that their nuts remain tight (or not).

6.8 Safety culture

The report of the Heathrow Inquiry includes the paragraph: 'those involved in projects with the potential for major accidents should ensure they have in place the culture, commitment, competence and health and safety management systems to secure the effective control of risk and the safe conclusion of the work'.

To help prevent accidents, organisations should have a 'safety culture'. The idea of safety culture developed after the Chernobyl disaster (1986) and enquiries since then into major accidents (like the King's Cross fire, 1987 or Piper Alpha, 1988) have cited a lack of safety culture as a contributory factor. The HSE (Human Engineering, 2005) have produced a definition:

> The safety culture of an organisation is the product of individual and group values, attitudes, perceptions, competencies, and patterns of behaviour that determine the commitment to, and the style and proficiency of, an organisation's health and safety management.

This is made readily understandable by the failure of the Challenger Space Shuttle (Section 6.5). Beforehand, the culture had been one of continuing unless the shuttle could be shown to be unsafe. Afterwards, it was one of continuing only if the shuttle could be shown to be safe. NASA's shift in attitude could well be beneficially applied to many structural engineering projects. It will be noted that in many of the accidents cited in this chapter, an underlying theme has been for production to continue and for the needs of commerce to take precedence over the prudence of safety. That same theme is echoed in the report on the Boeing MAX crashes (US House Committee on Transportation and Infrastructure, 2020) and the report includes a criticism of a lack of safety culture.

Prioritising production or programme has been a factor in many failures. In 2011, part of the new Dutch FC Twente stadium roof collapsed during erection. The causes were several (see Chapter 8, Box 8.36) but the official government report noted that a contributory factor was undue haste and pressure. 'It appears that overly-tight completion dates were allowed to drive the programme in an unsafe manner. The programme, whilst important, must never over-rule safety.' As for the Herald of Free Enterprise or Piper Alpha disasters, there was tension between the desire to maintain production and progress, and the need to be sure that continuation was safe. In the Grenfell Tower disaster (Chapter 9), there seems to have been commercial pressure to choose a cheap envelope cladding. The one selected had lower fire resistance as compared to the more expensive option. The CDM regulations in the UK place responsibilities on the client, not least the need to ensure that sufficient resources and time are allocated to all steps in the design and construction sequence.

6.9 Chapter summary

Human beings are fallible. They make mistakes, sometimes innocently, sometimes culpably. Humans get tired and distracted. Indeed, fatigue is cited as one of the principle underlying causes in many road accidents. It is not for nothing that moving parts on machines are guarded. The evidence from past disasters is that mistakes are easily possible, including gross mistakes, and these can go undetected until it is too late. Many studies suggest that human failings are a prime underlying cause of engineering failures.

- Humans have different abilities, different temperaments, different attitudes and different priories and this has to be recognised in any plan that seeks to ensure overall safety. Likewise, organisations have different methods of working and different attitudes and different cultures. Such differences also occur between different countries. The safe execution of any project should recognise those facts.
- The presumption on any engineering project should be that a human error is likely. That may be in the design, it may be in the execution of the work or in the subsequent management of the product in service. Consequently, in assessing any project's safety, as well as looking at all the technical hazards that might be relevant, it is equally imperative to look at the hazard, risk and consequence of human error.
- If the consequence is severe, then efforts need to be made to design out the risk. After the Croydon tram crash (Rail Accident Investigation Branch, 2017), design changes were made to minimise the risk of drivers falling asleep and to minimise reliance on reacting to a slow-down signal.
- What we should not do is just shrug and assign blame. The challenge is to devise a design and to devise a project management system that recognise the possibility of error but seek to eradicate it or at least mitigate the chances of it occurring, especially if the consequences are adverse and safety related. If someone can inadvertently make an undetected operational mistake, the design is at fault.
- Discussing the possibility of human error might appear to concede incompetence, but that would be unfair. The task of producing a major project from conception to opening is immensely complex; there is a huge amount of technical information to assimilate and the challenges of doing that are immense. Considering the possibility of human error is merely facing up to that reality (Carpenter, 2008; Mann, 2011).

This chapter has highlighted various themes:

- It is painfully easy 'to miss the obvious'. In such cases, it is of little value to be able to assign blame: the damage will have been done. A better approach is to presume someone will err. So even where responsibilities might be properly defined, overall management should seek to minimise the risk of those responsible making an undetected error (see Chapter 13).
- There is a danger on any large project of poor allocation of responsibilities. As the Queen found out: 'At every stage, someone was relying on somebody else and everyone thought they were doing the right thing.' Safety management should be an integral part of any project's plan. There need to be clear lines of responsibility and proper management with checks and balances, i.e. proper control of risk.
- Communication is an ever-present problem. It is bad enough in a community that purports to speak one language. In a world of international contracts with different languages, the risks of misunderstanding are even greater. The UK construction workforce now includes a very significant proportion of overseas workers whose mother tongue is not English; that raises the risks of misunderstanding.
- A recurrent theme is a mismatch between what designers thought was going to be built and what was actually built. That might be a consequence of poor communication or simply of misinterpretation, or it might be due to lack of care. In all cases, it is prudent to verify that the construction is matched to design intent. This involves some form of QA plan and some form of on-site inspection.
- With any innovative design there should be a degree of prototyping and checking matched to the confidence of the design team (and regulators if appropriate). The objective should be to ensure that enough technical knowledge is available to eliminate the possibility of critical error. What is to be hoped is that there is no fundamental design flaw, including any linked to error or human ignorance, of the technical demands. For this reason, some major projects adopt a strategy of design based on 'proven technology'. Whilst undoubtedly less risky, if all designs were based on that policy, no progress would ever be made. Nonetheless, no one should assume that the 'best' design teams are infallible nor that 'modern' knowledge is good enough to prevent failure. In recent history, the failure of the Charles de Gaulle Airport terminal and the failure of the Millennium Footbridge are examples. On any innovative designs, third-party checking is highly desirable (see Chapter 13).
- In service, nothing should be changed unless under the control of a 'design authority' who can be confident of foreseeing all implications (a corollary of this is that proper engineering records should be kept of all projects).
- A safety culture should be adopted. One aspect of that is to be very careful that commercial pressures do not override safety concerns.

SCOSS has promoted the principles of the 3Ps: People, Products and Processes (Chapter 13). People are first in the ranking simply because people, and their behaviour, are so key to achieving safety.

References

BBC News (2017), 'Failed satellite programmed with "wrong co-ordinates"', www.bbc.co.uk/news/technology-42502571, 28 December.

Brady, S. (2014), 'Citicorp Center Tower: failure', *The Structural Engineer*, **94**(2).

Brady, S. (2015), 'Hyatt Regency: the human price of failure', *The Structural Engineer*, **93**(5).

Brady, S. (2016), 'West Gate Bridge collapse – the story of the box girders', *The Structural Engineer*, **94**(10).

Brady, S. (2017), 'Beyond the limits of imagination: what do the Comet aircraft failures teach us?', *The Structural Engineer*, **95**(9).

Carpenter, J. (2008), 'Safety risk and failure: the management of uncertainty', *The Structural Engineer*, **86**(21).

COMSAFCIVENG and H. Shirley Smith (1969), 'Report of the Committee on Safety in Civil Engineering', *Proceedings of the Institution of Civil Engineers*, **42**(1), 143–152.

CROSS (undated), 'Confidential Reporting on Structural Safety', www.structural-safety.org.

Cullen, W. D. (1990), *The Public Inquiry into the Piper Alpha Disaster*, Vols 1 and 2, HMSO.

Dallard, P., T. Fitzpatrick, A. Flint, A. Low, and R. Ridsill Smith (2001), 'The Millennium Bridge, London: problems and solutions', *The Structural Engineer*, **79**(8).

Department of Employment (1975), *The Flixborough Disaster*, Report of Court of Inquiry, HMSO, www.icheme.org/media/13689/the-flixborough-disaster-report-of-the-court-of-inquiry_repaired.pdf.

Hackitt, Dame Judith (2018), *Building a Safer Future. Independent Review of Building Regulations and Fire Safety: Final Report*, CM 9607.

HSE (1994), 'Collapse of NATM tunnels at Heathrow Airport', HSE.

HSE (2003), 'Train derailment at Potters Bar 10th May 2002. A Progress Report by the HSE Investigation Board'.

Human Engineering (2005), 'A review of safety culture and safety climate literature for the development of the safety culture inspection toolkit', Research Report 367, prepared by Human Engineering for the HSE, www.hse.gov.uk/research/rrpdf/rr367.pdf.

MacLeod, I. A. (2005), *Modern Structural Analysis: Modelling Process and Guidance*, ICE Publishing.

Mann, A. P. (2011), 'Passing on knowledge', *The Structural Engineer*, **89**(10).

Presidential Commission on the Space Shuttle Challenger Accident (1986), 'Report of the Presidential Commission on the Space Shuttle Challenger Accident', Washington DC, https://history.nasa.gov/rogersrep/genindex.htm.

Rail Accident Investigation Branch (RAIB) (2017), 'Derailment at Grayrigg 23 February 2007', Report 20/2008 v5, www.gov.uk/raib-reports/derailment-at-grayrigg.

Rail Accident Investigation Branch (RAIB) (2017), 'Overturning of a tram at Sandilands junction, Croydon 9 November 2016', Report 18/2017, https://assets.publishing.service.gov.uk/media/5a2a6294ed915d458b922f27/R182017_171207_Sandilands.pdf.

Royal Commission of Inquiry into the Failure of West Gate Bridge (1971), *Report of Royal Commission into the Failure of West Gate Bridge*, www.parliament.vic.gov.au/papers/govpub/VPARL1971-72No2.pdf.

SCOSS Alert (2017), 'Inquiry into the construction of Edinburgh Schools', www.structural-safety.org/media/397456/scoss-alert-inquiry-into-the-construction-of-edinburgh-schools-final-20-february-.pdf.

SCOSS Alert (2018), 'Summary of *Building a Safer Future. Independent Review of Building Regulations and Fire Safety: Final Report*'.

Temporary Works Forum (TWF) (1994), 'Collapse of NATM tunnels at Heathrow Airport (1994)', www.twforum.org.uk/about-us/famous-failures/heathrow-tunnels-1994/.

US House Committee on Transportation and Infrastructure (2020), 'Final Committee Report on the Design, Development and Certification of the Boeing 737 Max'.

Wood, J. G. M. (2005), 'Paris airport terminal collapse: lessons for the future', *The Structural Engineer*, **83**(5).

7 Material and product failures

7.1 Introduction

7.1.1 Introduction

With care and proper detailing, building materials can last; there are wooden structures hundreds of years old and brick structures thousands of years old. But even tough stone erodes, and the norm is that all materials degrade over time. Safety requires an understanding of what may happen, plus an ability to understand relevant degradation processes and thereby predict rates of property change. Although long experience has taught designers many skills in traditional detailing, for example, on how to mitigate the persistent effects of water, experience also shows that material degradation can surprise and cause failure (Hewlett, 2021). Hence designing for durability is as important as designing for initial strength. When assessing an existing structure's safety, a key aspect is to consider its material state: a numerical value of safety only exists at a point in time linked to the material properties that coexist. Nowadays there is much pressure on sustainability, plus reuse of components; so, designing for durability, and being able to predict life, are increasingly important skills.

There are many estimates of corrosion costs to national budgets and suggestions that total costs are a noticeable percentage of the world's GDP. In the US, there are reports of huge annual sums to replace structurally defective bridges and maintain decaying ones (the importance is so high it features in State of the Union addresses[1]). As in the UK, one cause of such decay is poor drainage and the use of winter de-icing salts. In the UK, the Highways Agency has a significant budget just to maintain its bridge stock. The acknowledgement of such repair, and its scale, is evidence of structural weakening, and provides an alert to account for corrosion in any periodic

1 Every four years, the American Society of Civil Engineers' 'Report Card for America's Infrastructure' depicts the condition and performance of American infrastructure in the familiar form of a school report card – assigning letter grades based on the physical condition and investment needed for improvement. Their assessment for 2021 was C– (not good), with huge sums being cited for essential repairs (see www.infrastructurereportcard.org).

assessment of a structure's safety (see Section 7.4 generally and Section 7.4.2 in particular).

On many projects, issues of corrosion and maintenance can become driving factors in the overall design. The legendary demands of 'Painting the Forth Bridge' were such that the idiom became part of the English language, indicating a never-ending task. For safety, maintenance demands must be anticipated, which requires provision of a plan and adequate means of safe access, frequently via dedicated routes or moveable gantries.

Routine causes of decay are rain, ice, wind and thermal change. Wetness decays timber and corrodes steel, and dampness causes fungal attack on many products. In some materials, water content has a direct influence on strength. Freeze/thaw cycles affect even tough brickwork with the expansion of water at low temperatures causing bursting. Frost damage can be extensive. Sunlight causes degradation through UV attack. Repeated cycling through hot and cold temperatures is a source of cracking and gradual weakening. Even the atmosphere is a hazard. In the 1960s, before the Clean Air Acts, acid rain was a national concern in respect of buildings and forests. Rainwater absorbing industrial pollution turned into a weak acid, which has a corrosive effect on the fabric of buildings. The national costs for the repair of ancient cathedrals remains high. Acid rain remains a source of contamination and degradation in parts of the world to this day. Local to our coasts, salt-laden spray and air significantly accelerate corrosion and ought to be a design consideration for any coastal or offshore structure (even at a domestic level, metal parts, such as stays on plastic windows, degrade in coastal homes).

Protection against corrosion/degradation should be a key step in the design of all civil engineering/building projects. Partly this is achieved by specification of durable products; partly by controlling the design of members to achieve durability and partly by detailing to lessen the risk. Thereafter, in-service inspections to detect the onset of damage ought to be a key part of the overall safety strategy. Every householder with timber window frames knows that routine preventative painting pays dividends. For certain structures, protection against corrosion is a dominating part of the design. In industrial structures, local pollution (say in chimneys and flue liners) can be so intense that thickness protection governs their design. In coal-fired power stations, the degradation rate of concrete in precipitators was formidable. Bearing in mind that power station output stops if the chimney fails, it is readily appreciated just how important attention to durability can be. That concern is heightened once it is realised that opportunities for inspection and repair are often restricted, even during plant outages.

7.1.2 Historical development

Product quality, as opposed to material quality, is an allied concern. Chapter 1 discussed the difficulties early engineers experienced in creating sound cast iron beams. There were difficulties both in dimensional control and in ensuring beams were relatively blemish free. Hence for a long time, industry has been continually evolving improvements in manufacturing technique, furthering the aim of ensuring that products purchased have consistent and reliable properties. Experience reveals that this is not without setbacks. Chapter 6 referred to the promotion by SCOSS of the principles of the 3Ps: People, Products and Processes. Chapter 6 dealt with 'People'. This chapter deals with 'Products'.

Material property control is a specialism. Over many years, industry has evolved specifications to define and assure product quality so as to rid designers of the need to spend time on that task themselves. The standardisation movement was originally driven by the needs of mass production. It took two major forms; the standardisation of size and the standardisation of properties. We might argue that another key factor was the standardisation of measurement itself, which has been a demand since historical times, yet the introduction of the metric system is still incomplete in the UK. The metric system was 'invented' in France in 1799 but only adopted in the UK in 1965 and even then, only partially. Popular trade often retains imperial units; long distances being measured in miles, office space in square feet and land areas in acres. In the US, the metric system has still not been adopted. That divergence is a real source of confusion and potential safety weakness. Even between European countries there are those that use the newton (N) and those that use the decanewton (daN), a symbol barely understood in Britain. Then there are tons, tonnes and American short tons (2,000 pounds rather than 2,240 pounds). There is also the short hundredweight, not to mention American gallons and British gallons (ratio 1 : 1.2, the American gallon being smaller). Or even in words, with the UK (old) billion meaning 10^{12} and the US billion meaning 10^{9}, though usage now is to standardise on 10^{12} as a trillion. In continental European languages, 10^{9} may be a milliard whereas 10^{12} is a billion.

Having such diversity raises a real chance of confusion and human error (Chapter 6). Within any one country that risk ought nowadays to be low for new work. However, much of the building industry's workload on refurbishment will refer back to calculations and drawings set out in imperial units, for which later-educated generations will have no 'feel'. It is all too easy to make an error and the case of the Mars Climate Orbiter is reported in Chapter 6. There is also a case of a Boeing jet which ran out of fuel due to a miscalculation of volume based on unfamiliarity with metric units. Chapter 6 (Section 6.5) records a mix-up in units which contributed to the sinking of the *Vasa* warship in 1628.

Early industrial societies used their personal prestige to create a quality image, such as 'Toledo steel' (best in Europe for making swords) or 'Sheffield steel' or 'Belfast linen'. However, such vagaries were of little use as industry developed. Standardisation started to be significant after the Industrial Revolution, driven by the need to interchange parts and later promoted after numerous material mishaps. The difference between Brunel's broad-gauge and Stephenson's narrow-gauge railway track is well known with Brunel's perhaps technically superior gauge being discarded for the sake of uniformity between

1864 and 1892. Whitworth's thread, the world's first national screw thread standard, was devised by Joseph Whitworth in 1841 and widely adopted until metrication. In America, early progress was driven by the railway companies who required huge amounts of steel for tracks. Likewise, in Great Britain, the BSI founded in 1901 (claiming to be the world's first national standards body) issued its first standards to reduce the number of sizes of tramway rails: goals of standardisation were not just to assure quality, but to reduce the number of parts being made. That incentive continues in sizing (modular co-ordination) and product dimensioning. Hence, for example, rebar bending shapes have been harmonised and reduced in the UK by BS 8666: 2005 (and its predecessor). Before that, many companies had their own shapes. Universal beams (structural steel) were introduced about 1957. Before that date, there was a proliferation of different beam types, all with different properties.

For obvious reasons, standardisation became pre-eminent in industrialised societies such Britain (BSI) Germany (DIN, introduced in 1917) and the USA (National Bureau of Standards, 1901 and ASTM, 1898). Later these were joined by other industrialised countries such as France (1926) and Japan (JES, 1921, now JIS). Because these few countries dominated world supply, their standards became de facto norms and BS/ASTM/DIN/JIS[2] became used throughout the world. More recently, in Europe, this proliferation of standards has been seen as a weakness in the sense of imposing unnecessary costs on manufacturers as they need to produce or test to show compliance with national standards. Hence since 1975 there has been a project to harmonise across Europe (EN standards) and this is leading to the demise of BS and DIN and should reduce costs and reduce error consequent on product unfamiliarity. There will also be a demise of familiar BS building design codes, but in the short term that is seen as source of concern in terms of potential to permit mistakes (Chapter 6).

7.1.3 Repetition

A nightmare linked to product failure is that of making a mistake that repeats multiple times. This occurs even in mass manufacturing, exemplified by occasional product recall. It has also occurred in civil engineering. In Manchester, UK, in 1994, large parts of the track laid for the new metro system had to be taken up since there appeared to be a generic fault with the form of polymer material used to bed the rails. After 12 months' use, cracks were occurring, perhaps created by traffic and freeze/thaw cycles. In 2010, there were press reports that a generic fault had been discovered in the grout which bonded upper sections of offshore wind turbine shafts into lower portions via a sleeved form of joint. Degradation was allowing the upper shafts to slip. Projected repair costs were substantial.

2 Interestingly, the ASTM website quotes one of the first known specifications from the Book of Genesis: 'Make thee an ark of gopher wood; rooms shalt thou make in the ark, and shalt pitch it within and without with pitch.' And we might say that the extract from Genesis 11:3–4, 'Come, let's make bricks and bake them thoroughly', is an early example of a manufacturing specification.

7.1.4 Current issues

As design engineers, we nowadays rely very heavily on having confidence in the products we specify, particularly in their material properties. A relatively recent problem linked to market globalisation has been the falsification of records such that substandard products have been inadvertently purchased (see reference below to SCOSS Alert, 2013). In 2013, there was a national scandal initiated in Ireland when spot testing found that supermarket 'beef' burgers contained traces of horse meat (or in some cases significant percentages of horse meat). Apart from the cultural problems of eating horse (all health risks were denied) the basic issue was lack of confidence by the public as to knowing what they were eating. The major supermarkets exercise quality control and registration of suppliers; if they cannot control what we eat (and miss a major fraud like this), it's hardly surprising that within the myriad of building producers we use there are cases of substandard products arising from weaknesses further back in the supply chain. A problem with the food scandal was that the distribution methods for the contaminated products made traceability virtually impossible. Likewise, a serious consequence of product fault or record falsification is an inability to trace that product once it has been sold on. The problem is clear if the product is a much-used commodity like bolts, which will have been widely distributed without sales records. In these circumstances, combating safety concerns is virtually impossible.

Falsification of records might seem shocking, but there have been major scandals involving reputable companies. NASA was a victim of this around 2011 when it was discovered that an aluminium supplier had been faking quality certificates for years. NASA alleged that this led to the loss of two satellites. More recently, around 2017, Volkswagen was found guilty of faking diesel emission tests on its engines, and was penalised with huge fines. Alleged falsification of tests has been a feature of the 2021/2022 Grenfell Fire Inquiry.

Perhaps none of this is so surprising. Use of shoddy materials and shoddy building techniques is captured in the expression 'jerry-built', which has been around since the mid nineteenth century.

SCOSS has published an 'Alert' on 'Anomalous documentation for proprietary products' (SCOSS Alert, 2013; see also CROSS Reports 254, 259, 284, 299, 326, 331, 338, 474, 5132) prompted by feedback that in a few instances, certification accompanying proprietary products has stated compliance with standards or specified requirements when in fact this has been untrue.

Material product reliability depends on the quality of the basic product but may also be linked to how that product is shaped or adapted for use in a building and whether that conforms with the manufacturer's intent. There have been cases of failure linked to faulty fabrication or conditions of in-service use. Hence, overall, the safety principles about materials are:

- Do we have confidence in the product's provenance?
- Do we know that the product is suitable for its design application? (Cf. Grenfell Fire queries in Chapter 9.)
- Do we know that fabrication processes have not degraded properties? (Cf. examples in this chapter.)
- Do we know the product has been installed correctly? (Cf. comments on the Edinburgh schools' failures in Chapter 6.)
- Have all potential degradation mechanisms been foreseen? (Cf. examples in this chapter)
- Is the product inspectable in location? (Cf. comments on the Edinburgh schools' failures in Chapter 6 and in this chapter on bridges.)
- Are there provisions for maintenance? (Cf. examples in this chapter, especially on bridges.)

In practical terms, the fundamental properties of most concern to designers are strength and member size. Strictly speaking, both are time dependent. Some of main building materials can change strength with time and some, like steelwork, can thin with time to reduce the area available for taking load. It is also possible that stiffness varies with time. Chapter 13 includes comments on QA/QC techniques to assure material/product quality.

7.2 Brickwork and blockwork

Brickwork can be extremely durable. Many of our most historic buildings are brick built and have lasted well for centuries. 'Design Note 7' from the Brick Development Association (2011) provides general guidance on brickwork durability. Nevertheless, there are potential problems and brick surface spalling from freeze/thaw is a common sight in winter months. Frequently that is indicative either of poor detailing or the mis-ordering of brick type based on ignorance of potential in-service demand. Over time, and with thermal movement, the pointing of brickwork will weaken and need renewal. Failure to do this might lead to ingress of rain with the risks that entails, or it might lead to weakening of the brickwork as a panel.

The most commonly reported structural problems with brickwork (as a product) relate to failure in side-loading either from wind loading or soil pressure. There are many instances of garden walls and other boundary walls failing and it is easy to spot these potential hazards on any suburban walk. Whilst these hazards might sound trivial, there are several cases where boundary walls have collapsed onto children and killed them (CROSS Reports 59, 94, 116, 6001). Such failures are not caused by material deficiencies so much as by faulty design.

One weakness in older houses with cavity walls relates to brick tie failure. Formerly, ties were poorly galvanised so they corroded, permitting walls to bulge or be sucked outwards by wind (houses most at risk are those built in the 1920s and 1930s). Ties tend to corrode as the cavity void is moist or they might corrode if the outer leaf has mortar with high sulphur content. In that form of corrosion, the weakness might be detected by rust expansion opening up the bed joints. Conversely, internal tie corrosion might remain undetected until it is too late. Any metal embedded in the outer leaf of brickwork forms a potential problem. In older buildings, lintels were sometimes made of poorly protected steel and once corroded, the general expansive forces would be enough to create horizontal cracking along whole bed joints sufficient to lift complete upper brickwork panels. Any such opening-up of course weakened the panel as whole against lateral wind pressure. There have been cases of large panels falling (CROSS Reports 92, 186, 5082, 5098).

Anchorage corrosion is a generic hazard for all large-panel structures and for all non-load-bearing brick panels used as infill. Modern specifications and local building regulations will nowadays demand use of more durable anchorages, such as stainless steel. In any assessment of an older building's safety, a key need is to investigate the durability status of any metal restraints (although not relating to brickwork, gradual degradation of roofing slate nails is enough to permit slates to come off, often in high winds, and they are deadly missiles). In more modern houses, lack of adequate strapping against wind suction has often resulted in non-load-bearing panels being sucked out, especially the triangle of brickwork at a gable end under the roof.

A building work problem exists between bricks and concrete as the former expands with time and the latter contracts. This differential movement has been the cause of severe crushing of brickwork infill panels in concrete frames (i.e. as used to create shear walls). A common problem between older steelwork and close proximity facing brickwork is the cracking of the brickwork caused by rust expansion on the steelwork face, often called 'Regent Street Disease'.

7.3 Timber

As with brickwork, well-protected good quality timber can last for centuries. Equally, there are a range of degradation sources that can be sufficient to weaken or destroy timber rather quickly. Whilst frequently such degradation will be 'non-structural', the same degradation sources can apply to structural members and cause collapse. This is obviously an issue for historic structures and the demands of conservation require ingenious assessment and strengthening techniques. The hazard of timber degradation changes with time. When investigating older homes or buildings, it is common to find timber quality that is simply unobtainable today and that decline in timber quality is one reason for the growth in plastic window frames. On the positive side, nowadays we have many quality timber products, such as plywood (though that can delaminate) and engineered trusses with proprietary connectors, which did not exist before modern times.

Standard degradation mechanisms are wetness of any kind, or fungal or insect attack. Dry or wet rot severely weakens timber and is often consequent on poor construction allowing persistent damp. Beetle attack is primarily a specific hazard for older buildings. Water per se is not necessarily a degrader; the Romans used elm for water pipes and Venice is built on oak and pine piles (protected in airless mud). It is certainly possible to build robust and durable marine structures out of appropriate timber. Another source of degradation and risk comes under the category of 'unauthorised change' (see Chapter 6). It is not at all unusual to find floor joists notched to permit pipe runs without much thought being given to the weakening effects of reduced joist size (CROSS Report 209).

For many engineered timber products, the source of weakness is, as with many structures, the joints (CROSS Reports 21, 62, 249, 273). In timber engineering, designing joints to transmit load has always been the challenge. Indeed, it is the size of timber required to make the joints that often drives main member sizing. In timber, capacity is affected by material quality and by moisture content. If that moisture content changes beyond design expectation, so does capacity. Two examples are included below (Boxes 7.1 and 7.2).

Box 7.1 Roof failures

Fair centre roof, Jyväskylä, Finland (2003)

This was a complete collapse of 2,000 m^2 of a fair centre roof. The roof was made of laminated trusses with 55 metres spans. Collapse was precipitated by failure of the dowel joints in a mode not anticipated by the code (shear block failure). (A comparable earlier failure had been that of the glulam velodrome roof at Ballerup arena (Siemens Arena) in Denmark in 2003.)

Supermarket inner ceiling failure, Sysmä, Finland (2005)

In another Finnish failure, a supermarket roof collapsed (400 m^2). The cause again was joint failure. The joints had never been designed and there were no details provided for construction. Joints mostly had two nails only and the investigation found that when the connections dried out after nailing, nail withdrawal strengths decreased to about 43% of code predictions.

These failures occurred under roof snow loading but that was not the prime cause; the snow was merely the load that highlighted the weakness.

More details can be found in Finnish Safety Investigation Authority (2005).

Comment: A lesson about the importance of joints and of matching design assumptions to material properties. (These reports are from the Accident Investigation Board of Finland.)

Box 7.2 Ice rink, Bad Reichenhall, Germany (2006)

This roof collapsed under 300–400 mm of snow; 34 people were injured and 15 died. It was determined afterwards that water damage had weakened the roof, causing the failure. Design parties were charged with manslaughter.

The building had been constructed in 1972 and used box girders at 7.5 metre centres as the structural form. A particular issue of this disaster is that the leaks were known about and strong concern had been expressed about safety. But such concerns were ignored, and the public were allowed to use the rink. At the time of failure, snow depths were well below predicted limits.

Comment: There are several lessons. First, a predictable degradation hazard existed. Second, the failure took place after 34 years of use, hence supporting the common observation that safety degrades over time. Third, for safety, never ignore warning signs.

As with all materials, a generic issue is to ensure that the as-constructed design is compatible with design intent (see Chapters 4 and 5). Timber can be more vulnerable in this context since it is a natural material with dimensions not so easy to control. An example of this consequence can be seen in Box 7.3.

Box 7.3 The Rosemont Horizon Arena, Chicago, USA (1979)

This was a major collapse occurring during construction. The roof was supported on 16 glulam arches each spanning 88 metres. These were subdivided into three trusses intended to be laterally braced via the purlins. But the arches were misaligned and varied in height as much as 300 mm in one span. Consequently, it was impossible to fix all the purlins. According to the reports, about 53% of fixings were missing. The whole structure collapsed due to instability under wind loading, killing five people and injuring 16 others.

Comment: Although this was a timber failure, there is a clear link to lack of reality over design and construction. Even at a 'tight' tolerance of L/1000, it would be possible for any pair of arches to differ in height by say 2 × 88 mm, i.e. about 180 mm (and so not align). One lesson of Chapters 4 and 6 was to be sure that what was intended to be built was actually built. Another lesson is to account for tolerances.

Timber as a material can be vulnerable to fire and Chapter 9 gives examples.

7.4 Steel

7.4.1 General

Steel comes in various grades, but there is no significant difference in corrosion rates between them. However, if thinner sections are used because of using material with higher permissible stresses, then the effect of thickness loss will be proportionately greater. Clearly a 1 mm loss from 5 mm thick plate will be much more significant than a 1 mm loss off a 50 mm plate. For that reason, design codes specify minimum steel thicknesses in certain members to guard against a significant loss of safety should excess corrosion develop. It is of course possible to select special weather-resisting steels (commonly under the trade name Corten) or stainless steels, and those are essential for certain applications, particularly where in-service inspection is impossible. In other cases, steel can be protected by painting or galvanising, but neither of these treatments is a panacea.

Degradation mechanisms are corrosion of various kinds, but also time-dependent phenomena such as relaxation, creep and fatigue. In some circumstances, stress and corrosion combine to exacerbate the cracking rates. Corrosion alone has been enough to cause structural failure, as also have cracking and fast fracture. These might all be classed under 'material failures', whereas failure under load is more properly looked at under another heading. Certain fabrication processes can increase or even cause material failure and the chief culprit is usually the introduction of inadvertent hardness in the parent metal from the application of heat followed by rapid cooling.

7.4.2 Failure by corrosion

The development of rust is well known. In older cars, life was more likely determined by the body's resistance to corrosion than it was by wear of mechanical components. There are different forms of corrosion: creation of a general surface film, pitting and overall surface wastage. There is also bi-metallic corrosion created by the electrical cell set up between different metals (Box 7.4 gives an example). This also exists in normal welded steel as the weld metal is nobler than the parent metal, and in severe cases it is common to see a line of enhanced corrosion parallel to weld seams. It is especially important to avoid bi-metallic effects if the attack is towards the fasteners when these are less noble than the sheets they connect. Corrosion rates are generally exacerbated in alternating wet/dry conditions, by dirt (which retains moisture) and in thin gaps (crevice corrosion). A key factor is that rust expands to many times the original metal volume, so the development of rust in crevices creates a bursting force. In the 1960s a gantry over the railway lines near London's Clapham Junction collapsed with the cause being rust expansion pushing riveted plates apart (undetected) until the rivet seam 'unzipped'. The original code rules for minimum spacing on fasteners between plates derive from empirical rivet spacings targeted at keeping connected plates in close contact, so limiting the risk of crevice corrosion. Corrosion is accelerated in salty marine environments by chloride ions and that same chemical effect can corrode embedded rebar.

A safety requirement is to be able to inspect structures in service to detect the onset of corrosion before it progresses too far. Many failures have resulted from a combination of lack of inspection and lack of inspection capability. The blast destroying the Stockline

Box 7.4 Lighting fixtures, Boston Tunnel, USA (2011)

A problem was discovered with lighting fixtures in the Boston road tunnel when some were found lying in the roadway. They weighed 54 kg each. Investigations revealed that their mountings had failed with the cause being bi-metallic corrosion between aluminium and stainless steel in the presence of salt water. Reports stated that 25,000 fixtures would have to be replaced at a cost of $54 million.

Comment: First, this is an example of bi-metallic corrosion having serious consequences both for costs and safety. Second, it is an example of the risks associated with one error being repeated multiple times (see Chapter 5 and Moskowitz, 2011).

Plastics factory in Glasgow in 2004 is one such example (Chapter 11). Lack of inspection was a feature in each of the US Mianus River Bridge failure (Box 7.5); the US Silver Bridge failure (Box 7.6); and the Scottish Stewarton Bridge failure (Box 7.7). Many other examples exist.

Corrosion protection requires an appropriate choice of protected materials, plus attention to detailing primarily to ensure drainage. In structural components known to be vulnerable, regular inspection is required to detect the onset of damage

Box 7.5 Mianus River Bridge, USA (1983)

There are reports of US bridges collapsing due to corrosion. One example is that of the Mianus River Bridge. In this incident, a 30 metre suspended section fell completely when the bridge was about 30 years old. Cars were plunged into the water and three drivers died. The bridge's structural form had one section supported on each of its four corners by hangers. Each hanger consisted of a plate strap with single pins top and bottom with the pin retained in place by a capping plate and nut. The bridge spanned a salty tidal estuary and was subjected to winter de-icing salts from above. Failure began by corrosion expansion apparently forcing a keep plate off one lower pin end so allowing the pin to come out and thereafter all local support was lost. The hanger support on the opposite side then had to carry twice its nominal load yet was not designed for that contingency. Moreover, the hanger plates had also displaced such that the pin was subjected to both shear and eccentric bending. Stresses were high and eventually the supporting pin failed in fatigue and in so doing, the supported span fell catastrophically.

Apparently, the original design had a drainage system that diverted run-off away from the hangers, but this had been blocked off. There also appear to have been few inspections so allowing corrosion to build up without detection. Even if inspection had been carried out, the corrosion might not have been obvious. In this example, the risk from salt corrosion ought to have been anticipated; the structural form was vulnerable to a single point failure and prediction of failure by corrosion could easily have been made. Mitigation would have been possible by having a 'catch system' (as was applied in the retrofit) and by inspection and maintenance to detect the onset of corrosion and remove it; but this was not apparently done.

Comment: The lessons are the reality of corrosion and how it can degrade safety with time. Corrosion ought to have been obvious here as a design consideration, and the structure should have been designed to be durable and 'inspectable' (Chapter 5). This is also an example of the safety concern of 'event chains': corrosion leads to loss of support, then stress increases, so fatigue damage becomes more serious. Additionally, the structure was demonstrably not 'fail-safe' (Chapter 5).

Box 7.6 Silver Bridge collapse, USA (1967)

An earlier bridge disaster in the USA was also partly caused by corrosion. The Silver Bridge over the Ohio River collapsed more or less instantaneously in 1967, when it was around 40 years old. This was a suspension bridge with chain links as its main support system. The links were fabricated from high strength steel (520/mm^2 yield). Rather unusually, these links were only arranged in pairs such that if one failed, the load on its partner had to double (historically, eyebar bridge links such as the ones used on Telford's Conwy Bridge (North Wales) had several bars making up each link thus giving multiple redundancy). Failure cause was put down to fracture of one link which then overloaded its partner, precipitating a total failure. The fracture cause was fatigue growth from an originally small defect which had grown by a mixture of stress/corrosion/fatigue. The investigation also concluded that initial defects could not have been detected by known technology nor could they have been detected in service. The bridge was also carrying heavier loads than designed for.

Comment: As with the Mianus Bridge, we might say the design concept was basically flawed in not being robust against a single component failure. It can further be said that degradation mechanisms were foreseeable but both bridges suffered from an inability to detect defects likely to cause failure and in any case, there was a lack of inspection in service.

Two points stand out. Cables are made up of multiple isolated strands so a failure in any one strand is not an immediate cause of concern and if detected, gives much advance warning of degradation. To that extent, cables offer safety advantages over linked bars.

Second, the Silver Bridge bars were high-strength steel. The endurance limit of steel is the same whatever its grade so when fatigue is a design criterion, high strength offers no advantage. It might be thought it equally conveys no disadvantage and that would be true if the stress range were the same. But high-strength steel will naturally be used so that it can sustain higher stress. So, the chances are that numerically it will endure a higher stress range than would have been the case had an alternative weaker steel grade been used. In that light, higher-strength steel structures are going to be more prone to fatigue damage.

and thereafter repair to prevent damage progressing before safety is jeopardised. Quite clearly cable damage in major suspension bridges is to be avoided on both safety and cost grounds and great efforts are made in such bridges to protect the main cables. Nevertheless, damage does occur, and Stahl and Gagnon's 1996 book, Cable Corrosion in Bridges and Other Structures, depicts many pictures of failures, particularly at anchorages (Stahl and Gagnon, 1996). Proper design requires that anchorages are both protected and detailed to be 'inspectable'. In 2005, inspections revealed the onset of corrosion in the main cables of the Scottish Forth Road Bridge. Over the following years, major precautionary works were carried out. Safety was ensured because the main cables consist of multiple strands such that the loss of any one would not precipitate the type of failure seen at, say, Silver Bridge (Box 7.6). Moreover, the onset of corrosion was detected before much damage had taken place and detected in time to initiate action. Corrosion of cables is not unique to the Forth Road Bridge and has been discovered on other major bridges across the world; preventative action consists of wrapping the cables in waterproof elastomeric material and fitting a dehumidification system which blows very dry air through the cable cluster (dehumidification is also used inside box girders).

Box 7.7 Stewarton Bridge, Scotland (2009)

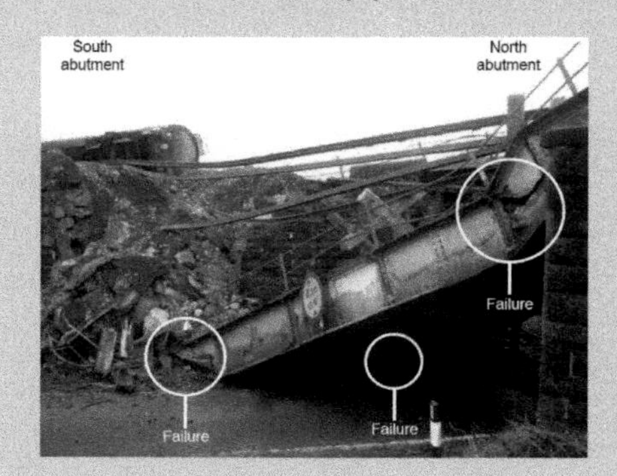

Figure 7.1 The collapse of Stewarton Bridge (courtesy of Rail Accident Investigation Branch, 2010)

This was a wrought iron bridge that collapsed as a fuel train was passing over, leading to its derailment. The bridge's centre and side girder webs had been thinned by heavy corrosion and so failed in shear. The corrosion site was hidden by timber boards retaining ballast so was not picked up on inspection. On top of that, calculation assessments made no allowance for loss of web material over time (see Rail Accident Investigation Branch, 2010).

Comment: The effects of corrosion are dramatic but the lesson, as in so many other cases, is to beware of hidden corrosion. If elements cannot be inspected, suspect the worst.

7.4.3 Fabrication effects

Certain steel properties will be affected by methods of fabrication and these may alter the risks of failure by either fracture or fatigue. Failure by fast fracture is the more dangerous type since there might be no warning of an event with consequences that can be catastrophic. The risk of fast fracture is affected by each of the choice of material, the fabrication process and the structural detailing, whereas the risks of fatigue failure are not normally directly affected by choice of steel, but only by fabrication and detailing. The influence of detailing is the extent to which it introduces or controls stress concentrations. However, in practical structures it is certainly not possible to eliminate all stress concentrations; they exist at every re-entrant corner and around every bolt hole.

Probably the most common risk factor is the application of heat by uncontrolled welding or flame cutting. It is not the heat per se that affects the properties adversely, but the rate of cooling afterwards that governs the local steel microstructure. Essentially, if cooling is too rapid, the steel is left with a hard and brittle microstructure which, in the worst cases, shifts the mode of steel failure from a ductile to a brittle one and hence may lead to fracture. Since safety is fundamentally undermined if failure modes are brittle, any process which might inadvertently degrade steel properties towards brittleness needs to be understood. The design objective is to have a tough steel and not to degrade intrinsic toughness values via the fabrication process.

One example of what can go wrong can be seen in Figure 7.2 which depicts re-entrant corners flame cut to form standard end connections in I-beams. Flame cutting followed by rapid cooling leaves the edge in a hardened condition. On top of that, the flame-cutting process will probably leave notches to act as stress concentrations. In the figure, it will be obvious that loading on the beam will create a local tensile stress at the re-entrant corner and that stress will be intensified by the re-entrant shape, made even worse if there is a notch from the flame-cutting process. Thus, there is a very high stress on a brittle surface with ideal conditions for fracture creation (see also Box 7.8 on the Sea Gem failure).

Steels can be purchased in a variety of strength grades and a variety of sub-grades that offer enhanced toughness such that material can be safely used in low temperatures. The first protection against low temperature fracture (very rapid crack propagation) is to specify a steel with the right properties. Codes guide as to the grade required,

Figure 7.2 Fracture starting from flame-cut re-entrant corner

and selected Charpy values are a function of service conditions, temperature, material thickness and whether or not the steel is welded. Thicker steel is more prone to fracture because there is a better chance of it having a less favourable microstructure (Section 7.4.4). Welded steel is always more at risk (Section 7.4.7) and in most classic brittle fractures, the steel has been welded.

In much weld testing, the focus is on strength, but for safety there needs to be assurance that the parent metal is left in a ductile condition. Specifications for structural design in earthquake areas usually prohibit fabrication processes that undermine that key demand of ensuring ductility: hence notches and so on are to be avoided in any locations where plastic hinges might develop. Equally, hole punching can be forbidden since that procedure risks leaving the steel around the hole in a hardened state, perhaps with micro cracking from the punching operation. Under high imposed strains, such cracks can propagate to destroy connection strength.

Casting poses obvious risks exemplified by very early historical failures (Section 7.4.6). Big Ben cracked after casting in 1856 as did the Kremlin Tsar Bell in 1735, when it was drenched with water to extinguish a fire. The risks of cast metal in structures still remain, illustrated by the failure of certain cast cable anchorages on the Glasgow Clyde Arc Bridge in 2008 (Box 7.9). CROSS has issued a topic paper on castings (CROSS Topic Paper, 2010).

Box 7.8 Sea Gem oil rig collapse, UK (1965)

In 1965, Britain was just starting out on the extraction of oil and gas from North Sea reserves. The Sea Gem was a converted barge oil rig which collapsed catastrophically killing, 13 people. A public inquiry was convened to establish the cause.

The rig's deck was supported on ten independent legs and leg extension was accomplished by jacking down with the deck suspended from each leg via four tie bars onto a jacking frame. At the time of failure, the deck was being raised but was lifted unevenly and whilst being returned to level, there was loud bang whereupon the rig collapsed. Two legs had given way allowing the rig to tilt sideways and sink.

The inquiry found that the whole structure was notch brittle and that failure had been precipitated by tie bar fracture. These tie bars had been fabricated from plate which had been flame cut and the flame cutting was irregular. There had also been patch weld repairs. It was found that cracks had originated from weld defects and from sharp radiused corners between the tie spade and its shank. It was also thought that loading from the lifting operation, perhaps combined with a change in temperature, had precipitated the failure (it was about 3 °C at the time).

Comment: Clearly there are common features in this failure with the inevitability of uneven loading, with the fabrication effects of flame-cut edges and the lack of comprehension of brittle fracture cause. In terms of the structure overall, there is the common theme of gross failure being precipitated by local failure.

Box 7.9 Clyde Arc Bridge, Glasgow, UK (2008)

The Clyde Arc Bridge was opened for traffic in 2006. Its deck is hung from cables attached overhead to an arch spanning the river. Around mid 2008, one upper hanger connection failed, and cracks were found in a second. Published photographs show the hanger end completely fractured through across the net section with the pin hole in the centre. As a consequence, it was determined that all the cast steel hanger ends had to be replaced. There has been no official report on causation but, whatever the cause, the mode of failure was a highly undesirable fracture.

Comment: The mode of failure was certainly fracture so it looks likely either that the design failed to identify that mode of failure or that the material had insufficient fracture toughness. It will be noted that the steel used for the deck hanger ends was 'thick'.

7.4.4 The role of thickness

In moderately thick steels loaded in a through thickness direction, lamellar tearing can be a design issue and special plate can be purchased as a precaution. In thicker steels the risk shifts to one of fracture. The risk of cracking in 'thick' steel is higher than in 'thin' steel for a variety of reasons (Burdekin, 1999). In the late 1980s there were some spectacular fracture problems in the US when welded jumbo sections were used in tension (flanges 125 mm, webs 75 mm). In one truss, the tension boom spontaneously cracked right through whilst in service. Forensic examinations proved the material to have low toughness and the investigation concluded that there were inherent problems with the way jumbo sections were produced, leading to excessive grain size and especial weakness in the centre of the thick sections (where the cooling was too slow) and at the flange/web junction (where grain refinement by rolling was weakest). The initiation point (certainly for one jumbo fracture) was the hardened edge of a flame-cut mouse hole utilised for completing the butt weld across the flange. These failures led to recommendations for not welding jumbo sections and for controlling the cooling rate when flame cutting. One example of precautions taken when fabricating with thick steel can be found in Berenbak et al., 2001.

7.4.5 Punching

Punching is commonly employed as a cheap means of making holes. But cold working raises both the steel strength and its hardness and if the working is too great, as say in trying to punch through thicker steel, micro cracks can be generated around the hole edge. Such cracks have been known to propagate explosively in zones of plastic stress (hence seismic codes forbid punching in regions of anticipated ductility demand). Likewise, trying to bend steel through too tight a radius can create cracking, hence avoidance requires the specification of minimum radii and/or the controlled application of heat.

7.4.6 Product failure by fracture

The danger of cast iron cracking / brittle fracture was always understood, at least empirically, as most early bridges such as that at Ironbridge (1779) used arches as their structural form to keep steel in compression (albeit that bridge still has cracks within it). Longer bridges supported by chains such as the Menai Suspension Bridge (1826) used wrought iron (at least for the chains) for better reliability. Nevertheless, cast iron remained in use for beams, with Robert Stephenson suffering a severe setback in 1847 when the cast iron girder bridge he had designed over the River Dee near Chester collapsed as a train crossed over; the cracking perhaps precipitated by fracture of the girder's bottom flange. Consequently, cast iron was phased out of use, with the Norwood Junction railway crash of 1891 being a deciding event; the cast iron bridge there collapsed by fracture emanating from a large blowhole in the flange. Thereafter steel was the preferred material.

That situation of using steel in building continued in the same style more or less up to the Second World War. Up to that date, most structures and their connections were made by riveting, where fairly thin plates of steel were bonded together. This gave two benefits: first, the steel was 'thin', so the material's tensile reliability was better (than for thicker steel) and second, failure of one part in a multiple-riveted assembly did not necessarily spell disaster for the whole unit.

Around 1920–1930, a series of papers on welding were published in the journal of the Institution of Structural Engineers, heralding the start of a new era. In 1932 the journal had published a first report from the Steel Structures Research Committee, which highlighted some welding tests with the comment: 'A discussion and description of tests on the welding of steel structure, a subject which is of the greatest interest as offering perhaps one method whereby the intolerable noise associated with riveting may be avoided.' (Apart from the health dangers of hot handling, the health risks of deafening were grim.) Thereafter the demand for rapid production throughout the Second World War gave welding technology a strong boost. Alas, the new technology brought new problems of which the most spectacular were the full fractures of a few American Liberty ships (Figure 7.3).

The Liberty ship failures offer some classic images. These were all-welded vessels and some of them just split in two in calm waters more or less instantly for no apparent cause. The boats were mass-produced with plates joined by welding and this created one of the problems in the sense that once a crack started, there was then no natural arrest point. In older ships, joined by riveting and with riveted-on stiffeners, there were at least opportunities for crack lengths to be contained. Although the property of toughness was known about in the 1940s, the full extent of how it could be degraded was not properly understood. Toughness as a material property had been formally investigated by the French scientist Charpy (1905) who gave his name to the eponymous test which defines

Figure 7.3 Classic photo of Liberty ship fracture

toughness by a measure of the energy a steel specimen can absorb whilst fracturing. The test can be used to demonstrate changes in steel toughness with temperature, the toughness magnitude dropping off dramatically below a value known as the transition temperature. Fracture risks increase as temperatures drop. But although the Charpy test demonstrates toughness variation, it did not in itself explain the fast fractures observed in ships. This was investigated by navy engineers and eventually led to greater understanding of cracking and fracture mechanics.

Underlying all this, a puzzling problem in the strength of materials had been identified by Inglis in about 1914, when he looked at stress concentrations and showed that theoretically, the localised stress which must exist linked to flaws/notches could be many times the known tensile capacity of steel. Yet patently, despite these localised concentrations, most steel structures were able to cope with such stresses in service and not only from flaws introduced by design (such as sharp re-entrant corners), but also from flaws which were intrinsic to the basic manufactured material. The plausible explanation for survival relied on the capability of a steel's ductility to yield locally without the high stress generating a fracture.

A later milestone in the understanding of cracking came about in 1920s when Griffiths pondered the phenomenon, speculated on cause and deduced that there was such a thing as a tolerable crack length that could exist in a tensile stress field, and this led onto the definition of a critical crack length, marking the transition point between stability on the one hand and rapid instability via crack growth on the other. Griffiths thought in terms of energy rather than stress. This was the birth of fracture mechanics, a complex subject seeking to address questions of crack stability of a given length in a given material in a given stress field. In advanced studies on safety, fracture mechanics are used to assess the length of tolerable cracks and one caveat is that to be tolerable, such lengths must be detectable by NDE (non-destructive examination); otherwise there can be no confidence in the structure's safety.

7.4.7 Effects of welding

Observed failures under service conditions have often been associated with welding, which is clearly one of our most common joining technologies. Whilst most welds perform perfectly well, it remains important to comprehend what problems welding may introduce. The reason that welding is one root cause of failures can be explained as follows: first, many welds contain a defect of some sort (albeit these might be at microscopic level and have no effect on static strength) and those defects might be sharp, thence given the right orientation in the stress field they may generate extremely high stresses at their tips. In effect such defects might be the seed for future crack development, either via fatigue or fast fracture. Note: this condition applies to all welds, even those nominally not load bearing: the very act of welding can introduce crack initiation conditions. Second, welding requires the application of great heat and subsequent cooling. If the cooling is too rapid, it may generate a hard and therefore brittle microstructure. Conversely, if cooling is too slow, grain size might be too large, so lowering inherent toughness. Thus, even with sound parent metal, an inappropriate welding procedure may degrade the basic material properties and those may coincide with a crack-like feature. On top of this, the differential cooling, post welding, leaves a residual stress field in the material which will be of yield value. In short, welding can simultaneously add the three risk factors most feared: a crack, a susceptible microstructure and a built-in high tensile stress. For all these reasons, it is vitally important to make all welds to a proper qualified procedure and to carry out some post-weld NDE, to a degree proportionate with the weld's service function.

Equally it is those same factors which explain the Liberty ships' failure: first, the basic material had a poor toughness; second, the welds probably had defects, and third, the driving energy for the crack (in apparently low external stress conditions) was the internal residual stress left over from welding. Finally, the continuity provided by the all-welded structure allowed the crack to propagate a considerable distance without arrest. Nowadays these risk factors can be controlled by appropriate steel selection, by application of proven weld procedures and by post-weld examination to eliminate serious flaws. In unusual circumstances, post-weld heat treatment can additionally be used to enhance fracture toughness, since this relaxes the residual stresses (Berenbak et al., 2001).

In the aftermath of the Northridge, California earthquake of 1994, it was observed that a great many welded joints of the steel rigid frames used in the region had cracked (Maranian, 1997). Similar failures were observed after the Kobe earthquake in 1995. This was put down partly to poor material and partly to faulty welding procedures whereby large welds had been laid down (to enhance production rates) but allowed to cool too quickly. Whilst the welds might have possessed adequate strength, they lacked ductility, thus in the plastic distortion imposed under seismic conditions the welds cracked and the cracks propagated into parent material.

7.4.8 Fabrication control: welding

Although there are risks in welding and flame cutting, proper procedures can reduce these to tolerable levels. Codified welding procedures take account of a material's weldability and control the amounts of heat to ensure that the weld and its local heat affected zone (HAZ) have appropriate engineering properties. Post-weld NDE can ensure that welds only go into service in their intended state. But designers need to be cautious when steel is thick or is to be used in very cold temperatures and if there is fatigue loading. The weld quality standards for welds sustaining fatigue loading need to be higher than they would be for welds only sustaining static loading, simply because imperfections are that much more important.

Process control is equally required when there is to be thermal (flame) cutting, because if the cutting is uncontrolled then, just as in welding, the local microstructure next to the cut can be embrittled. Standards normally impose process control to ensure edge hardness remains within a safe range.

7.4.9 Product failure by fatigue

General

Fatigue is a pernicious form of failure which most often arises when the possibility has not been anticipated. Fatigue damage is crack growth from an initial defect, the growth being driven by fluctuating stress. The classic perception is one of crack growth under many thousands of cycles. Whilst that frequently is the case, fatigue life is inversely proportional to (stress range)^3 so that means if the fluctuating stress range is very high, life can be reached in relatively few numbers of cycles. A second contributor to the risk is that the fluctuating stress to drive crack growth is the true elastic stress. Unfortunately (refer to Chapter 3), this is rather hard to estimate, partly because all secondary stresses that must exist, but are normally ignored in static structural design, have a real and adverse effect on fatigue life. Collectively that means that there is fundamental uncertainty in life prediction since the stress range is never known with accuracy and life is sensitive to its magnitude.

Examples

A good example is shown in Figure 7.4. This is the end of a long-span beam underneath a dance floor. The beam was nominally 'pin ended' so its connection was designed for shear only, but obviously there is significant end restraint via the deep end plate, so a large secondary moment developed at the connection. The main girder was 'long span' so had a low natural frequency and the first cause of concern was excessive flexibility and bounce of the floor caused by dancers jumping up and down at the floor's frequency to cause resonance. The result was beam vibration and hence frequent application of the secondary moment at the restrained end with eventual cracking from a point of stress concentration. Figure 7.5 shows a running tube in a roller coaster track. As the wheels move over the joint (which has a slight lip) they thump down on the trailing edge and exacerbate the lozenging effect on the tube, eventually driving a fatigue crack. Again, predicting the stress from dynamic wheel impact over (an uncertain) lack of fit is unreliable. Another classic case of fatigue damage is that which can occur at the top of a welded crane girder. In normal plate girder design, the welds (flange to web) are sized solely to carry the shear between them as from QAy/I. But in a crane girder, there is always a fluctuating vertical stress directly under the wheels which might be carried by the welds if the web has uncertain fit up to the flange (which is probable in parts). This stress can be high, creating short lives, and there are several reports of crack development along the welds (Senior and Gurney, 1963; Mann, 2011) and Figure 7.6. Unfortunately, this web/flange junction is also a location that might be masked by the crane rail aligned overhead and so cracks may propagate undetected, eventually leading to separation between flange and web.

Poor fit up has historically also been the cause of fatigue damage in bridge decks; Figure 7.7 shows the addition of stiffeners to the underside of the bridge deck and their joints by welds. As with the crane flange / web intersection described above, unless the stiffener is forced into contact there may well be gaps between the stiffener and the deck and thus the weld may have to carry longitudinal shear plus vertical varying load from wheels crossing the deck. This has led to fatigue failures.

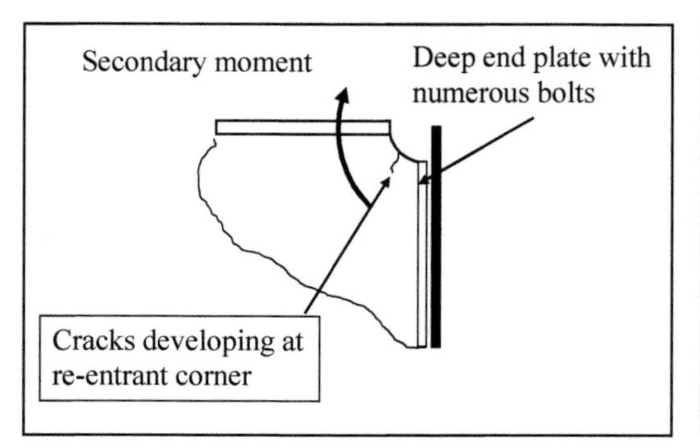

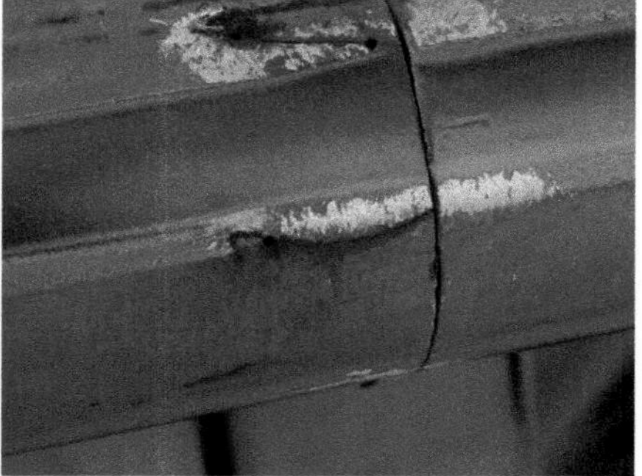

[Left]Figure 7.4. Fatigue crack propagating from corner stress linked to a high secondary moment

[Right]Figure 7.5 Fatigue cracks propagating

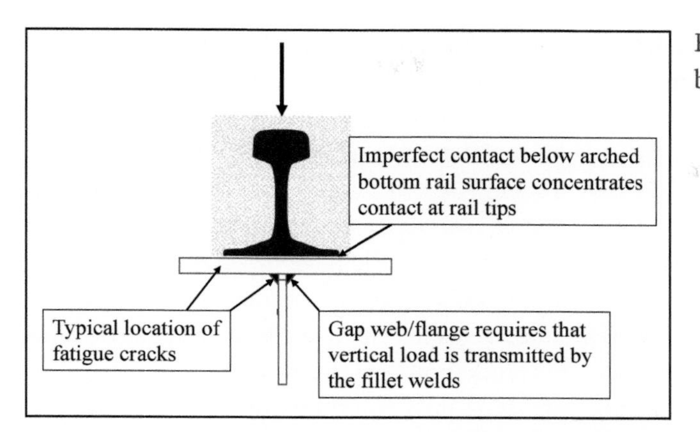

Figure 7.6 Fatigue cracks below a crane running rail

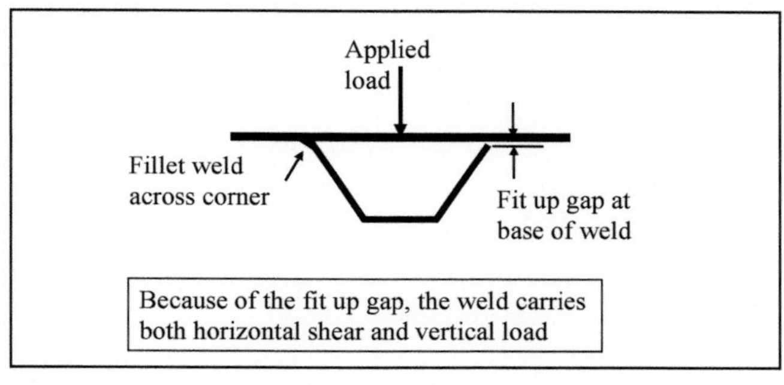

Figure 7.7 Fatigue cracks below a bridge deck

Figure 7.8 Fatigue crack growth in a large pin

Strategy

Where fatigue loading exists, safety must be ensured by design which includes proper detailing and by regular in-service inspections using NDE techniques as appropriate. Really safe design should ensure that potential cracks start on the outside of a component's surface and work inwards, rather than in the undetectable interior and work out. This is the 'leak before break' argument discussed in Chapter 5. This desirable state is not always possible, and Figure 7.8 shows a large pin which has failed in fatigue. Elastically, the peak stress is on the pin centre and fatigue cracking has progressed from the centre outwards. It can be seen from the fracture surface that a significant part of the inner core was lost before the stress on the outer annulus became large enough to cause a straightforward fracture on the remaining steel. At that point, the entire pin failed.

Given the fundamental uncertainties, a further design objective is to ensure that components are tolerant to crack growth in the sense of being able to survive with cracks of reasonable length. That length should be detectable before elongation to total failure and the length should be such that cracks can be detected either visually or by NDE via an inspection regime. It is, of course, highly desirable to have redundancy in the system for although prediction of fatigue crack progression is uncertain, it is equally very unlikely that two parts nominally loaded equally will in fact be loaded equally, and therefore unlikely to suffer critical fatigue crack growth at the same rate and time.

The initial design should identify points of likely crack initiation and the structure should be arranged such that these locations are 'inspectable'. There has been at least one accident on a 'ride' linked to fatigue cracks propagating too far in a location hidden by seat padding. Hence in any structure where fatigue is considered a hazard, the design process ought to include making sure that inspections are possible by providing both access and visibility. In any structure required to sustain fatigue loading, it is vital that fabrication quality is matched to design intent. This will always include post-fabrication weld inspection with the weld acceptance criteria being more demanding than those applicable for welds only required to sustain static loading.

Risk

A problem with fatigue is that the worst failures have often occurred when the designer has not spotted that a fatigue condition exists (as in the dance club beam example above). In the case of oil rig failures like that of the Alexander Kielland (Box 7.10) (fatigue not identified) the varying loads were linked to varying sea states. Oscillation from wind-induced vibration might be quite hard to anticipate, but several failures have occurred in this way. It is not the frequent application of wind that counts so much as the possibility of a large number of cycles being imposed from harmonic oscillations, such as may be generated by vortex shedding or other wind dynamic response phenomena (examples are given in Chapter 8).

Technical issues

Technical issues that increase the risks from fatigue loading (and thence cracking) are:

a. There is a low stress limit below which fatigue cracking will not occur. But at higher stresses, fatigue life is finite. Life is determined not so much by the maximum stress but by the stress range acting on the component. Furthermore, life is inversely proportional to (applied stress range)^3. Thus, any enlargement of the stress range potentially has a dramatic effect on life, shortening it considerably. In fatigue-designed structures, the stress range must be kept low. It should be

Box 7.10 Alexander Kielland rig collapse, UK (1980)

The Alexander Kielland rig failed catastrophically in 1980 and 123 people died. The loss of life was greater than might otherwise have been expected as the rig was being used as a worker's floating hotel. At the time of the incident, those on board heard a loud crack and this was followed by rig tilt, a short period of stabilisation and then total capsize.

The Alexander Kielland had a different structural form to the Sea Gem (Box 7.8). Its deck was supported on three giant tubular legs which were mutually cross braced. The failure inquiry determined that collapse was initiated by fatigue of one of the main bracings. The initiation site for crack progression was identified as a small 6 mm fillet weld used to join onto a non-load-bearing flange (this was found to have lamellar tearing in it). Collectively it was thought that there were cold cracks in the welds, increased stress concentrations due to the weakened flange plate, a poor weld profile, and cyclical stresses from wave action all creating conditions for a fatigue crack to grow. It was also concluded that weld faults had been present since fabrication.

Another conclusion of the inquiry was that the rig had a poor command and control structure in place to respond to crises, and that had also contributed to the loss of life.

Comment: The technology that caused this disaster was not novel. When fatigue is an issue, all welds (non-load bearing and load bearing) are significant in their effect on fatigue life (see Section 7.4.7). Another feature covered in Section 7.4.7 is that as part of the overall safety assessment, is it vital to check that fabrication quality achieved is what is required by the design. There was apparently no reason why the Kielland weld defects could not have been picked up in post-fabrication inspection. What this failure also has in common with the Sea Gem is the adoption of a structural form where total collapse was consequent on the failure of a single member (from a foreseeable cause) and that violates the principle of robustness (Chapter 4).

As with many failures, human error played a part and the need to have an effective emergency response system on complex engineering projects is a part of overall safety (Chapters 5, 9 and 13).

obvious that if secondary stresses or stress concentrations are neglected, a structure can easily be subjected to a high stress range when under fatigue loading. The consequences then are that structural lives might be measured in hundreds of cycles rather than thousands or millions.

b. It is unfortunate that secondary stresses, which can be mostly ignored in 'statically' loaded structures, really count in fatigue-loaded structures and such stresses are almost impossible to compute with accuracy, yet fatigue cracks respond to the true stress state. Consequently, cracks can develop in all sorts of surprising circumstances/locations.

c. There must be a tensile stress state. But if there is a residual tensile stress in the structure yet that part is in the compressive range of external loading, then an applied compressive stress will merely cycle the tensile stress from high to low rather than the other way around. Thus, fatigue failures can occur in apparently compressive zones.
d. An added complication occurs when corrosion and fatigue combine. Where there is corrosion (or fatigue in a corrosive environment) lives can be shortened even further and there may be no endurance limit, even at low stress.
e. Checking for fatigue is an obvious condition on structures carrying traffic or alternatively in any moving structure (see Chapter 8 on roller coaster track).

Avoidance of fatigue damage

The first line of defence is to be aware that a fatigue-loading regime exists and to design joints to codes accordingly. Second, there must be appropriate NDE before joints are put into service, to ensure weld quality. Lastly there must be an inspection/monitoring regime throughout service whose objective is to detect crack initiation before cracks progress far enough to become dangerous. In advanced structures such as aircraft components, certain parts are just time limited in use because of the dangers of fatigue crack propagation (but even so there have been fatigue-related failures in aircraft: see reports on the Comet and the Airbus in Chapter 6).

7.4.10 Product failure from galvanising

Galvanising is a very common system for protecting steel from corrosion and in most circumstances it is used safely and successfully. In rare circumstances it can cause problems. One form of failure is damage to fresh zinc coating by the formation of 'white rust'. This is a powdery white substance that can form on fresh zinc surfaces. Zinc provides protection by transforming into surface stable oxides but these need time to develop and storage of freshly coated zinc in stacks (no access to atmosphere and in damp conditions) is a risk factor inhibiting that desirable stable surface condition.

Not all steels are easy to galvanise, and this has not always been appreciated. High strength steel can be susceptible to hydrogen embrittlement arising from the pickling process used prior to dipping. In 1968, the spontaneous fracture of four galvanised Macalloy bars occurred on a bridge (the bars were being used to post-tension brickwork) and this caused a major safety alert.

More recently, a worldwide investigation has been carried out on the causes of liquid metal assisted cracking (LMAC) (Figure 7.9). This is an apparently rare phenomenon or

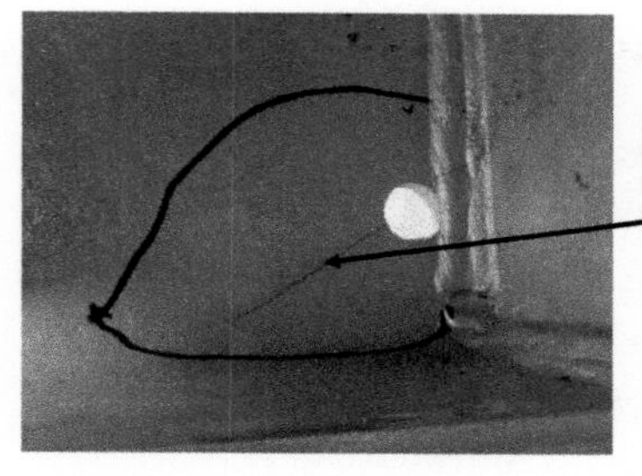

[Left] Figure 7.9 Liquid metal assisted cracking

[Right] Figure 7.10 Galvanised rails corroding

at least one hitherto not widely recognised, where for some reason steelwork cracking occurs whilst the part is in the galvanising bath. A significant worry is that the cracks may become filled with zinc and thus not be apparent until the load applied in service opens the cracks up. SCOSS has issued advisory notes about this danger (SCOSS Update Note, 2006) and the BCSA, in conjunction with the Galvaniser's Association, has issued guidelines for inspection and these identify the risk factors (BCSA and GA, 2005). A risk mitigation factor is always to inspect the steel once it has been galvanised and when just put into service with some loading.

Galvanising does not last for ever, and its life is linked to the environment and degree of pollution (Box 7.11). In salty marine environments, its life can be relatively short (Figure 7.10).

Box 7.11 Galvanised lamp posts, UK (1990s)

SCOSS expressed concern in 1999 over the large number of UK columns in need of replacement (SCOSS Bulletin, 1999; CROSS Reports 98, 114, 610, 5101). One death resulted from the collapse of a 35-year-old tubular steel column in Gateshead in 1995 and a pedestrian was seriously injured when a steel column fell in Westminster in January 1998.

Comment: Lamp posts present an interesting example of safety choice where failure is concerned. Driving into a light steel lamp post may shear it off with the post falling back over the car, but the post is light so will probably be supported by the car's roof. In contrast, hitting a 'stronger' concrete post may well cause more injury either via the impact and rapid deceleration or by the heavy post falling on the car and crushing it. Lamp posts also highlight the quandary in multiple items: one wrong, all wrong.

7.4.11 Fastening failures

A variety of fastenings are used and they have several modes of failure (CIRIA, 2019). CROSS reports several failures linked to the use of blind bolts (CROSS Reports 71, 86, 113, 170). More widely, given a pattern of failures linked to failed fixings, SCOSS issued an Alert in 2010 on 'The selection and installation of construction fixings' (SCOSS Alert, 2010) with the opening words: 'SCOSS has been concerned for some time at the use of structural fixings where these so called "minor" items have not received the attention they deserve given their safety critical nature. The Committee is aware of a few failures which had the potential for multiple fatalities.' The Alert also stated: 'The Construction Fixings Association (CFA) reported in 2004 that a significant percentage of specified fixings were changed without due authority.' This warning comes under the generic heading of failure cause due to unauthorised change, as raised in Chapter 6.

One failure mode is of fastenings loosening due to vibration. An extreme example of this occurred in two tunnels in the UK. The tunnels were lined with panels secured by self-tapping screws. One day, one of the panels fell off onto a passing lorry. On investigation it was found that the holding screws were loose, and the cause was that design envisaged use of self-tapping screws into a pre-fixed backing strip; both the screws and backing strip were stainless steel (to guard against corrosion). Unfortunately, stainless steel as a material strain hardens very quickly and so it is impossible to make a self-tap type of fixing: the fixing just jams up or burns out. On site, the installer had clearly found themselves with a problem, so to resolve it had drilled large pilot holes to get the screws in. The fixing was totally ineffective for the screws just unwound under the fluctuations from air pressure generated by traffic passing through the tunnel. CROSS Reports 148 and 149 both concern ceiling collapses where fixings loosening via vibration were an issue. In any structure that is moving, vibration will be a potential issue and measures need to be taken to secure items against coming loose.

Vibration also causes stress fluctuation so it is naturally linked to fatigue. The most severe danger arises when some initiating force acts at resonance so as to cause structural vibrations at natural frequency, since then a large number of cycles of high stress can rapidly build up. In the dance floor case mentioned earlier, the floor was being forced to vibrate by clubbers dancing; other causes are from reciprocating machinery or wind and those are described in Chapter 8.

CROSS has reports of epoxied-in fittings pulling out or mechanical anchors pulling out below intended capacity (CROSS Reports 44, 182, 244, 422, 429, 433, 615; SCOSS Alert, 2010; SCOSS Alert, 2014). Anything that depends on correct installation should be conducted under a procedure with supervision, and capacity should thereafter be verified by on-site testing.

CROSS has reports of weaknesses with grade 10.9 bolts, although they are widely used. The use of bolts with strengths greater than 10.9 is considered poor practice in general structural engineering as higher grade material may lack essential ductility, plus modes of fastener failure may shift from shank yield to thread stripping, which is undesirable. A rare form of bolt failure can occur through hydrogen embrittlement. These failures are especially dangerous as they are often delayed to occur spontaneously some considerable time after installation. Higher material strength is a risk factor, as is scale, in that unusually large diameter bolts may be more at risk (CROSS Reports 490, 499, 585, 600; CROSS, 2015).

Cases are known of procurers just ordering the wrong material. In one case of temporary works failure, the contractor mistakenly procured grade 4.6 studs instead of the grade 8.8 ones intended (the two appear visually identical). The connection failed and a death occurred.

7.5 Stainless steels

It might be thought that using 'stainless steel' is a panacea to all durability problems and generally it is, provided the right grade is used. Certainly, in earlier times, corrosion of galvanised ties in brick cavity walls was a problem and the substitution of the more durable stainless-steel variety should eliminate that concern. There are, however, many forms of stainless steel with some being more durable than others, and some grades are totally unsuitable in certain atmospheres. In marine atmospheres, lower-grade stainless steel can corrode with pitting, suffer from crevice corrosion, or fail by stress corrosion cracking.

In structural engineering, several cases of stress corrosion cracking have occurred with devastating consequences. These have mostly occurred over swimming pools where the atmosphere is warm and laden with chloride ions. Failures have been reported in many countries. Problems have been known in the UK (in wire hangers) but Page and Anchor (1988) report a failure (with fatalities) in Switzerland in 1985 when, after 13 years of service, a heavy concrete ceiling fell over a pool (Box 7.12). The collapse

Box 7.12 Uster swimming pool, Switzerland (1985)

Twelve people were killed when the concrete roof over this pool collapsed after 13 years of use. The roof was hung from stainless steel rods which failed. An investigation concluded that rod failure had been brought about by chloride-induced stress corrosion cracking. Overall, the form of failure was a progressive collapse. Similar failures have occurred over other pools (Steenwijk, the Netherlands (2001), and Kuopio, Finland (2003)). In each incident, the causes and form of collapse were the same. Chloride ions can attack certain grades of stainless steel, especially in warm conditions and the combination of stress and corrosion develop the phenomenon known as stress corrosion cracking. The development of cracking is insensitive to the source or magnitude of stress. In one incident in the UK, failures came about simply from the residual stresses due to twisting wires together. In the Dutch incident, some of the stress was residual stress induced by plate bending.

Comment: A clear safety danger here is the risk of lack of technical knowledge when using a relatively novel material.

Box 7.13 Thelwall Viaduct bearings, UK (2002)

Thelwall Viaduct is long, carrying the M6 over the Manchester Ship Canal, and its main girders were supported on rollers made of a strong (martensitic grade) stainless steel. In July 2002, inspections revealed a failed roller. Further problems were found with the form of the failure being cracking right through the roller on its vertical plane. As a measure of seriousness, replacement costs of over £50 million have been suggested. Examination of the cracked surface showed that cracking had been progressive under fatigue. The investigators concluded that cracking started from the roller ends where there were intergranular corrosion fissures which in turn caused hydrogen-induced stress corrosion and the general level of tensile stress across the centres of the rollers was sufficient to drive crack growth (a cylinder loaded vertically will develop tensile stress in the horizontal direction).

Comment: This failure again might be classified under risk as a small multiple item whereby if there is one failure, all will be affected, at great cost.

resulted from brittle failure of austenitic stainless steel ceiling supports. SCOSS issued a fresh alert about this type of failure (SCOSS Alert, 2005).

Stainless steels offer designers many valuable properties. Some grades can be very ductile and suitable for use at very low temperatures. Others offer the best possible corrosion resistance. Others offer high strength. But caution is required when using any material in unusual applications and unless there has been a long history of usage demonstrating reliability. A good example of this is the failure of roller bearings under Thelwall Viaduct (Box 7.13).

7.6 Concrete

7.6.1 General

When modern reinforced concrete (RC) was first 'invented' around the beginning of the 20th century, it was thought to be a wonder material capable of lasting forever. After all, certain Roman concrete structures were still in existence. Reinforced concrete can last, but it can also suffer badly from cracking and durability problems. The first national code for RC was issued in 1934. Construction output was slow until the late 1940s, but immediately after the Second World War, RC structures received a real boost due to the scarcity of steel and the huge demand for new building, not least in housing. In the period after that, there were many technical innovations in understanding performance; in the development of better concrete and in product innovation. In overall terms, structures became lighter and more 'efficient'. However, up until the 1960s, issues of concrete durability were not widely appreciated by the design and construction community with the result that the national stock of buildings deteriorated badly and left a huge workload of repair and replacement. It has been observed that many of the industrialised processes of the 1950s and 1960s left a legacy of deterioration. Somerville, provides an overall review (Somerville, 1986). As one example, he points out that calcium chloride admixture, which had been commonly used as an accelerator, was only banned in 1977 because of its adverse influence on reinforcement corrosion. General guidance on concrete durability is given by the Concrete Society (2004, 2009, 2015).

A dramatic example of concrete degradation on strength is given by the collapse of the Piper's Row multi-storey car park (UK) in 1997 (Figure 7.11). One night, the top deck failed when supporting columns punched through, resulting in a 120 tonne section falling onto the floor below. One cause was degradation exacerbated by poor repair and inadequate inspection. As the investigation states (Wood, undated) the collapse 'highlighted the need to identify in appraisal and inspection the risks from shortcomings of design and construction, and the effects of deterioration, before safety is compromised'.

It is not in the national interest that investment in infrastructure should be squandered by premature failure. Moreover, putting investment aside, a structure deteriorating is one that is also eroding its safety margins. Hence for safety, designers need to address durability problems and understand cause. Likewise, for safety in through-life assessment, there needs to be comprehension of potential failure mechanisms. Alas in

Figure 7.11 Collapse of Piper's Row car park (courtesy of Jonathan Wood)

concrete there have been numerous degradation causes. Hence there is a need to be able to predict how long a structure might last and to predict its rate of degradation. Perceived lack of durability is a major issue in safety-related structures such as containment structures for hazardous materials and those used for nuclear containment (in Canada and the US, regulators have established that some reactor buildings are suffering from alkali–silica reaction). Establishing rates and degradation modes is critical: instant failures (such as that at La Concorde Bridge, see Section 7.6.8) or the Genoa motorway bridge (see Section 7.6.8) are intolerable, but where degradation is progressive, it at least gives time for authorities to make judgements about either repair or replacement: it is simply not possible to replace major structures instantly and assessing safety under less than perfect status is a necessity. An example of the challenges can be seen in Jackson and Moore (2017).

Somerville's paper (Somerville, 1986) provides a very good overall description of the essentials. Key is preventing embedded rebar from corroding since once this starts, the metal expands, bursting off its cover, and the rate of corrosion will then accelerate. Embedded rebar is ideally kept passivated by the surrounding alkalinity of the concrete which in turn is linked to the cement content and other factors. It is also linked to cover, concrete's life being sensitive to quite small reductions in cover thickness. Over time, concrete reacts with carbon dioxide in the atmosphere and this gradually reduces the alkalinity of the cover depth until it reaches as far as the embedded rebar. Codes ought to minimise the risk by ensuring minimum cement content in mixes by enforcing adequate cover and by enforcing certain construction practices such as compaction. Nonetheless, and realistically, failures such as those in Edinburgh schools in 2016 (Chapter 6, Section 6.3.4) highlight that site inspection and enforcement of site quality standards are not what they should be. It is not hard to observe instances of corroded rebar on concrete surfaces where cover has been lacking.

Certain chemicals are known to cause degradation; calcium chloride and carbon dioxide have been mentioned above. Groundwater with excessive sulphates is another and in certain parts of the world aggressive chemicals within soils can lead to demands to embed foundations in protective material. Precautions always include carrying out adequate site investigation to assess the risks. Good concrete can only be made with proper aggregate and codes ensure testing to eliminate aggregates that cause long-term problems. Frost damage can occur, and surfaces that are badly exposed (such as roads) may be protected by specifying air-entrainment. In certain industrial processes (say, chimneys) acid attack from fumes can cause severe localised erosion and degradation. Salt generally is a problem and the use of de-icing salts on roadways is a major cause of highway structure deterioration. In vulnerable areas such as bridge parapets, the problem can be so bad that stainless steel rebar has to be used.

Over the years, other degradation mechanisms have come to light. Principal amongst these are:

Box 7.14 Sir John Cass School swimming pool and gymnasium roofs, UK (1973)

In this collapse, the roof over the swimming fell completely and suddenly. In the same period, there were two equally serious failures, one of the Assembly Hall at Camden School and one at the Bennett Building in Leicester University. In all three cases, the failing beams were made from high alumina cement (HAC) so there was a common factor. It was found that, under certain conditions, the original concrete lost strength dramatically. Reports from the Sir John Cass School incident quote a strength reduction from an original of 65 / 54 N/mm^2 down to 30.5 / 12.7 N/mm^2 (Barclay, 1974; Institution of Structural Engineers, 1974), i.e. down to 20–50% of the original strength. Neville (Neville, 2011) quotes that at Leicester, strengths dropped from 55.2 N/mm^2 to 6.9 N/mm^2 (12% of the original strength). These failures created a nationwide scare and a need to investigate many buildings. HAC was banned from use in 1976.

Comment: This story amply illustrates the dangers in having a material which exhibits degradation or safety issues no one anticipated. Industry needs to beware in the introduction of all new materials. There is also a lesson about record keeping. At the time, many owners would have been quite unaware of whether HAC was present or not in their buildings and inadequate records compounded the identification demand.

- HAC conversion;
- alkali–aggregate reaction (AAR) / alkali–silica reaction (ASR);
- mundic;
- sulphate attack and thaumasite;
- chloride attack.

7.6.2 High alumina cement (HAC) conversion

HAC was originally marketed as a strong cement that gained strength very rapidly, with obvious uses. It was formerly much used in making precast concrete products. Alas, it was in practice found to suffer 'conversion': it reverted to a much weaker strength over time. Conversion took place if the original water/cement ratio was not controlled, if the product was not properly cured and if the product was in warm, moist environments (such as over swimming pools). HAC may still be encountered in demolition or building alterations. Box 7.14 describes some collapses.

7.6.3 Alkali-aggregate reaction (AAR) and alkali-silica reaction (ASR)

AAR and ASR are chemical reactions between alkaline components in cement and the active silica-based minerals of some aggregates. The reaction forms a gel which absorbs water and the ensuing expansion creates internal pressures which can be large enough to

Box 7.15 Chambon dam, France (2012)

Chambon is a 90 metre tall × 239 metre long concrete gravity dam completed in 1934. Over the years it has experienced AAR, causing the concrete to expand and crack sufficiently as to require major repair.

generate micro cracking in a patchwork pattern. ASR is a more widespread phenomenon and the more harmful to concrete's mechanical properties, the general trend being of significant reductions in compressive strength. Internal expansion will enforce noticeable global movements in large masses of concrete. However, the action takes place over years so does not pose immediate safety threats. Nevertheless, it appears an issue of concern in some structures including dams (Box 7.15). The Concrete Society (1999) and the Institution of Structural Engineers (1992) give guidance on minimising the risk.

7.6.4 Mundic

There are a range of concrete durability problems linked to reactive aggregates. AAR and ASR are two. Mundic is another of concern in the UK. Mundic is a mineral of chemical pyrites embedded in the mining waste locally used in south-west England as aggregate. When water penetrates the concrete, sulphuric acid can be created. This acid then causes the concrete to crumble. It has been a concern largely to housing in the south-west of England from where the aggregate originates. This form of failure is one of several linked to unsuitable aggregates.

7.6.5 Sulphate attack and thaumasite

Sulphate attack on concrete can be very serious, eating away a large proportion of concrete sections. Attack can be external or internal. The external type comes from burying concrete in ground with aggressive chemical- or sulphate-bearing water. Attacks cause cracking, expansion and general degradation of the cement paste leading to loss of strength and/or loss of concrete. Thaumasite is a calcium carbonate silicate sulphate hydrate which forms within concrete from the hydrates in the matrix, causing softening. Attacks are uncommon but since it is the attacked hydrates that provide most of the strength, thaumasite formation results in severe weakening. Affected concrete might have virtually no strength. In the UK, attacks have been discovered on M5 bridge foundations and other buried structures.

7.6.6 Chloride attack

If chlorides from salty atmosphere or road salts penetrate concrete, they will in time cause corrosion of embedded steel with subsequent expansion, spalling and general weakening. Perhaps a more dangerous form occurs in dense, strong concrete, since then rebar corrosion might continue without the spalling warning signs.

7.6.7 General cracking

All concrete cracks to some extent under stress. That stress might be generated by external load or it might be generated by internal actions. Such internal action is high during casting of thick sections when heat generated by the setting process creates expansion which is subsequently restrained. There are several other forms of 'non-structural cracking' and guidance on causes is available (Concrete Society, 2010). Cracking may not affect safety per se but cracking and crack growth may be indicative of some design deficiency or progressive failure, so should always be investigated.

7.6.8 Reinforcement corrosion

The capacity of reinforced concrete as a material clearly arises from the combination of concrete strength with the added rebar. The factors that protect embedded bar from rusting are several, including chemical and mechanical factors. Essentially the alkaline environment created by enveloping bars with cement-rich concrete provides protection (but see Somerville, 1986). It is worth pointing out that variations of cover are a good example of the test of sensitivity (Chapter 4), i.e. a small reduction in cover is linked to a big reduction in corrosion protection. Over time, atmospheric CO_2 interacts with outer layers of concrete to carbonate it. This results in loss of alkalinity to the concrete around embedded steel (and in short lives if the cover is thin) and hence the onset of corrosion. Other factors, such as the initial water/cement ratio and the compaction during placing, are also influences. Boxes 7.16 to 7.19 give examples of failure due to reinforcement rusting.

Box 7.16 Kongress Halle, Berlin, Germany (1980)

The Kongress Halle was a show case 'architectural' building. It had a shell roof restrained in perimeter tension by a ring beam. The whole structure suddenly collapsed without any warning.

The failure cause was implicit in a poor structural design that failed to consider corrosion prevention. The tendons in the ring element became corroded because there was poor detailing and inattention to movements that allowed water to enter tendon ducts. When the tendons failed, so did the whole roof. It was also reported that the roof joint concrete was very porous and had a high chlorine content.

Comment: The obvious concern is corrosion and in this case that concern led to more widespread concerns about bridges (see Box 7.17). Additionally, in safety terms, three other principles were violated:

a. There were no warning signs (or if there were, they were not picked up by inspection).
b. There was probably no means of monitoring the tendon condition.
c. The collapse was sudden (in what might be argued a Class 3 structure) (see robustness in Chapter 4).

Box 7.17 Post-tensioned bridges, several countries (1970–1990)

SCOSS raised concerns about the durability of ducted tendons in bridges as early as 1979. A number of failure reports exist, such as for Angel Road Bridge on the North Circular Road in London (1980); Taf Fawr bridges in Merthyr Tydfil (1982); Folly New Bridge in Oxfordshire (1988); Blackburn Road Bridge on the M1 in Sheffield (1990), and Botley Road flyover in Oxfordshire (1992). All these were found to have tendons corroded or broken following corrosion.

Abroad, the Melle Bridge, constructed in Belgium during the 1950s, collapsed in 1992 due to corrosion failure of its post-tensioned tie-down members.

The Ynys-y-Gwas Bridge in West Glamorgan collapsed without warning in 1985 due to corrosion of its tendons. This failure led to increased concern about the risk of tendon corrosion, about the adverse effects of chlorides and the need to improve methods of detection, and the essential need to provide effective methods of corrosion prevention.

Comment: A major concern of all these failures was the lack of warning of reduced safety with potential for sudden failure (as in Box 7.16).

Box 7.18 Charlotte Motor Speedway Bridge, North Carolina, USA (2000)

This bridge was part of a motor sports complex. At the time of the incident, fans were crossing the bridge when a 24 metre section suddenly failed and dropped about 6 metres. The bridge was a post-tensioned precast unit about five years old. The failure cause was said to be corrosion of the bridge tendons brought about by using calcium chloride as an accelerator.

A spectacular failure of a major motorway bridge occurred in Genoa in 2018. The high-level Morandi Bridge collapsed whilst in use, leading to 43 deaths. A whole centre deck fell out, initiated by failure of one of the concrete stays, which then set off a chain reaction of deck failures. The cause is suspected to be degradation since press reports claim this had been known about for some time but not acted upon. Dramatic pictures of this collapse can be seen on the web, as also can a video of another real-time failure of another Italian bridge (Lecco Notizie, 2016). The cause of this failure is a likely to be the familiar

Box 7.19 De la Concorde overpass collapse, Montreal, Canada (2006)

This shocking failure caused headlines worldwide. A complete 20 metre span of the bridge suddenly dropped out, crushing vehicles below. Five people were killed and six were injured. Three cars were on the deck at the time of its fall and two cars were below. Apart from these immediate costs, consequential costs on traffic disruption were substantial.

The span that fell had been supported at either end on half joints. The investigation revealed that these cantilever joints had failed with cause being put as shear failure on a horizontal plane that had been progressively growing. Causes were poor fixing of rebar (not in proper locations), low quality of concrete deteriorating in freeze/thaw conditions; lack of shear rebar in thick slabs (permitting brittle failure) and no waterproofing, allowing water degradation. Additionally, concrete removal and rebar exposure during earlier repairs had caused weakness.

There was also criticism of authorities for putting off repairs to balance budgets. Moreover, a similar problem had occurred elsewhere in 2000 so giving warning that aging structures were weakening.

Comment: This tragic failure illustrates many of the usual causes; first, a mixture of human and technical error. Human error in not responding to generic risk, and in putting short-term costs before safety. There are also errors in concept, permitting a mode of failure that was undesirable (brittle shear failure) and failure of detailing. The initial construct error falls under the category of making sure what was built matched design intent. The problems of error during earlier repair seem to fall under the category of unauthorised change. Proper inspection and maintenance should have prevented this incident, although in-service inspection was virtually impossible (SCOSS Topic Paper, 2008).

problem of deck half joints as described in Box 7.19. Failure of cantilevers where water can easily percolate into wider top gaps has been a feature in domestic balcony collapses.

7.6.9 Concrete product failure

Another way that rebar may affect safety adversely is by it being either missing or misplaced. The report in Box 7.19 cited missing/misplaced rebar as one of the causes. Likewise, in the collapse of beams in Sir John Cass School, the fact that some bars finished short of beam ends was cited as a contributor. In 2013, newspaper reports suggested that a tall building in the US (Wikipedia, undated) had to be built shorter than planned (28 storeys rather than 49) as rebar was missing from lower levels (the building was subsequently demolished). Other reports of other buildings cite rebar missing simply because the bars were impossible to fit in. An issue with concrete design/detailing is the skill of being able to see that bars will fit into the planned section with enough clear space to permit placing and compaction of planned concrete. If this is not done, predicted member capacities are meaningless. The demand on detailing is particularly

Box 7.20 Short piles, Hong Kong, China (1998–2000)

A series of scams related to substandard works were discovered over this period. In 1999, monitoring revealed excessive settlement at two buildings being constructed on a housing estate. Investigators found that of the specified 36 large-diameter bored piles only 4 met requirements. A total of 21 were shorter by lengths between 2 metres and 15 metres and 11 were resting on soft mud instead of bedrock. Other cases were found on other projects and for safety reasons, two blocks had to be demolished. There were 142 prosecutions and estimates suggest HK$650 million had been squandered (£55 million).

Comment: There have been cases reported to CROSS of fraudulent material and product certification.

significant when using large diameter bars since bend radii can be so large that enclosed bars are displaced significantly from where the designer might have intended. A feature of a Liverpool tower crane collapse, described in Chapter 6, was that dowelled-in bars to the pile cap were epoxied in at a length much less than intended. In 1998–2000 there was a scandal in Hong Kong when it was found that rebar cages in piles had purposely been cut short to save money and this was intended to be hidden from inspectors, who were only shown required bond lengths projecting from the top (Box 7.20). Around 15 sites were thought to be affected and criminal charges were laid.

The possibility of gross error can never be discounted. There are several reports of global progressive failures consequent on punching shear failures of flat slabs around column heads. In a report on the Harbour Cay Condominium failure in the US in March 1981 (National Institute of Standards and Technology, 2017) the structure collapsed during construction due to design error coupled with construction errors of using the wrong sizes of rebar and chair height. The designer never performed any calculations to check for punching shear, the most common form of failure for this structural form. In 2003, a casino parking garage collapsed in Atlantic City (Chapter 8). Four workers were killed and 30 were injured. The simple cause was that floors were just not connected to their supporting walls with the required bars and the whole structure just fell apart (US Department of Labor, Occupational Safety and Health Administration, 2004).

Corrosion mechanisms that affect structural steel equally apply to steel embedded in concrete. CROSS Report 147 describes a tank that failed due to stress corrosion (Box 7.21).

In 2019 widespread failures of reinforced autoclaved aerated concrete (RAAC) planks were discovered in the UK. This product dated back to the 1960s and the failures were revealed by sudden collapses. SCOSS Alert (2019) gives more details. It has recently been revealed that numerous UK hospital roofs are afflicted.

Box 7.21 Concrete tank failure, UK (2009)

This report was about the sudden and catastrophic failure of a tank formed of precast panels prestressed together using unbonded cables in grease-filled sheaths. For whatever reason, the tendons failed, and the failure was put down to stress corrosion.

Comment: Issues are the suddenness of the failure and the cause of corrosion.

Another recent problem (post 2017) has occurred in Ireland, where mica in concrete blocks has led to extensive dramatic cracking and weakening affecting thousands of homes.

7.7 Soil

Soil of various kinds is mostly what we build on. The whole science of soil mechanics is about understanding soil properties and capability; the loads soils can support, and the pressures soils might exert. The role of water either as a water table or as it moves through the ground or over surfaces is crucial, not least in its power to be a corroding medium, or via its erosion potential. Many building failures relate to inadequate anticipation of soil movement and to poor judgement about the capability or otherwise of supported structures to accommodate differential displacement (Chapter 3 on the meaning of failure; see also Burland, 2008). Such building faults rarely pose a direct safety hazard except in exceptional circumstances such as in the Leaning Tower of Pisa.

Perhaps the most dangerous of soil failures from a safety perspective is when soil gets inadvertently washed away. This is a common occurrence via coastal erosion undermining UK seaside properties. The power of water to move soil was shown up dramatically in Newcastle-upon-Tyne in September 2012, during a severe storm. In one place, a culvert collapsed permitting a local torrent to run beside a block of flats. The torrent was so severe it removed all the material around the pile foundations rendering the whole building unstable: the flats had to be demolished. It's possible to see videos on the Web of incidents of cars or pedestrians suddenly disappearing through roads or pavements as a result of collapse following washout from faulty drains below. SCOSS has issued an Alert on sudden loss of ground support (SCOSS Alert, 2017).

There have been many dramatic failures due to bridge scour that has either undermined or washed away foundations, leading to complete collapse. In safety terms there is a good deal of uncertainty associated with the risk of scour; inspection and early detection might be hard and the form of failure might be instantaneous. In the severe Cumbrian floods of 2009, the Northside Bridge at Workington (UK) was swept away. The failure was so fast that a policeman on the bridge at the time (trying to keep traffic off) was swept to his death. Scour is a large cause of bridge failures (see Box 7.22).

Box 7.22 Malahide Bridge, Dublin, Ireland (2009)

This bridge is a 176 metre long viaduct built over an estuary and it carries the Dublin to Belfast railway. The bridge has a concrete deck resting on masonry piers dating from 1860. In 2009, a 20 metre section gave way just after commuter trains had passed over. The cause was the collapse of a masonry pier precipitated by foundation scour.

Comment: This was a 'near miss' and loss of life could have been very great. As with many serious collapses, the suddenness of the event was critical. In many ways a more alarming fact was that the bridge had recently passed an inspection and been declared safe, which reinforces the notion that scour is not always easily detectable. In safety terms, that suggests a need for added assurance over the stability of bridge foundations within rivers because their failure precipitates a particularly undesirable consequence in both speed and extent of collapse (Brady, 2017).

Water inundation when working below ground is always a risk, as Brunel found out to his cost when trying to complete the Rotherhithe Tunnel in 1828. But it can occur in ordinary excavations when sheet piling or coffer dams fail. In 2002, part of the Union Canal bank through Edinburgh gave way, flooding some local areas to a depth of 1.5 metres. Several buildings had to be evacuated. When sub-ground water pours in, the flow might be at a constrained rate, whereas if a canal bank bursts and there is a large volume of water retained in the canal sector, the flows will potentially be rapid and enormous (in the First World War, the Belgian government deliberately opened sluice gates of a canal to flood an area a mile wide as part of their defences).

Landslips are a frequent problem, particularly alongside railway embankments where the safety risk is line blockage. Slips can occur anywhere, and longevity is no real proof of safety or security. As an example, in 2008 it was reported that Abraham Darby's Iron Bridge (the world's first) was being distorted by land slips at its abutments.

As with all other materials, danger lies in the material itself and in the techniques for construction. CROSS reports that undermining existing foundations is a problem frequently encountered by building control bodies, and there are numerous examples of resultant substantial failures. There are incidents where incompetent undermining has resulted in gross collapse of the wall overhead and death of those involved in the excavation (Box 7.23) An extreme example would be the widely reported toppling of a whole block of flats in Shanghai in 2009 (Box 7.24).

7.8 Glass

Glass is a common material with widespread use in windows, but also increasingly used in more architectural and structural applications. Consequently, glass is available in many different forms: plate glass, toughened glass, laminated glass and so on. Thus,

Box 7.23 Orkney sea wall, Scotland (1997)

Two volunteers were working under the foundations of a sea wall on North Ronaldsay when it collapsed, burying them under a 5 tonne concrete section. The controlling organisation, the Princes Trust, was fined £10,000.

Comment: This is perhaps mostly an example of failing to verify stability during phases of construction.

Box 7.24 Shanghai flats, Shanghai, China (2009)

Figure 7.12 Complete overturn of the Lotus Riverside apartment block, Shanghai Alarmy 2D2P8E5

A newly built 13 storey block of flats completely overturned in Shanghai (the foundations were piled). Amazingly only one worker was killed. The failure seems to have been caused by undermining on one side exacerbated by heaping spoil on the other.

Comment: Few failures are more spectacular than this one.

almost uniquely, selection of type is expressly linked to an anticipated mode of failure. Car windscreens cannot utilise glass which breaks into shards but must instead use a type that disintegrates into less risky fragments on impact. Periodically there are reports of spontaneous glass cracking, especially in windows with nickel sulphide inclusions, albeit there appears some lack of consensus that such inclusions are the cause of all the problems. The main safety issue is of glass falling from height.

The 12th SCOSS Report (SCOSS, 1999) referred to façade treatment and included:

> Failures of such glass constructions have occurred. The principal causes of glass failure appear to be poor design and workmanship. Stress concentrations arising from lack of fit, edge damage during construction, incorrect installation of fixing and seals, and lack of allowance for in-service movement or temperature change may each lead to glass failure, often at an early age. Failures sometimes arise due to manufacturing defects in the glass itself.

And thereafter:

> A SCOSS concern is that designers should ensure that they fully understand the behaviour of structural glazing systems before using them in innovative applications. Another concern is that glass constructions and glazing systems including fixings, should be inspected carefully at the time of construction and subsequently during the lifetime of the building.

The Institution of Structural Engineers' Structural Use of Glass in Buildings (Institution of Structural Engineers, 2016) will provide overall design guidance. In 2019, SCOSS issued an Alert on the structural safety of glass in balustrades (SCOSS Alert, 2019).

7.9 Chapter summary

- No construction material is immune to degradation and many failures arise from that material degradation. These cost society a great deal of money.
- Failure risk in service is linked to the material itself, to its operating environment plus the fabrication/construction techniques used.
- Safety requires thought be given to degradation potential at the design and assessment stages.
- Safety requires that designers appreciate common degradation mechanisms (and the rate at which degradation takes place) and design accordingly to reduce the risk of failure.
- Safety requires that in-service inspection is both possible and takes place. The objective must be to identity the onset of weakening before it progresses too far.
- Safety requires that provisions for maintenance are made (which may also demand provision of dedicated access).
- Numerical safety margins must be looked at in the light of material state.
- Designers routinely rely on material and product provenance. A safe design should always verify this by requiring appropriate certification and perhaps additional testing.
- Designers routinely presume that construction will match their design intent. A safe design ought to verify this.

References

Barclay, C. F. R. (1974), 'Collapse of roof beams Sir John Cass School, Stepney', *The Structural Engineer*, **52**(7).

Berenbak, J., A. Lancer and A. P. Mann (2001), 'The British Airways London Eye. Part 2, Structure', *The Structural Engineer*, **79**(2).

BCSA and GA (2005), *Galvanising Structural Steelwork: An Approach to the Management of Liquid Metal Assisted Cracking*, British Constructional Steelwork Association Ltd and Galvanizers Association.

Brady, S. (2017), 'Lessons from the Malahide Viaduct collapse', *The Structural Engineer*, **91**(7).

Brick Development Association (BDA) (2011), Design Note 7, 'Brickwork durability', www.brick.org.uk/admin/resources/g-brickwork-durability.pdf.

Burdekin, F. M. (1999), 'Size matters for structural engineers', *The Structural Engineer*, **77**(21).

Burland, J. B. (2008), 'Foundation engineering', *The Structural Engineer*, **86**(14).

CIRIA (2019), 'General fixings. Guidance on selection on whole life management', C777; 'Management of safety critical fixings', C778.

Concrete Society (1999), 'Alkali–Silica Reaction – Minimizing the Risk of Damage to Concrete, TR30, Concrete Society.

Concrete Society (2004), *Enhancing Reinforced Concrete Durability*, TR61, Concrete Society.

Concrete Society (2009), *Repair of Concrete Structures with Reference to BS EN 1504*, TR69, Concrete Society.

Concrete Society (2010), *Non-Structural Cracks in Concrete*, 4th edn, TR22, Concrete Society.

Concrete Society (2015), *The Relevance of Cracking in Concrete to Corrosion of Reinforcement*, 2nd edn, TR44, Concrete Society.

CROSS (2015), Newsletter 38, www.cross-safety.org/uk/news-and-events/cross-uk-newsletter-38.

CROSS Report (March 2014), Tension systems and post-drilled fixings.

CROSS Report (September 2010), The selection and installation of construction fixings.

CROSS Report 21, Timber designs.

CROSS Report 44, Fixings to steelwork.

CROSS Report 59, Fatality from free standing wall collapse.

CROSS Report 62, Timber truss.

CROSS Report 71, Blind bolts.

CROSS Report 86, Blind bolt fixings.

CROSS Report 92, Collapse of a gable wall.

CROSS Report 94, Serious injury from free standing wall collapse.

CROSS Report 98, Unsafe street lighting columns.

CROSS Report 113, Blind bolt query.

CROSS Report 114, Unsafe lamp posts (news).

CROSS Report 116, Death from wall collapse (news).

CROSS Report 147, Post-tensioned precast concrete tank failure.

CROSS Report 148, Suspended granite ceiling collapse.

CROSS Report 149, Rendered ceiling failure.

CROSS Report 170, Blind bolt failure on offshore platform.

CROSS Report 182, Glass panel fixings failure.

CROSS Report 186, Collapse of large panel system (LPS) buildings during demolition.

CROSS Report 209, Risk in notching timber studs.

CROSS Report 244, Failure of epoxy fixings due to high temperature.

CROSS Report 249, School roof timber truss collapse.

CROSS Report 254, Steel connector failures and forged certificates.

CROSS Report 259, Quality of some imported steel components.

CROSS Report 273, Collapse of proprietary timber roof.

CROSS Report 284, False CE certificates.

CROSS Report 299, Documentation for imported large diameter steel pins.

CROSS Report 326, RAPEX notification on some imported RHS stock (news).

CROSS Report 331, Certification of steel sheet piling.

CROSS Report 338, Concern about CE marking for reinforcing steels.

CROSS Report 422, Fixings related to 395 partial roof collapse at shopping centre.

CROSS Report 429, Failure of epoxy resin bonded anchors in concrete.

CROSS Report 443, Post-fixed RC anchors – erroneous assumptions leading to unsafe design.

CROSS Report 474, Imported steel.

CROSS Report 490, Bolt failures due to hydrogen embrittlement.

CROSS Report 499, Failure of high strength studs.

CROSS Report 585, Failure of supposedly grade 10.9 bolts.

CROSS Report 600, Failure of tower crane studs.

CROSS Report 610, High mast light poles at all UK sites.

CROSS Report 615, Inadequate bolted connections supporting stairs.

CROSS Report 5082, Cladding and glazing.

CROSS Report 5098, Cladding on buildings.

CROSS Report 5101, Lighting columns.

CROSS Report 5132, Certification, use of products and other associated matters.

CROSS Report 6001, Two-year-old boy has died after a wall collapsed (news).

CROSS Topic Paper (2010), 'Major cast metal components', www.cross-safety.org/uk/safety-information/cross-topic-paper/major-cast-metal-components.

Finnish Safety Investigation Authority (2005), https://turvallisuustutkinta.fi/en/index/tutkintaselostukset/other/tutkintaselostuksetvuosittain/muutonnettomuudet2005/b22005ymarketinsisakatonputoaminensysmas.html.

Hewlett, B. (2021), 'A checklist of water damage', *Proceedings of the Institution of Civil Engineers – Forensic Engineering*, **174**(2).

Institution of Structural Engineers (1974), 'Failure of roof beams at Sir John Cass's Foundation and Red Coat Church of England School, Stepney', *The Structural Engineer*, **52**(8).

Institution of Structural Engineers (1992), *Structural Effects of Alkali–Silica Reaction: Technical Guidance on the Appraisal of Existing Structures*, IStructE Ltd.

Institution of Structural Engineers (2016), *Structural Use of Glass in Buildings*, 2nd edn, IStructE Ltd.

Jackson, P. and S. Moore (2017), 'Strengthening of the Hammersmith Flyover, London (Phase 2) – strengthening project focus', *The Structural Engineer*, **97**(8).

Lecco Notizie (2016), https://m.youtube.com/watch?v=akS6Tp5x8Y4.

Mann, A. P. (2011), 'Cracks in steel structures', *Proceedings of the Institution of Civil Engineers*, **164**(FE1).

Maranian, P. (1997), 'Vulnerability of existing steel framed buildings following the 1994 Northridge (California, USA) earthquake: considerations for their repair and strengthening', *The Structural Engineer*, **75**(10).

Moskowitz, E. (2011), 'Big Dig woes still pose threat', Archive.boston.com/news/local/massachusetts/articles/2011/03/17/some_big_dig_light_fixtures_found_damaged_by_corrosion/.

National Institute of Standards and Technology (NIST) (2017), 'Harbour Cay condominium collapse Florida 1981', www.nist.gov/el/harbour-cay-condominium-collapse-florida-1981.

Neville, A. M. (2011), *Properties of Concrete*, 5th edn, Pearson.

Page, C. and R. D. Anchor (1988), 'Stress corrosion cracking of stainless steel in swimming pools', *The Structural Engineer*, **66**(24).

Rail Accident Investigation Branch (RAIB) (2010), 'Derailment of a freight train near Stewarton, Ayrshire, 27 January 2009', Report 02/2010.

SCOSS (1999), *The Twelfth SCOSS Report, Structural Safety 1997–99: Review and Recommendations*, SETO Ltd.

SCOSS Alert (2005), 'Stainless steel: a reminder of the risk of failure due to stress corrosion cracking in swimming pool buildings', www.cross-safety.org/sites/default/files/2005-10/stainless-steel-stress-corrosion-cracking.pdf.

SCOSS Alert (2010), 'The selection and installation of construction fixings', www.cross-safety.org/sites/default/files/2010-09/selection-installation-construction-fixings.pdf.

SCOSS Alert (2013), 'Anomalous documentation for proprietary products', www.cross-safety.org/sites/default/files/2013-02/anomalous-documentation-proprietary-products.pdf.

SCOSS Alert (2014), 'Tension systems and post-drilled resin fixings', www.cross-safety.org/sites/default/files/2014-03/tension-systems-post-drilled-fixings.pdf.

SCOSS Alert (2017), 'Sudden loss of ground support', www.cross-safety.org/sites/default/files/2017-07/sudden-loss-ground-support.pdf.

SCOSS Alert (2019), 'Failure of RAAC planks', www.cross-safety.org/sites/default/files/2019-05/failure-reinforced-autoclaved-aerated-concrete-planks.pdf.

SCOSS Alert (2019), 'Structural safety of glass in balustrades', www.cross-safety.org/uk/safety-information/cross-safety-alert/structural-safety-glass-balustrades.

SCOSS Bulletin (1999), 'Bulletin 3 – February 1999', www.cross-safety.org/sites/default/files/1999-02/summary-scoss-review-1997-99.pdf.

SCOSS Topic Paper (2008), 'The partial collapse of the "de la Concorde" overpass bridge: Laval, Canada', Briefing Note SC/08/009, www.cross-safety.org/sites/default/files/2008-06/de-la-concorde-bridge-collapse.pdf.

SCOSS Update Note (2006), 'Liquid metal assisted cracking', SC/06/59, www.cross-safety.org/sites/default/files/2006-03/liquid-metal-assisted-cracking-update.pdf.

Senior, A. G. and T. R. Gurney (1963), 'The design and service life of the upper part of welded crane girders', *The Structural Engineer*, **41**(10).

Somerville, G. (1986), 'The design life of concrete structures', *The Structural Engineer*, **64**(2).

Stahl, F. L. and C. P. Gagnon (1996), *Cable Corrosion in Bridges and Other Structures*, ASCE Press.

US Department of Labor, Occupational Safety and Health Administration (2004), 'Investigation of the October 30, 2003, fatal parking garage collapse at Tropicana Casino Resort, Atlantic City, NJ', www.osha.gov/sites/default/files/2019-12/2004_r_03.pdf.

Wikipedia (undated), https://en.wikipedia.org/wiki/The_Harmon.

Wood, J. G. M. (undated), 'Pipers Row Car Park, Wolverhampton: Quantitative Study of the Causes of the Partial Collapse on 20th March 1997', www.hse.gov.uk/research/misc/pipersrowpt1.pdf.

8 Design and construction failures

8.1 Introduction

The time between project conception and completion can be long and the whole process complicated, often involving large teams and maybe teams of discipline specialists all interacting with each other. Even the smallest project can involve architects, civil, structural and services engineers, and contractors. Opportunities for confusion are rife and in any conference on failures, an underlying theme of 'miscommunication' is one that invariably recurs. It is the norm for changes to be made and every time that happens, the possibility of some unfortunate consequence is potentially overlooked.

Significant irritations exist because different teams do not understand the demands of others. Any practising contractor will complain about designs that are 'impossible to build': maybe steelwork connections that are unachievable onto member sizes defined by others or reinforced concrete that is too congested. Most designers will tell tales of clashes on interfaces: steel columns not fitting onto misplaced holding-down bolts or items constructed incorrectly. Every practising engineer will have anecdotes of late demand for penetrations through key structural items or 'minor modifications' that create significant delay or have structural implications. Such issues have recurred throughout construction history and are unlikely to disappear.

Assessing safety involves coping with uncertainty, not just in the civil structure but in the project overall. Chapter 10 deals with the demands on structures due to natural catastrophes and describes potential extreme loading cases that might have to be accounted for. Chapter 11 describes some man made incidents and some of the consequences of failures when perhaps not enough thought has gone into what might go wrong.

In looking at design and construct failures, we should not forget the risks to site workers who have the task of translating paper designs into reality. There are many hazards/risks and some of these are described in Chapter 12 on health and safety. Statistics from around the world vary but there remain many avoidable deaths and injuries. There are direct and indirect financial costs from failures alongside human tragedies. It is instructive to examine each of the major project stages and look at failure examples within each. These stages might be:

- feasibility;
- concept;
- safety concept;
- architectural design;
- structural design;
- civil design;
- detailing;
- site construction;
- commissioning;
- in use;
- demolition.

8.2 Why failures occur

8.2.1 Introduction

Design is the whole process of bringing a project to reality. It is not limited to structural calculation alone but embraces the concepts of structural form, load path, sizing of structural elements and how these activities interface with each other. It includes the selection of materials and their specification, the production of drawings showing general arrangements and details, and the whole business of interaction and form evolution with other construction professionals such as architects and services engineers. Architects create vision and impose limits on member sizing, disposition and available space for stability systems. Engineers need to understand architectural aspirations so that they can detail support structures appropriately and transfer applied loads back to ground. Services engineers apply loads both global and local and require holes through members in the most undesirable of locations. There is always conflict between the ideals of structural form and economy and the needs of the other parties.

An especial interaction is that which exists between structural form and construction. Put simply, most structures are assembled piecemeal sequentially and there are thus inherent strength and stability problems to be tackled whilst structures are in their incomplete state. Many historical failures have occurred during this stage.

Engineers, constructors and others are human: they make mistakes. It is foolish to assume that mistakes will not be made (a topic dealt with in Chapter 6), so starting from that premise, management processes need to be set up to catch these techniques and strategies are covered in Chapter 13. The stories of Concorde, the Space Shuttle, the Hubble Telescope and others show that civil engineering is not alone in this (see Chapters 6 and 11).

Not all 'failures' make headlines, but occur nonetheless often in bewildering circumstances. In recent times, a significant high-rise building has not been able to open because a contractor omitted rebar from large support beams, because if he had put the bars in, congestion would have been so great that he would not have been able to concrete (but why he failed to advise the design team of this remains a mystery). On another project, there were circular columns with welded hoop links. When the building height was advanced, someone dropped a hammer down the cage and all the steel links shattered. The cause was poor welding onto rebar and the welding process had embrittled the steel. The project team was then faced with the headache of what to do about all the lower columns which had already been constructed in substandard material. These two examples encapsulate the generic issues of:

- poor design and an inability to construct;
- human error in communication;
- severe consequence;
- poor detailing and lack of coordination between details and construction.

8.2.2 Design and analysis

It is interesting to study the history of structural analysis as a tool for predicting structural performance (MacLeod, 2008). Techniques only became available gradually and construction sizing by science, rather than intuition or empiricism, was uncommon until late in the 19th century; and it was only the last quarter of the 20th century that tools for advanced analysis and performance prediction became both easy to use and affordable (even now our Building Regulations have 'deemed to satisfy' sizing to aid those who have no facilities for computation). Before such analytical methods were developed, failures occurred regularly, but they decreased as more rational methodologies emerged.

8.3 Historical failures

In our own times failures recur and the professions struggle to learn and disseminate the lessons. Over the last 50 years there have been some landmark catastrophes.

It's easy to be pessimistic about failures, but in many cases, enlightenment has followed and led to improvements in understanding for the future. There are many

Box 8.1 Pemberton Mill, Massachusetts, USA (1860)

The headline of the New York Times on 11 January 1860 read 'HORRIBLE CALAMITY; Falling of the Pemberton Mills at Lawrence, Massachusetts. Five Hundred Persons in the Ruins. TWO HUNDRED OPERATIVES KILLED. ONE HUNDRED AND FIFTY WOUNDED. The Ruins on Fire Probably a Hundred Persons Burned to Death.'

As headlines go, this was dramatic and deservedly so and the event has since been described as the worst industrial accident in Massachusetts' history. The building was five storeys high, 85 metres long and 26 metres wide. It had brick walls (but these were slender and heavily perforated for windows), cast iron columns and timber supported floors. According to contemporary reports, the mill collapsed more or less instantly, trapping about 600 workers (perhaps 120 died). The cause seems to have been adding heavy machinery to upper floors. This overloading, allied with poor construction and iron columns which were brittle, came in for criticism. Some hours after the structural failure, a fire started and burned some of the survivors, adding to the death toll. Such contemporary mills were required to be as cheap as possible, there was no regulation and little regard for safety. The outcry was significant, with a demand for increased safety standards, but little was done.

Box 8.2 Knickerbocker Theatre, Washington DC, USA (1922)

This was a bad collapse of the entire roof. The roof had been in place for five years but collapsed completely during a film screening, killing 98 people in the process and injuring 133 more. The initiating cause looks to have been a heavy snow build-up. The incident is of interest since it was described in one of the very first technical papers published in The Structural Engineer (Godfrey, 1923), with the author putting the incident down to 'miserable design and construction, and a failure to understand the principles of stability'. See also Chapter 1.

failure cases which offer eternal lessons, but for understanding, it is better to concentrate on those of more recent times and examples are described in later sections. Some older ones of interest to set the scene are described in Boxes 8.1 and 8.2.

Other 'classic' failures are described in Chapters 1 and 2 (such as the 1879 Tay Bridge disaster).

8.4 Feasibility stage

8.4.1 Introduction

Self-evidently, no project should start unless it is feasible: yet examples testify that that objective is not always realised. Feasibility is the stage in any project when enough design ought to be carried out to check that the conceived project can be designed and constructed within the required time and budget. It is a stage when all key functional demands and all key uncertainties and risks ought to be anticipated and provided for. The task requires some considerable experience and judgement in trying to pick out all the project drivers and allow for them. Sometimes projects fail spectacularly because all along they were not feasible. In 2013, the BBC was forced to abandon its Digital Media Initiative: this scheme was supposed to transform the way staff developed, used and shared the corporation's video and audio resource. The technology apparently failed and £98 million was wasted. The NHS has also had some serious problems with IT projects that just did not work, as have the defence industries with some of their equipment procurement. In civil engineering, every project is more or less unique, so invariably carries risk, especially when 'innovation' is demanded. Cost overruns on major projects are legendary and often arise consequent on technical setback or 'unforeseen circumstances', for example, in the ground conditions. The original budget for the Channel Tunnel was overrun by about 80%. In recent years, project teams have tried to take a more mature attitude to risk by embracing its reality, recognising there are all sorts of uncertainties. Thereafter, rather than ignoring them and trusting they will not arise, the approach is to try and register what might go wrong and set boundaries to try and ensure a better probability of out-turn cost and programme. Many of the reported real-life failures could have been prevented, and their huge consequent costs contained, if a better attitude to risk and uncertainty had prevailed from the outset and if it had been convincingly demonstrated that the projects were feasible. This applies to projects small and large. But spotting what might go wrong is not easy.

8.4.2 Examples

In one reported case a large tank was being constructed below ground. The team thought that a formal check against flotation was required at detail stage so failed to assess the risks at feasibility stage. Actually, the water table was much higher than they had guessed, thus the upward forces were immense and significant amounts of ballast were required to hold the tank down. This project, as conceived, was not feasible against the assumed ground conditions. Other cases are known about in which uplift on buried structures had just been overlooked and the buried components either floated or their bases cracked upwards from water pressure. These are indicative of governing load cases that had not been properly checked out.

A significant amount of the industry's workload is refurbishment. It's common for there to be great uncertainty about the condition of older buildings. Records often don't exist, materials have deteriorated, and unrecorded changes have been made which might have weakened the structure or added load. Clearly, advance survey is prudent but frequently impossible before the building is 'opened up'. Safety requires that inspections are carried out at that stage and appropriate decisions made. In one example, a plan was made to add floors to an old building which was thought to have the requisite strength because it had originally been designed for more floors, which were never built. However, on opening up the roof (on which the further structure was to be supported) it was found that the filler joists embedded in the old concrete had corroded away to such an extent that it was no longer feasible to proceed with the scheme. There was then a dispute over blame: the project had proceeded too far before its feasibility had been verified.

The example cited above shows how essential it is to carry out surveys to verify information. Site investigations are standard but even those have their risks: drilling through buried services and so on (even unexploded bombs might be encountered (Institution of Structural Engineers' Health and Safety Panel, 2017)). In any project to extend older structures some condition survey is required.

Any feasibility stage ought to include an assessment of the practicalities of building the project. This includes gaining access and sequencing the works, delivering materials to site and so on. Those studies should ensure that surrounding structures can support the weights of material deliveries and cranage; it is not unknown for such heavy equipment to collapse existing structures below. Studies should include an assessment whether the structure will be stable in intermediate stages and not overstressed in any partially completed state. For complicated structures, an assessment of buildability ought to be mandatory. In practice, the term 'buildability' is not easily defined and means different things to different parties. The failure of some box girder bridges in the 1960s is illustrative of the issue. In 1868, Sir Benjamin Baker wrote: 'Of the numerous practical considerations and contingences to be duly weighed and carefully estimated before the fitness of a design for a long-span bridge could be satisfactorily determined, none is more important than those affecting the facility of erection' – wise words.

8.4.3 Interaction with other disciplines

Many structures support moving loads: bridges, crane support structures, rail track and whole moving structures. The interaction between the applied loading and its structure below includes uncertainty. There is an interface of maybe dynamic interaction and maybe fatigue and there is an interface between 'mechanical engineering' and 'structural engineering'. Historically the interface has proved problematical. Brunel's scheme for an atmospheric railway project failed, at least partly due to his inability to engineer the sealing interface between the moving train and its static track. Other large civil structures are supported on bearings and there have been failures. The structure on which bearings sit has to be rigid enough to support those bearings and minor deviations in bearing shape can lead to premature failure. Great care has to be applied to ensure that bearings are not subject to strain-controlled loads, which may develop if there are minor distortions of the support: press reports on the Wind Tower at the Glasgow Science Centre (2002) claim that the base thrust bearing, on which the tower is supposed to rotate, failed and had to be replaced: the cause has not been published. Two bascule-type bridges are known to have suffered vibration on opening and closing due to mismatch of a complicated support system (involving hydraulics) and the application of load. Assumptions were made about the load distribution that were just incorrect and came about because of a lack of understanding of how a hydraulic support system works, an example of failure across a mechanical/civil interface. Likewise, there was criticism after the Ramsgate link bridge failure (Chapman, 1998 and Box 8.11) that inadequate attention had been paid to bearing lubrication (a mechanical skill) and certainly the possibility of additional forces due to seizure or excess friction ought to be considered in any design.

Another interaction lies between structural engineers and services engineers with the issue being the demand for penetrations. In one major international project, a tender design was created that looked excellent. However, the designers had failed to use enough judgement to allow for later considerable penetration demands for services through slabs and through the shear walls providing overall stability. In practice, significant modifications and cost claims followed.

8.4.4 Other design issues

As well as resolving key strength and stability issues at the feasibility stage (and resolving construction matters) it is also necessary to define other targets such as durability and fire resistance. These might seem a matter of too much detail, but in one major dispute, the concrete mix was specified with long durability in mind to cope with the aggressive site conditions. This required considerable cement replacement with GGBS (ground granulated blast-furnace slag). Whilst that might seem sensible, the mix was very hard to place and quite incompatible with the very dense reinforcement quantities also required. It proved impossible to get reasonable finishes and impossible to guarantee surrounding the rebar with dense compacted mix and thus the durability specification was essentially self-defeating. If insufficient member size is not allocated at the project start, rebar covers may be inadequate, and the structure will not last (see examples in Chapter 7). Equally, low cover might provide insufficient fire resistance (Chapter 9). Likewise, if the need for extra-resistant materials such as stainless steel is not recognised early (say on bridge parapets), the project will not last (again see examples in Chapter 7).

Several failures have occurred through the design team failing to spot appropriate modes of failure and allow for them. A failure to spot fatigue as an issue will always run the risk of bad failures (Figure 8.1) and expensive remedial action (Chapter 7).

Figure 8.1 Fatigue failure

8.5 Concept stage

8.5.1 Introduction

Lots of disputes arise that have to be settled by legal action. Their existence is proof enough of misconception between the aspirations of those who conceive projects and those entrusted with carrying them out. Such disputes may be about scope, about performance or achieved quality or they might be about delay. Any case getting as far as legal proceedings is suggestive of a potential for two sides to have quite different views, and presumably firmly held views, about original design intent.

Over the years, the construction industry has invested huge amounts of time in drafting standardised conditions of contract (in an effort to allocate risk clearly); huge amounts on specifications; so that all parties understand what is being asked for (as in the National Structural Steelwork Specification: British Constructional Steelwork Association, 2010) and huge amounts of development time on design standards (say as for steel and concrete) so all parties can work to the same set of criteria. Despite best endeavours, all such documents undergo revisions as areas of weakness are uncovered and opportunities for confusion still exist. Chapter 6 described a few cases created by gross confusion, not least just by mixing up units.

In 2003 the ACEC (American Council of Engineering Companies) issued 'A guideline addressing co-ordination and completeness of structural construction documents' (ACEC, 2003). In that document, they wrote of concerns about communication especially relating to the quality of information produced for construction documents, criticising their frequent incompleteness with implications for structural safety:

> Inadequate communication results in budget and schedule over runs, disappointed owners and a potential risk to the safety of the building occupants and the public. Successful communication is critical; for the protection of the public safety which is a structural engineer's first priority as a professional.
>
> During the early 1980s professional liability insurance claims increased substantially because of increased claims.
>
> Furthermore, as computer-based analysis has proliferated so has the complexity of both analysis and code requirements making it that much harder for the less experienced to get a feel for the accuracy and adequacy of their design. The progression towards more efficient use of materials has further increased this complexity.
>
> In summary, there has been a widening gap between the ability of the construction design profession to adequately describe its design,

> and the ability of the construction industry to adequately develop a bid and schedule representative of that which it ultimately requires to construct projects.

That position probably remains unaltered by the passage of time.

8.5.2 Stiffness

A fundamental error can exist when proper attention is not given to stiffness early enough (Box 8.3). This is particularly problematical since stiffness depends on 'E' (Young's modulus) which cannot be changed. It is very hard to increase either global or local stiffness if insufficient is provided for at concept stage. Errors on stiffness might manifest themselves as problems via their impact on architectural finishes or via displacement of floors or in extreme cases, via excess dynamic response. When New York's Empire State Building was erected, its reported sway (or more correctly its oscillation) was such that occupants complained of motion sickness. Nowadays, tall buildings are checked not just for sway but for their vibration characteristics and the effects of oscillatory motion on occupants. Sometimes positive damping is added to improve acceptability. A good example of this is the damper added on the top of the Taipei Tower (Taiwan). The damper there is large hanging mass; a pendulum, tuned to sway out of synchronisation with the tower's global sway.

An excessive concentration on strength rather than stiffness can lead to stiffness being overlooked. Problems might also arise through presumption of modelling in the sense that steel beams are almost uniformly assumed to be end-fixed or pinned. Neither pre-

Box 8.3 John Hancock Tower, Boston, USA (1986)

The tower is 60 storeys high with its entire facade sheathed in glass. For architectural reasons the tower is deliberately slender and in fact won prizes for its appearance. Unfortunately, its design was beset with problems and many windows just detached and fell off (both a commercial disaster and a huge hazard to anyone below). Surrounding streets were closed when winds reached a certain speed and the steel frame was shown to twist and sway. As at the Empire State (built 40 years earlier), occupants on the upper floors complained of the motion. That issue had to be resolved by the addition of damping. The sway may have caused the window problem though other reports suggest that it was a combination of oscillation and thermal stress generated within the glazing system. For whatever reason, all window panes had to be replaced at huge cost. Obviously, there were flaws in the concept. This incident is also a good example of the risks that lie across discipline boundaries where the consequences of structural performance on other disciplines may not be appreciated by either side.

Box 8.4 Palau Bridge, Palau, Western Pacific (1996)

This bridge spanned between two islands in the group, linking the main populated areas with the airport, and it also carried a fresh water supply. The bridge was a pre-stressed, long-span (240 metres), concrete construction. At the time it was the longest concrete girder bridge in the world. The two sides were cantilevers joined in the middle by a hinge. After about 18 years in service, the cantilevers had deflected by creep, shrinkage and pre-stress loss until the centre sag was around 1.2 metres down from its initial position. The bridge then collapsed utterly, causing the government to declare a state of emergency.

Full details of the cause have never been released, but Burgoyne and Scantlebury (2006) suggest reasons.

sumption is true; connections of whatever form are more truly semi-rigid with the consequence that beams assumed to be fixed-ended will deflect rather more than predicted and beams presumed to be pin-ended rather less. It is of course not so easy to predict concrete member deflections with accuracy and so the starting point is usually some empirical ratio of span/depth that is known to provide satisfactory results. Concrete member deflections are also complicated by longer-term creep. The consequences of excess deflection might just be commercial but are potentially disastrous. Excess deflection was one of the first indicators of trouble on the Palau Bridge, which collapsed in 1996 (Box 8.4).

Excess deflection is a potential indicator of other problems, especially dynamic response. This has only emerged as a truly common problem in relatively recent times. The reason is that, not so long ago, floor spans of 6–8 metres were the norm and longer spans rarer. However, clients began demanding longer spans and these became technically possible through more efficient design and stronger materials. However, natural frequencies are a function of $\sqrt{(\text{stiffness/mass})}$ and this tends to be a function of $(L)^2$, and if the arithmetic is followed through with longer spans and lighter mass, it turns out the spans of 8–9 metres and above have frequencies that coincide with those of normal footfall traffic. Thus, such spans get excited by footfall and cause a nuisance (see example of a dance floor beam in Chapter 7, Section 7.4.9). In that one case, it might be said that the cause was a fundamental error in concept. Other problems of vibration and excess deflection have developed in grandstands and bridges (see Chapter 6, Box 6.6) and retrofitting can be very expensive.

8.5.3 Stability and load path

Self-evidently buildings should be stable 'as a whole'. But in many structures, there are mixtures of construction disciplines where, say, steel frames rely on concrete cores for global stability with lateral load transmission (e.g. wind) relying on the floors acting as diaphragms. There should be no ambiguity on what holds what up at what stage of the job. Problems arise in the construction phase due to sequencing and perhaps in later life during alterations when a vital component on the stability load path is removed; Chapter 4, Figures 4.13 and 4.14 show examples.

8.5.4 Safety and functionality

It is never too early to have clarity and definition of any key functional safety demands that the project might have to satisfy or of the key hazards that might apply. Is a nuclear power station or a waste dump even feasible at concept stage if the public cannot be assured over safety? Currently, the UK is considering how to extract gas from shale reserves by fracking, yet there is much public opposition, partly on environmental grounds and partly on perceived safety grounds. Knowledge of these drivers can determine the entire project. In the UK in the 1980s, there was keen debate over connecting the UK to mainland Europe. There were bridge champions and there were tunnel champions. Those arguing against the bridge were concerned about placing obstacles in one of the world's busiest sea lanes and one frequently fogbound. In the event, the best option was considered to be a tunnel and that is what was built. Having regard to various severe bridge impacts that have taken place, allied with environmental disaster consequent on tanker wreck (see Chapter 11 and Box 8.5) perhaps that was the wisest choice; but tunnels too have their problems (see Chapter 9). Any very tall structure built from now will consider the possibility of terrorist attack (Chapter 11). Even at concept stage, the key targets of functionality must be set, thus both the Thames Barrier and the corresponding Maeslantkering (Netherlands) (Chapter 10) were both required to withstand very extreme flood heights, bearing in mind the consequences of failure. Noting what happened in New Orleans in 2005 (Chapter 10), those targets were a wise choice.

> **Box 8.5 Sunshine Skyway Bridge disaster, USA (1980)**
>
> This was a bridge in Florida. It was destroyed in 1980 when a freighter hit one of its piers. As a consequence, about 366 metres of the deck fell 46 metres into the water below, taking cars and a bus with it: 35 people died (Sunshine Bridge, 1980).

8.5.5 Global Stability

The generalised dangers of instability can be no better shown than in the failure of Aberfan tip (1966) described in Chapter 10. A structure's stability system needs to be defined from the outset. This applies to both global stability and local stability, which is achieved by restraining parts in compression via provision of a stiff load path back to a point of strength. In 1986, a school sports hall collapsed at Rock Ferry near Liverpool.

Box 8.6 Rock Ferry Comprehensive School, Birkenhead, collapse of sports hall, UK, (1986) and Camden School, London UK (1973)

A sports hall collapsed at Rock Ferry School (Menzies and Grainger, 1976). The structural form included standard timber trussed rafters supported on masonry side walls. The roof completely collapsed, primarily because it did not contain any diagonal bracing in its rafter plane nor in its ceiling tie plane or normal to the plane of the trusses. Looking at the structure as a whole, there were no clear load paths for transfer of lateral wind loads to ground. It is unknown if the wall designers assumed their walls were propped via the roof or the roof designers assumed their roof was propped via the walls. In either case, the roof was an inadequate diaphragm, not stiff enough in its own right to transfer any wind loads onto the walls and not stiff enough to transfer any wall wind load to neighbouring side walls.

A feature of the Camden School failure is reproduced in (Figure 8.2). See also Beeby (1999).

The support of the roof beams was on a very narrow ledge such that any beam rotation moved the point of support right to the edge and there was thereafter little continuity to hold the wall and roof beam together. There was a flaw in concept.

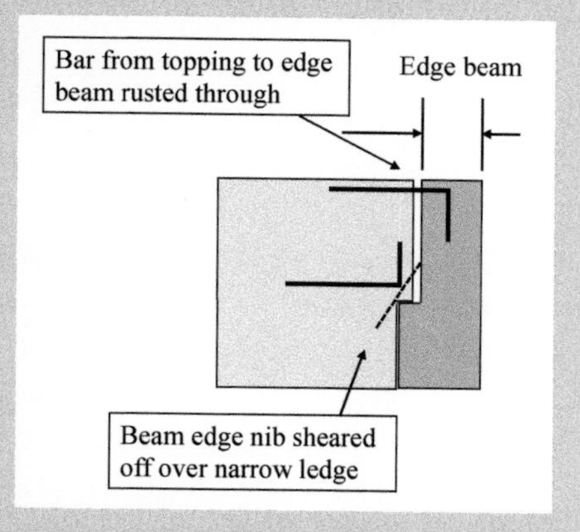

Figure 8.2 Failure of the edge support

Box 8.7 Hartford Civic Centre roof failure, USA (1978)

The roof over the arena was a large, deep, space frame structure (two-way spanning) (plan dimensions about 100 m × 110 m). One night, after five years of use, the roof collapsed completely, shortly after a major basketball match had finished. Loading from heavy snow seems to have been the final trigger.

The roof was complicated, being based on a series of inverted pyramids but it was essentially a two-way spanning truss. Because of its complexity, it had been analysed by computer and indeed seems to have been an early example of computer usage. Nonetheless, there was a basic and fatal flaw in the structural concept. Purlins that ought to have braced the truss compression booms were not in the boom plane but were located some distance above, and thus the only restraint to boom buckling was via some stub members separating the purlin plane from the boom plane below. This was not enough; boom members buckled, and when collapse started, it was progressive and catastrophic. Presumably the computer model assumed all members were restrained and only assessed stress, so did not 'warn' of the buckling failure mode.

As usual, other factors also contributed to the collapse. Even some of the restraints present were too slender to do their job properly. Dead loads are reported to have been underestimated by 20%. Whilst the reason for this is unknown, it is not at all unusual for the percentage weight of fixings to be a significant additional percentage of member weight and rather difficult to estimate. There were warning signs even at construction stage, when deflections were far too large and more than anticipated (erection involved force fitting), and that fact alone should have raised doubts about the computer analysis validity. Other factors related to confusion over who was responsible for what, the ignoring of problems during erection, and not building what was intended to be built.

But essentially there was a fundamental flaw in structural concept (for more detail see Brady, 2015; MacLeod, 2005).

The structural form was of standard trussed rafters supported on side walls but there was a basic flaw in concept (Box 8.6). The similar failure of a school roof in Camden in 1973 (Building Research Advisory Service, 1973) is most notorious for HAC beam weakening (Chapter 7) but is also an example of a stability failure consequent on inadequate beam edge support (this might be said to have failed the sensitivity test of Chapter 4).

The failure of the civic centre roof at Hartford illustrates another failure in concept (Box 8.7).

8.6 Safety concept demands

8.6.1 Introduction

Many projects are safety related in the sense that some malfunction of the process can cause harm to people. An obvious example is a chemical plant or a nuclear power station. In both these examples, the civil structures may perform some function within the overall safety case, say to act as a barrier or to withstand leakage or blast. In other projects, the failure of the civil structure itself might be the cause of catastrophe: dam failures will result in devastation. Chapters 10 and 11 will give several examples of failures where harm has been caused and where environmental and commercial consequences have been traumatic.

For any project it therefore pays to be clear from the outset what the safety function of any component might be and to use appropriate measures to ensure commensurate safety in service. 'More safety' is required where risks to the community are high.

The failure of the civil entrance structure to a dry dock in Chatham's Royal Navy Dockyard is a good example of an unforeseen failure with disproportionate consequences – in this case on a submarine (Box 8.8).

Box 8.8 Chatham Dockyard caisson entrance failure, UK (1955)

Dry dock entrances are closed off by a barrier (known as a caisson) which is rather like a ship's hull. It is floated into place at the entrance of a dock and then 'sunk' against seals by pumping water ballast into an internal tank. As dock water is pumped out, external water pressure forces the caisson against the seals to make a watertight joint. Caissons look substantial (Figure 8.3) but this is deceptive, as vertical stability (called preponderance) is slight since the ballast weight only just exceeds hydrostatic up thrust. Consequently, there is a danger, if tides rise exceptionally, of changing the uplift thence causing the caisson to float. To prevent this, neutralising 'free flood chambers' are provided which allow tidal water into the chamber via a series of holes set above the normal caisson water line. At Chatham, some painters were working inside the caisson, but it was draughty because of the holes, so they blocked them up. The tide rose and the caisson floated. Water then rushed into the dock and a tidal wave, reflected off the back wall, picked up a docked down submarine and transported it across the River Medway, where it sank and was a write-off.

This might be an example where the precise safety function of the free flooding was not clear or was too easy to overcome, and where dramatic consequences followed from one error (Board of Inquiry, 1955).

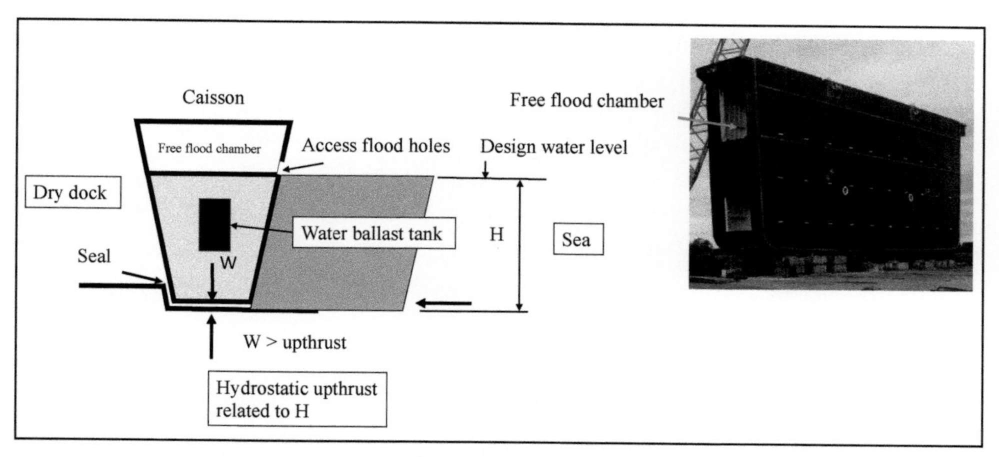

Figure 8.3 Caissons and the Chatham Dockyard failure

The failure of the Ronan Point block of flats (Box 8.9) might be attributed to a failure to spot the safety issues (the hazard) linked to supplying gas in such structures plus their inherent vulnerability to a gas explosion. Gas explosions are not common but not that rare either and given the number of tower blocks in existence, and the number of kitchens in such blocks, there is a reasonable probability of a gas leak and explosion in one of them at one point through life. Likewise, the Grenfell Tower fire (2017) was

> **Box 8.9 Ronan Point, London, UK (1968)**
>
> This block of flats in London failed in one of the most widely reported failures ever. A gas explosion blew out a wall panel and initiated a pack of cards-type failure: it was a classic stability failure where instability might be described as a gross change in state consequent on some minor event. Stability is a separate attribute to strength and to be safe, a structure must be both strong enough and stable enough (Chapter 4). Ronan Point also highlighted the need for another desirable attribute: robustness, which is a quality hard to define mathematically but conceptually it means that no structure should fail grossly, consequent on some inflicted minor damage (Chapter 4, Section 4.4). Although explosion was the prime cause of failure, the investigation revealed several other unsatisfactory features of the design, including poor workmanship and sensitivity to tolerances (Institution of Structural Engineers, 1969; Building Research Establishment, 1988; Bussell and Jones, 2010; Carpenter, 2008).

initiated in a kitchen by an electrical fault in a domestic device. Given the number of tower blocks in existence (and thence multiple kitchens) and given the frequency of electrical fault in some devices such as tumble driers, the probability of an initiating event at some stage in one kitchen in one tower has to be high.

Although the initiating Ronan Point event was a gas explosion, it was not one that would be regarded as that unusual. Quoting from the 15th SCOSS annual report (2005):

> Large panel structures first came to the attention of the wider structural engineering community (and the general public) in 1968, following the partial collapse of the Ronan Point flats in London. This type of construction exhibited fundamental flaws, both in its design and in its construction. The design relied on gravity forces and friction to hold all the panels together and this could lead to progressive collapse, as evidenced by Ronan Point, if one part failed.
>
> This occurrence was a seminal event and led to a major re-think of high density housing construction and a change to the building regulations in 1980 to regulate against disproportionate collapse.

Although the Ronan Point failure occurred in 1968, its repercussions were still being felt years later. In 2018, following structural checks, some tower blocks on the Broadwater Farm estate in London were evacuated and scheduled for demolition as they were deemed not sufficiently secure against collapse.

8.6.2 Control systems

Generically many modern engineering systems are automatically controlled by a system which is a collection of sensors, data transfer and an overall pre-programmed logical reaction via a PLC (programmable logic controller). Such systems can be used for industrial plant, transport systems (like driverless trains), railways, rides or moveable bridges and conceivably may become more common with 'intelligent buildings'. In structures supporting moveable loads having multiple lines of support, control systems can be used to maintain bogie alignment and so control crabbing forces on the supporting civil structure. Hence it is most important that the functionality of the control system is appraised. But before that, it is equally vital that the definition of what the system is supposed to do is set out and this implies dialogue between all engineering parties and comprehension of what the safety demands are, and that vision is not likely to be uniquely comprehended by any one party, and certainly not by the engineers charged with putting the software together.

Several control system failures have occurred. Control Systems Out of Control: 'Why Control Systems Go Wrong and How to Prevent Failure' (Health and Safety Executive, 2003) states that about 45% of all such failures resulted from an inadequate user requirement specification. Clearly civil engineers need not possess the technical skill of designing control systems themselves but when interacting, they should be aware of the concepts. As an example, the Herald of Free Enterprise disaster described in Chapter 6 (Box 6.1) could easily have been prevented by an adequate control system targeted at prohibiting sailing in advance of the bow doors being closed. Control systems are used on the London Eye and on the moving roofs of stadia, with potential to mitigate the loads and accident conditions imposed on the structure. The security design of systems can be adjusted to reflect the severity of the hazard and its consequences. This might be safe shutdown in a nuclear plant or it might be the automatic issuing of an emergency stop on moving plant. It is often hard to follow through the logic of system failures and identify what might happen. But there are techniques for this, which have been developed in the chemical industry (Chapter 13). The essential skills are transportable to civil engineering.

Moving structure design includes special problems because the safety of the entire project can depend on the reliability of the control system. Such structures can be anything from radio telescopes to dock gates to opening bridges to roller coasters. Generically they are systems made up of a combination of civil, structural, mechanical, electrical and control engineering. Risks occur with comprehension across the boundaries as between structural and mechanical (exemplified by bearings described earlier) but the interface with the control system is potentially a more dangerous one. Not only is there physical danger, but because the disciplines of control engineering and structural engineering are so diverse, there is a real danger of human lack of comprehension across the interface: one side not understanding the technology and the other side not fully comprehending what can go wrong and thus the demand on control system reliability ends up being ill-defined.

In any moving structure there are, for example, the twin hazards of 'not stopping' and 'starting up when not commanded'. Apart from the danger to humans, the forces involved in 'not stopping' may just be immense and as a worst case, the energy involved in trying to stop the heavy mass of a ship can be totally destructive (see Chapter 11). Figure 2.1 in Chapter 2 is quite a famous photograph from 1895, illustrating what can happen when a train fails to stop, but overshooting has occurred much more recently. The tragic Moorgate tube crash of 1975, in which about 43 people died was one such incident, and it led to the introduction of automatic control systems aimed at reducing the risk of human error (Chapter 6). But they are not in universal use, and in July 2013 Spain suffered a terrible rail crash at Santiago de Compostela as a train rounded a bend at twice the speed limit, derailed and crashed into a concrete retaining wall: 89 people were killed. The failure of an inspection gantry under the Severn Bridge in 1990 is another example that led to two deaths (the gantry was blown along and had no proper means of stopping and there were no end stops just before a gap in the track). In 2006, a civil contractor was convicted of manslaughter after a rail truck ran out of control down a track and killed four workers (Tebay, Cumbria). Such failures highlight the need for civil engineers to understand not only all the loads that might apply to their structure, but all the modes of failure that might exist as well (collectively 'non-stopping' is such a mode). This cause and effect can only be established by a systematic risk assessment (starting with hazard identification) based on some experience and inside knowledge of the entire system (Chapter 13).

Within 'rides' the attention given to control can be traced back to a failure at Battersea funfair in 1972 (Chapter 2) when a coaster started to run backwards and killed five riders. Following this incident, all cars have since normally been equipped with anti-rollback devices. The safety point is that in the generic risk assessment, a presumption is made that the lifting device can fail, and that implies an essential safety response.

The introduction of control systems to reduce the likelihood of human error does not eliminate problems, for it is always necessary to assess the risks of a faulty system and to address the consequences of component failure (Box 8.10). In the preliminary report of the Buncefield disaster (Chapter 9) it was reported that the supposed high integrity control system for preventing tank overfilling failed to work; the consequence was overfilling the tank by some 300 tonnes of petrol which flowed out, ignited and caused the largest civil fire since the Second World War; thereafter the nearby watercourse became polluted with the foam used to put out the fire. This is not new. In 1957, the Windscale fire occurred at Sellafield. This was Britain's worst nuclear disaster, releasing a large plume of radioactive contamination from the chimneys. Technicians initiated the failure by heating the graphite core to release Wigner energy, when actually, because of faulty temperature instrumentation, this was unnecessary and incorrect. As a result, the pile went out of control and it was nearly half a century later before the chimneys could be decommissioned.

A hazard and emerging risk is that of cyber-attack. The UK's National Health Service has been targeted with demands for huge ransoms. There is concern that

Box 8.10 Stadium roof

A football stadium was fitted with a moving roof. The two opposite sides were winched along by cables, each of which could stretch by different amounts. A control system was developed to assess the crabbing degree and control the drums on opposite sides to try and keep the roof square to its rails. The roof was operated remotely from a control room within the stadium. On one occasion, there was an urgent need to open the roof, but the system indicated a fault and issued a prohibition on opening. There had been a history of false signals, so the operator switched the system off. Unfortunately, on this occasion, the report was true and consequently when the winches were operated, the whole roof was pulled off its rails. Fortunately, it did not crash onto the crowds below.

In another incident, a moving telescope had been built and was designed to move along parallel rails (like a crane). During testing, the control system was not fully functional and the whole structure just ran off its rails and had to be scrapped.

a future attack could simply close down key parts of the nation's infrastructure. The UK government's National Cyber Security Centre can be consulted (National Cyber Security Centre, undated).

8.7 Architectural design stage

There is a constant tension between architects and engineers over aspirations on member size and so on, which all have the potential to create late problems if insufficient member space has been allocated. There have been disputes linked to floor zones being squeezed down too much in order to comply with planning restrictions on overall building height. Such disputes lead to pressure on designers to try and generate structural capacity from members that are too shallow.

Building design is essentially iterative, progressing in steps with compromises along the way. At an initial stage, one of the first engineering tasks is to establish a structural form and grid that offers adequate load paths and the promise of stability via shear wall, bracing or moment frames. Another early task is to lay claim to space, i.e. for beam and column sizes that are capable of accommodating members of adequate strength and stiffness. In concrete columns, for example, this might be assigning dimensions which give the engineer enough flexibility to incorporate a percentage of rebar and a concrete grade that together will provide enough strength. In slabs, a thickness has to be assigned that will achieve adequate stiffness. These early tasks require great skill in foreseeing future pressures on space demands. In all buildings, the interaction with services engineers is problematic as service design demands penetrations, often in inconvenient locations: bringing services up next to columns can demand weakening holes through critical floor-to-column shear interfaces. In structural steelwork, a skill is to size members with foresight, to ensure adequate connections can be matched to required member strength. Since forming connections is one of the most expensive parts of the steelwork package, it is no surprise that there have been innumerable disputes related to a demand to achieve over-stiffened and complicated connections simply because space has not been made available. In concrete work, no amount of analysis will compensate for an inability to get concrete into overly congested rebar cages.

8.8 Structural design

8.8.1 Introduction

Structural design embraces everything from choosing the structural form, applying load, analysing it, designing and detailing the members, and then terminating in construction. Failures may occur at any stage of the process due to ignorance in understanding, lack of communication or just error. There are uncertainties in everything and the history of code development from the fairly simple texts of half a century ago to now, is one of increasing content and complexities as codifiers have tried to incorporate technology progress. Unfortunately, that very process, and the ensuing complexity, is itself a potential risk.

Designers need to be educated to understand the basics of codes, not least so they understand their limitations. Concrete codes permit concrete slab design by yield line methods and, providing slabs are under-reinforced to ensure they have appropriate ductility, the method works well and has been verified by tests. But the original tests included slabs of only 'normal' floor dimensions. In one case, a two-way slab spanning 15 metres each way was designed by this method. There was nothing wrong with the arithmetic and no limitation of span in the code. But, of course, the slab deflected enormously under its self-weight and continued to creep. A dispute then arose over whether the slab was 'strong' enough. When looked at in detail, by the time the slab had reached a cracking limit load (indicating the limit of useful life) it was still way below its predicted yield line capacity, so whether the ultimate load capacity was sufficient was of little practical interest. Moreover, there was no proof that the postulated yield lines could sustain the plastic angles expected of them and the rotation at support points was such as to invalidate the presumptions of load transfer onto the supports. In short, the design approach was significantly at variance from code intent and thus invalid.

All designers need to judge when their design procedures are appropriate. Much of steel work design remains notional. Long experience shows that although such approaches are 'inaccurate', they are usually perfectly valid for producing safe structures: but not always. Chapter 7, Section 7.4.9 describes one negative consequence of inadvertent restraint at a 'pinned' connection. On the other hand, 'accurate' design that predicts high stress is no automatic indicator of concern either (Chapter 3). In buildings of normal height, elastic shortening effects are ignored, as are trivial foundation settlements. In tall buildings such assumptions can start to become unsafe both in terms of strength and serviceability. When elastic shortening effects are included for, the effects on predicted moments can be disconcerting and foreign to everyday experience and some judgement is required as to which loading cases to account for and which to ignore. Overall, the decision in height over when to account for such effects is fuzzy. Similarly, linear analysis is only valid so long as P-Δ effects remain small. Again, some judgement is required on when to include for them. As structures become taller and sways become larger it is important to make sure that the analysis assumptions remain reasonable. As a generality, when any structure or structural element exceeds 'normal dimensions', the validity of standard design procedures should be questioned (SCOSS Alert, 2018).

8.8.2 Design error

Any of the problems discussed under the sections above might be thought of as types of design error. The full process of 'design', encompassing concept, safety and interaction with construction, can be immensely complex so it's hardly surprising that occasionally it goes wrong. Sometimes the causes can be isolated to more specific activities within the more traditional processes of civil/structural engineering, which start with analysis and move on to member design and detailing. Quite often failure cause has been linked to overlooking a relevant hazard or design consideration or to a later change in use. Sometimes the cause has been technical ignorance. Problems of human error were dealt with in Chapter 6 and problems of materials in Chapter 7. Examples of other forms of failure are given below.

8.8.3 Just getting it wrong

The process of bringing an engineering project from concept to fruition can be enormously complicated. Many separate companies and many separate technical disciplines interact, plus there is the broad difference between design and construction. The documentation on any job, let alone the drawings with all their numerous modifications and clarifications, is staggering. It's almost impossible for one human to have the broad vision to see the implications and interaction across the project. What is especially worrying is the possibility of significant consequence based on a trivial error, since the chances of spotting that seem remote. A key aspect of the Grenfell Inquiry which was on going through 2021 was the interaction between the myriad of parties who might have contributed to the over-cladding design that went so badly wrong. One reported failure linked to a design error is that of the Ramsgate walkway: see Box 8.11

> **Box 8.11 Ramsgate walkway failure, UK (1994)**
>
> The walkway was part of an articulated structure that passengers used to transfer from the dockside edge onto ferries. It collapsed, killing six people and seriously injuring more. Part of the cause was that an extension arm at the base, around which the walkway pivoted, was short and stubby and assumed to be in pure shear at its support whereas in reality the stubs were short cantilevers with the connection interfaces being in shear and bending. That difference clearly resulted in a huge difference in elastic stress and thence a dramatically reduced fatigue life. There was concept error: it was just a mistake, but one made both by the designers and the checkers (Chapman, 1998).

8.8.4 Technical progress

For today's generation of young engineers, it's easy to overlook the immense strides in knowledge made over the lifetime of engineers now at retirement age. The Centenary Edition of the Journal of the Institution of Structural Engineers included papers setting out the tremendous progress in knowledge acquisition made over the century (MacLeod, 2008; Rolton, 2008; Sutherland, 2008). Alongside knowledge, we have analytical tools and computational power that could not have been dreamed of until comparatively recently. Welcome though that progress is, a moment's reflection will raise a knowledge query of how it will all be disseminated and understood by those at work now. There is a vast amount to learn across many skill disciplines and it cannot all be learned on a short university course. It is therefore no surprise that historically, and currently, many failures can be put down either to technical ignorance or to designers simply not being aware of hazards that might apply to their design. For that reason, engineers need a broad education for awareness and the later ability to call up specialised expertise when that is demanded.

Towards the end of the 19th century there was a large demand for bridges and many suspension bridges were built, partly because of their efficiency and through the availability of wrought iron, but no doubt also because they could at least be analysed and scientifically sized. A number failed from the then unknown phenomenon of aerodynamic instability. The Union Bridge at Berwick-upon-Tweed (built in 1820, using eyebar chains, with an approximately 135 metre span) was destroyed during a storm only six months after completion. Even Telford failed to foresee issues on instability and his Menai Straits Bridge (1826) was damaged a few days after opening and again in 1836, and has been progressively stiffened over its life. Clearly the cause was not understood by the time of the infamous Tacoma Narrows Bridge collapse (1940) over a century later. Lack of understanding of dynamic effects still seems to be at the root of some problems as described in Sections 8.8.10 and 8.8.12.

Other failures that may be categorised under technical ignorance might be many of the material ones reported in Chapter 7.

8.8.5 Tolerances

A common category of problem is that relating to tolerances, and disputes occur perhaps due to a failure of achievement by contractors or a failure on the design side to anticipate tolerance implications. In terms of durability (Chapter 7), it will be recalled that minor variations in concrete cover have quite significant effects on concrete life. Figure 8.4 shows a design of a precast, pre-tensioned beam. The mathematics were not in contention but the required eccentricity of the pre-stressing wire from the centroid was just 10 mm. Clearly any sensible allowance for wire location tolerances would have had a dramatic effect on the eccentricity and therefore effectiveness of the pre-stress distribution and thereafter on subsequent beam capacity (see Sensitivity in Chapter 4). A feature of the box girder bridge collapses (Section 8.11.3) and the collapse at Ronan Point was the influence of tolerance on the presumed position of certain members. Rolling tolerances in struts are one reason why buckling is always going to be the mode of axial load failure and modern codes require that any strut mid height connection is sized accounting for a P- δ effect.

8.8.6 Loading

Loading might seem one of the easier parameters to deal with, but it has its own uncertainties. Of course, the dead load of materials should never vary but there are inaccuracies in take off with designers just missing components out (Boxes 8.12 and 8.13). In one large steel roof, an error was discovered when it was found that the allowance included for connections was quite inadequate. In another commercial dispute on an airport terminal, the take off by the contractor varied significantly from the allowance included in the budget. On investigation, the percentage addition just due to trimming angles and 'secondary steel' (particularly for a baggage handling system) was found to be very significant. Figure 8.5 shows alternative designs for a beam, one with a slab on the top and one to reduced depth with slabs supported on shelf angles.

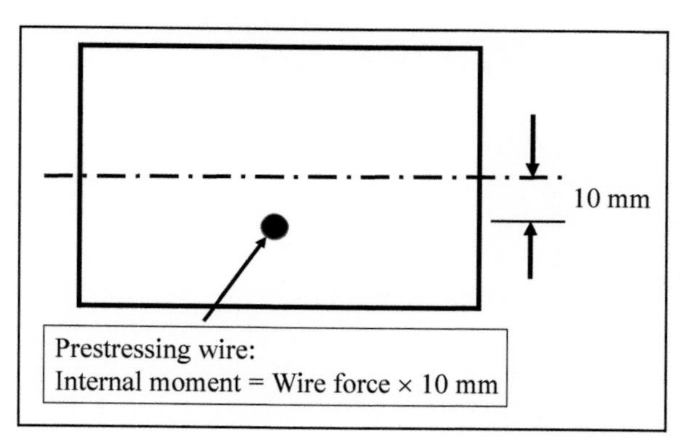

Figure 8.4 Design sensitivity

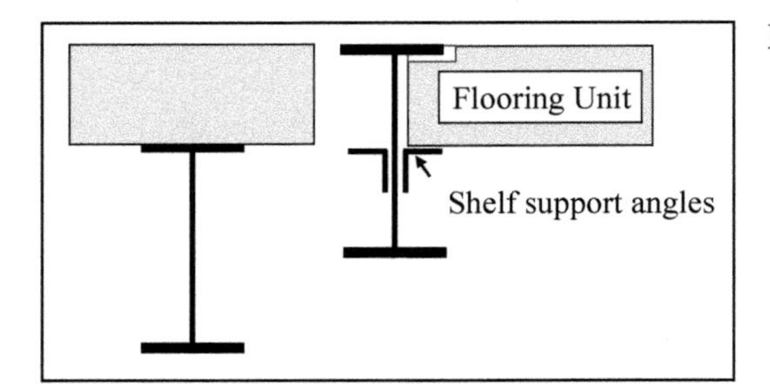

Figure 8.5 Alternative support systems

Whilst the reduction in floor depth might well be desirable, the addition of two 'non-structural' angles can easily add 50% to the beam weight. It has even been found that the extra weight due to paint can be high. Chapter 3, Section 3.3.2 describes an incident whereby the weight of concrete in a floor slab varied by a very large percentage just because tolerances had been overlooked.

Live load allowances are generally thought of as quite conservative, although cases are known in which gross overload has occurred, particularly from added plant loading. The collapses of the Rana Plaza building in Dhaka (2013) and the Sampoong department store (South Korea, 1995) both had reported excess machinery/plant load. In both those failures, machinery vibration was suggested as a contributory cause. Generally, a feature which renders live loading uncertain is the allowance included (or not) for dynamic

Box 8.12 Hotel New World disaster, Singapore (1986)

At the time, this was Singapore's deadliest civil disaster. The six-storey building collapsed in seconds, killing 33 people. Allegedly, the designer had not accounted for dead load in his calculations. Some columns are reported to have failed in the days before the final collapse.

Box 8.13 Versailles Wedding Hall, Jerusalem, Israel (2001)

This terrible event was filmed live and later shown on TV news. During a wedding, part of the floor suddenly gave way and 23 people plunged to their death. A further 380 people were injured. An initial inquiry blamed the collapse on defects in the lightweight coffered concrete floor system (which was a proprietary system). Whether this was the root cause of the failure or not, other factors played a part. There had been changes in building height (a floor was added) without proper checking, and this added loading. It seems this loading was partly dispersed by partitions below which were removed just before the collapse. Partition removal was followed by floor sagging which the owners levelled with grout so adding yet more load but no strength. It was this overstressed floor that failed.

effects linked to either people or plant activity (see Section 8.8.10). Of course, straightforward overload can occur and many UK bridges have had to be reassessed to cope with increases in traffic density and increases in individual lorry loads that have arisen out of changes in usage as lorry design has evolved (bridge overloading was found to be a factor in both the Silver Bridge failure (Chapter 7, Box 7.6) and in the I-35W Bridge failure described below (Box 8.33). Loading increased partly as bumper-to-bumper traffic jams became much more common with the huge growth in car usership, and in the case of the I-35 Bridge, because temporary construction materials were stored on the deck. Figure 8.6 shows typical 'live loading' on a roof access gantry. Three points stand out: (a) actual loading is considerably different from an idealised uniformly distributed load (UDL); (b) both the gross and local loading can be high; (c) probably the highest design load is that sustained during construction.

Figure 8.6 The reality of live load during construction

In very large roofs, the weight of rainwater can be temporarily high as it takes time for run-off to reach gutters. Excess deflection and thence ponding are potential issues. The risk is much worse if roof drainage is blocked, and worse still on flat roofs. When roofs are flexible, the potential load is not just a function of the weight of water or snow per square metre but also of its potential accumulation under roof sag. Snowmelt can be part of the blocking system. In 2013, part of Brazil's fabric-covered Salvador stadium roof collapsed after heavy rain. Press reports suggest that drainage was accidentally closed off. A rather spectacular collapse of the Metrodome sports stadium (an inflatable structure) in Minneapolis in 2010 can be seen live on the Web (Metrodome sports stadium collapse, 2010); it collapsed after heavy snow accumulation. The UK winter of 2012–2013 was severe and parts of Scotland received very heavy snowfalls. A spate of collapses in earlier years had led to a revision of the Scottish guidance for roof loading (CROSS Report 264). A potential linked issue on some roofs arises as their capacity has been slowly degraded, perhaps by water ingress, over the years and so it may be some time before the coincidence of excess load with weakened structure combines to produce disaster without warning (Boxes 8.14 and 8.15). In the interim, because the roof may have survived several winters, there may be a false sense of security.

Added weight of ice build-up in winter storms has been enough to cause failure and is a factor to be considered. Tall masts, transmission lines and cables are items potentially at risk. Chapter 10 gives examples of failures of pylons in Canada. In the UK,

Box 8.14 Katowice Trade Hall roof, Poland (2006)

This roof failed when there were about 800 people inside; 65 were killed and many injured. Those that survived the initial impact remained at risk through delays in rescue during the freezing weather. The roof was thought to have failed under heavy snow load.

Box 8.15 Ice rink, Bad Reichenhall, Germany (2006)

The ice rink was built in the 1980s. In 2006 it collapsed under heavy snow, trapping 50 people; at least 15 were killed. Accumulated snow was thought to be the trigger for failure, but the depth was not that unusual (see also Box 7.2).

ice build-up on cables was a factor in the failure of the Emley Moor TV tower described in Box 8.29.

Wind forms a key loading onto structures large and small. Only 50 years ago, wind loading practice amongst structural engineers was rather simple. It was known to be simple and unconservative and the collapse of Ronan Point prompted a significant upgrade in the pressure values. Since that time, the complexities have grown extensively and, as with many other branches of the profession, wind engineering has become a specialisation. Failures still occur, mostly explained by lack of appreciation of the way wind loads act (especially by imposing suction) rather more than by the magnitude. The extent of UK damage in storms of predicted 50-year intensity is such as to suggest that following code design approaches properly will result in safe structures (but see also Section 8.12 below on dynamic effects).

8.8.7 Analysis

Engineers now have at their disposal vast amounts of cheap computing power and a huge assortment of programmes that can numerically model ground and structural behaviour through both linear and non-linear ranges. Both static and dynamic analyses can be carried out and their graphical presentations are hugely beneficial in depicting performance. Finite element analysis (FEA) allows us to assess component stress in fine detail. Automatic design of concrete and steel members offers incredible productivity benefits. To a large extent, traditional methods of analysis and design by hand are now obsolete (see also Chapter 3, Section 3.4).

Yet with all those benefits comes risk (Carpenter et al., 2013 and Brady, 2015). No matter how much computing power is available, predictions still remain an approximation of real structural behaviour. There is much disquiet amongst more experienced engineers that too much reliance is placed on 'the model', with that reliance generating a lack of physical understanding on the part of the users. There is disquiet that much software can be used by designers who really lack experience to see the uncertainties attendant on software use. Sometimes this might be because the skills of computer analysis and modelling are now so advanced as to be carried out by specialists, perhaps mathematicians, who may not have the background to temper their predictions with engineering knowledge. Yet reliance on 'the model' is in many cases essential and one master model underpins many total projects. Looking at that in terms of risk ought to convey a need to manage the checking of the master model very carefully indeed, for much can be at stake. The cases of the Sleipner collapse (Chapter 6, Section 6.4) and Hartford Civic Centre roof failure (Box 8.7) are both classic examples illustrating the danger of inadequate analysis/modelling.

Having a structure designed by computer is no guarantee that it is going to be adequate. The interpretation of the output can be rather baffling as the following examples (Boxes 8.16 to 8.20 show).

Practical imperfections play their part in the behaviour of all struts and it is essential to assess their role in any model. Imperfections played their part in the box girder bridge collapses (discussed later) and in the failure of concrete cooling towers as at Ferry Bridge (Box 8.19).

Similar conceptual errors seem to have caused a water tower collapse at Seneffe

Box 8.16 Steel grillage

A large steel grillage was analysed by FEA (Figure 8.7).

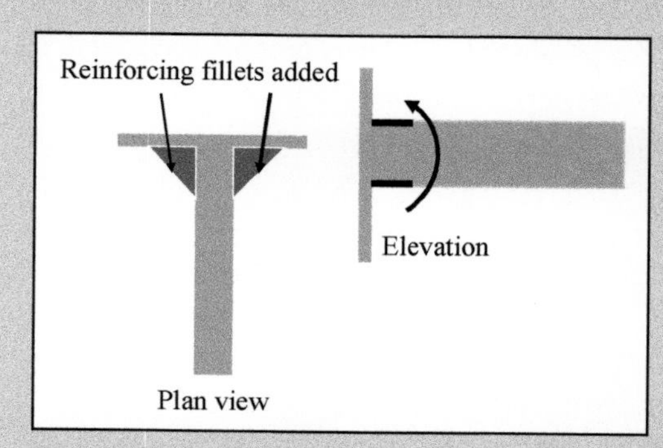

Figure 8.7 Incorrect interpretation of stress concentrations

On review there was puzzling addition of triangular stiffening fillets at the universal beam (UB) junctions (despite the UBs being connected around by full profile butt welds). The fabricator was dubious but made what he was told to make. On investigation, it was found that the modellers had created a mesh which picked up stress concentrations at the tips of the flange outstands and had considered these exceeded permissible limits. To reduce stresses, they added 'strengthening fillets'. On repeating model runs, the beams ends were now stiffer than before and so picked up more moment and the outstands picked up even more stress and the stiffeners were increased again. Such assessment of behaviour is quite at odds from notional design of steelwork, which treats whole cross sections as in uniform stress and relies on the ductility of the steel material to cause redistribution as required. In reality, once steel is welded there is always local stress at yield from weld shrinkage, but one which is, and can be, ignored. The concern of the modellers was quite unfounded and had just resulted from a lack of appreciation of how real structures behave.

In a similar incident, a large truss was fully modelled by FEA. The designers became concerned at certain junction stresses and increased member sizes only to find that the stresses increased some more. This went on until elastic criteria were satisfied. What the designers failed to realise was that they were picking up secondary stresses at the joints which they could have argued were of no consequence (a traditional design would have assumed the joints were 'pinned'). Furthermore, the magnitude of the secondary stress was entirely linked to their assumption of 'rigid' connections which engineering expertise ought to have told them was 'incorrect' and that any amount of joint semi-rigidity would have relaxed the stresses.

Although these two incidents were not structural failures as such, they do illustrate a potential for great confusion in modelling and assessing output.

Box 8.17 Posts and cables

A series of devices were carried on wires strung between lamp posts. The forces acting were self-weight, wind (in two directions) and temperature (from wire contraction). These were analysed on a plane frame program, producing alarming values for the bending moments at the base of the columns. The BM came from two sources; first, thermal contraction of the horizontal wires linking the heads of the posts producing tension on the outside of the posts at their base. Second, wind blowing along the line of the posts to produce a BM in the same direction. What the designers had overlooked was that wind in that direction would cause a sag in the wire, negating the thermal contraction effects completely, and the two effects were just not additive (certainly not enough to cause failure). Furthermore, the thermal stresses were themselves strain controlled and could not possibly have caused failure (Chapter 3, Section 3.3.9).

Box 8.18 Complex three-dimensional steel frame

An analysis was required of a complex 3D steel framework. The work was carried out on a standard FEA program and, after processing, was used to assess the stress limits. A large report was prepared. The team carrying out the work were not engineers and they implicitly assumed that if the (factored) stresses were less than yield the checks would prove the structure adequate. What the team had failed to grasp was that the members were slender and permissible stress limits would be lower than yield and governed by buckling. The analysis was thus fundamentally flawed.

> **Box 8.19 Collapse of Ferry Bridge cooling towers, UK (1965)**
>
> This failure was the dramatic collapse of a group of towers. The cause is often put down to wind loading and it is true that wind funnelling through the tower group was an issue. However, cooling towers are theoretically pure shell structures subject to plane membrane stress and these towers were analysed as such and reinforced as such. Unfortunately, towers cannot be constructed perfectly and any unevenness in their shell wall creates a local bending effect. If the shell has only a single layer of rebar (commensurate with presumed plane membrane response) it has no resistance against bending from imperfections. Modern shell walls now have two layers of rebar to give bending resistance.
>
> This failure might be thought of as a failure of modelling or a failure to appreciate the practical effects of imperfections, in which case it has similarities to Ronan Point and some box girder bridge collapses. Or it might be thought of as a failure consequent on detailing error or one that highlights the importance of detailing.

in Belgium (1972). This was a thin shell structure with inappropriate modelling and a failure to appreciate buckling as a failure mode. The initial tower design was made using membrane theory though some assessment of local shell bending was undertaken. Apparently, no checks of buckling were made for parts in compression and failure occurred by thin wall buckling during filling. Looking at the shape of the water tower, which is top heavy and defined as an inverted pendulum, there ought also to have been a load case of notional side load to account for the heavy mass on top and potential sloshing of the contained water side to side, perhaps generated by wind vibration (European Steel Design Education Programme, undated). These are potential uncertainties in the loading condition. It is not known if such factors were accounted for or were relevant, but other failures of top-heavy structures have occurred, such as the water tower in Goodland, Indiana (2011). This 300,000 gallon tank failed catastrophically and very quickly.

> **Box 8.20 C.W. Post Center auditorium dome roof collapse, USA (1988)**
>
> This dome collapsed after eight years of use, during a heavy snowstorm. Forensic investigators blamed the structural theory used for analysis and the application of loading. Domes are ideally strong under symmetrical loading but real loads from wind and snow are not applied symmetrically and create bending. In particular, the potential for ice and snow to build up on one side with superimposed asymmetrical wind suctions does not seem to have been appreciated (the analysis was based on simple membrane theory). Consequently, the roof failed by snap through buckling. To compound the conceptual faults, dead loads had been significantly underestimated and load sharing between members across the truss depth was inappropriate. The roof had been inspected shortly before its failure, but such inspection would only detect visible damage and not pick up conceptual errors in the original design.

The way loads are applied to the model may equally be a cause of error (Box 8.20). Geotechnical ground modelling in particular is fundamentally uncertain and the output needs treating with respect. After the Nicoll Highway collapse in Singapore in 2004 (Box 8.31), the official report found that the collapse partly occurred because the engineer responsible for the analysis did not understand the software, the model nor the local ground conditions (SCOSS Failure Data Sheet, 2004).

Summary

Given the amazing facility of today's computing power, it is easy to be seduced into believing the output is 'accurate'. However, it is always well to recall that whatever the modelling sophistication, it will remain an estimate of structural behaviour. With good modelling, that prediction will be a good estimate. But to be confident, the modelling always needs to be validated and the sensitivity to its assumptions verified (MacLeod, 2005 and Institution of Structural Engineers, 2002).

8.8.8 Statics: in service

Few failures occur under static loading apart from those due to error and perhaps those initiated by material corrosion (Chapter 7). Of course, mistakes do occur in analysis (Section 8.8.7) or because of errors in applied loading (Section 8.8.6). Errors also occur in execution. There are cases reported to CROSS of designers fundamentally failing to comprehend the needs of global and local stability in steel buildings or of specifying masonry walls far too slender.

Occasionally (Chapter 7, Section 7.6.9) mistakes are discovered too late. One example is the failure of a multi-storey car park in Atlantic City, USA, in 2003 (Bosela and Bosela, 2018). The results can be seen in Figure 8.8: there was a domino-type failure during construction with the main cause being a lack of connection between the floors and their supporting shear wall (on the right in the photo). Again, there were reports of concerns expressed by workers during construction, but these were ignored. Lack of connection and support were also a feature of the Camden School failure (Box 8.6).

Most failures under 'static' conditions are serviceability ones, but these might have safety consequences. In 2006, press reports of a multi-storey car park in Bournemouth (Alexander, 2014) talked of concrete spalling and the car park had to be closed just before the Christmas rush. It appears that the cause was lack of consideration of articulation of the long floor spans on their ledge supports, possibly due to differential thermal effects, possibly due to larger end rotations related to using long spans. There were suggestions that bearings should have been used to accommodate such support rotation. The car park is large and key lessons from this are to beware when any unit is 'long' and moreover when any unit is replicated lots of times, since if there is a generic error, the consequences will be widespread (SCOSS Alert, 2018).

Figure 8.8 Failure of Tropicana Casino parking garage (Alarmy 2D448AW)

There have been several in-service failures of structures primarily carrying tension loading. Chapter 4 showed how the safety of all structures is closely linked to their ability to redistribute local overstress via system ductility. Section 4.3 of Chapter 4 describes this and refers to Hambley's explanation of ductility demands in compression systems. Tension systems require the same ability to cope with overload, but there are key differences. Ductility of tension members themselves cannot be relied on unless their connections are stronger than the attached member and it is connection failures of various kinds that have frequently precipitated collapse. In a compression system, if the connection 'fails', strut displacement is likely to be limited but that is not true with tension systems and failure of any part can lead to catastrophic consequences (Mann, 2002). Boxes 8.21 and 8.22 provide examples. CROSS has several reports on ceiling collapses (CROSS Reports 6, 100, 101, 102, 103, 124, 130, 140, 148, 203, 251, 304, 426, 442) and although these might sound relatively trivial, many were heavy (due partly to carrying heavy services) and any could have caused fatalities. The Construction Fixings Association (2015) gives recommendations. There have also been some bad examples of failures, such as in the Boston tunnel failure (Box 8.21) and a failure in a Japanese tunnel (Box 8.22), when parts of the tunnel lining fell off onto cars below (Brady, 2013).

Box 8.21 Big Dig ceiling collapse, Boston, USA (2006)

A concrete ceiling panel, about 6 m × 12 m and weighing 2.85 tonnes, fell from the tunnel ceiling. The panel fell on a car, killing one passenger and injuring the driver. A section of the tunnel was closed for almost a year. Blame was put on bolts holding the panel up. These were epoxied in and it was reported that some bolts were too short and the epoxy was substandard.

The pulling out of the bolts can be seen in pictures of the incident, while post-failure inspections revealed several more bolts in the process of failing: partial pull-out. The exact support details are unknown, but reports say there was system of tie rods and flanges. Bolts were inserted into reinforced concrete and had curtailed length to avoid clashes. So, this failure might also be used as another example of the importance of detailing.

Box 8.22 Sasago Tunnel, Japan (2013)

As in Boston, concrete slabs fell from the ceiling onto cars, killing nine people. Press reports claimed that 'anchor bolts' used to secure concrete slabs to the ceiling were missing. Two vehicles caught fire in the accident and heavy smoke hindered rescue efforts.

8.8.9 Global stability failures

Global lack of stability might have been a factor in the collapses of the top-heavy Seneffe water tower referred to in Section 8.8.7. Potential danger also occurs when structures are being installed 'indoors' since then there is no lateral wind case and hence no obvious need to brace or stabilise against lateral load. It is for that reason that a notional load related to the mass being carried has to be defined. Real sway loads will invariably exist due to column lean and asymmetric gravity loading (Bragg Report, 1975; BS 5975). The latter case is significant when concreting on formwork: there is always a lateral sway case to consider and there have been several instances of bad formwork collapses during the concreting phase (see Chapter 4, Section 4.2.2, discussing stability during construction). One example is given in Box 8.23 and others are given in Section 8.11.

The Eurocodes recognise this potential failure mode in their requirements for robustness since the presumption is that in any multi-storey structure there is always a lean to one side defined at least by the plumb tolerance of columns and therefore always a tendency to sway sideways. Where shear resistance is light, P-Δ effects will exaggerate the sway. Critical cases must be those with heaviest mass high up, and some serious side sway progressive failures have occurred in high-bay warehouses:

> **Box 8.23 Barton Highway, Canberra, Australia (2010)**
>
> Nine workers were hospitalised when a concrete pour went wrong. The investigation found that formwork was not adequately braced to stop sideways movement when the concrete was poured. The bridge had to be demolished.

these have heavy mass supported by relatively light structures and, being indoors, no externally applied horizontal force to serve as an automatic demand for bracing. Moreover, parts of the racking are at risk of support removal or damage from forklift trucks (Taylor, 2010).

Recent experience has shown that temporary structures are at risk of global stability-type failures. This arises because they may be light, may be designed for low wind speeds commensurate with short-term usage, yet may have large surface areas (Box 8.24). A spate of stage failures at concerts has occurred, some with tragic consequences. Often the failures seem to have been precipitated by sudden gusts. Examples are: Rocklahoma Festival (2008); a collapsed temporary roof structure at the Indiana State Fair (2011) which landed among a crowd of spectators, killing seven people and injuring 58 others; Cheap Trick stage collapse (2011); the Belgian Pukkelpop stage collapses (2011) and the Ottawa Bluesfest Radiohead stage collapse at Downsview Park (2012), in which the main stage scaffolding collapsed, killing one performer. Other failures are known about, as for example, that shown in Figure 8.9.

Figure 8.9 Collapse of temporary structures (in this case a performance stage)

> **Box 8.24 Birmingham TV screen, UK (2006)**
>
> A very large TV screen put up as part of a temporary show and it overturned in gusty weather. Because the screen was temporary, it was designed for a low wind speed. There were deficiencies in the design but of interest here was its sensitivity to the presumption of wind speed. The screen's capability was to withstand around 12 metres/sec but the wind forecast on the day was uncertain. If, for example, the wind had gusted marginally higher, at say 15 metres/sec, the forces (and they were dominant forces, not some fraction of a dead + wind case) would increase by $(15/12)^2 = 1.56$, which was more than enough to negate any perceived safety margin. The lesson is to be alert because pressures are due to $(v)^2$. Hence, when a low speed is the starting point, the global forces are sensitive to errors in prediction (SCOSS Topic Paper, 2008).

The failure at Pukkelpop was especially dramatic as a sudden large squall arose with very high wind speed: five spectators died and 140 were injured. It is worth observing that the risks in such events are high because stages are surrounded by lots of people and opportunities for rapid escape are minimal. Because of the similarities, CROSS has issued Alerts (SCOSS Alert, 2010 and SCOSS Alert, 2012); see also Section 8.8.11, suggesting limitations on design.

One of the most definitive stability failures must be that of the Ronan Point block of flats which dramatically failed in 1968 (Section 8.6.1 and Box 8.9). The key issue this failure raises is not so much the danger of gas, although that is real enough, but in highlighting fundamental flaws in structural concept with respect to ensuring stability. On this occasion, catastrophe followed a gas leak, but other collapse causes are conceivable, and the issue then is to test the survivability of the structure, or its 'robustness'. This is not easy since structural engineers are more used to assessing adequacy by direct numerical verification of stress than by having recourse to non-numerical principles of good structural format. Defining a test for robustness is not straightforward. The Ronan Point failure also highlights the uncomfortable need to take balanced judgements over what is an acceptable risk: high-rise failures are more unacceptable than comparable failures in low-rise structures, though not to any individual who might be harmed. And the test relies on notions of survivability under arbitrary extreme loading conditions. SCOSS resurrected this issue in 2005 (SCOSS, 2005) since there was concern that many authorities had forgotten the lessons and were allowing gas back into high-rise dwellings, whereas an original risk mitigation measure had simply been to remove gas supplies. That the risk remained valid was illustrated by another quote in the report recording that 'Although they may be of low probability, they are high risk. In December 2004 an explosion at a block of flats in Mulhouse, France, caused partial collapse and killed 18 people' (see also the last paragraph of Section 8.6.1).

Additional risks to global stability occur when a building is made up of several parts. In those circumstances, it is essential that one party retains responsibility for overall stability and defines horizontal load paths and stability systems in all the orthogonal plan directions. The failure at Rock Ferry School is an example where this task appears to have gone wrong (Box 8.6).

8.8.10 Dynamics

The technical aspects of structural dynamic response are not that well appreciated, partly because most structures are only designed, and need only be designed, for static response. But therein lies a danger. Many classic failures under dynamic response have occurred simply because designers were just unaware that dynamics might be an issue and if they had been aware, they might not have understood the mathematics. A good number of people will be vaguely familiar with stories about a need for soldiers to break step when crossing a bridge (Box 8.25) but perhaps not know why. Technically it's quite simple: all structures have a natural frequency; they vibrate at one fundamental frequency if excited. If forces are applied to them at that same frequency, displacements build up with repeated force application and thus so also do stresses (i.e. stress is proportional to displacement). Common examples occur when structures support reciprocating machinery and when bridges had to support steam trains with eccentric masses on their wheels. Other examples relate to the rhythmic motion of people and a modern day example applies in stadia used for pop concerts, when large crowds are coordinated by the music beat to bounce or jump up and down at a very similar frequency. If that frequency happens to match the structure's natural frequency, large and alarming displacements can build up. Pop music's basic beats are typically in the 2–3 Hz range and thus also in the 4–6 Hz range at the first harmonic. Many structures, especially cantilevers, have frequencies well within those ranges.

Chapter 11 describes how vibration from machinery may have been a contributory factor to the Rana Plaza building failure in Dhaka (2013) or the earlier Sampoong department store (Korea) (1995). A total disaster consequent on turbine vibration was that which destroyed the Sayono Power Station (Russia, 2009) (Chapter 12).

Chapter 7, Section 7.4.9 describes some vibration problems with a long-span floor. Other cases have occurred on beams supporting gym treadmills and the like. In all these types of problems, predicting the degree of response is not easy. It is not enough that the input frequency and the structure's natural frequency coincide. After all, a flea jumping up and down on top of a suspension bridge at the bridge's natural frequency will do no harm; the ratio of masses needs to be such that measurable deformations take place. This can occur, and the well-publicised excess response of the London Millennium Footbridge on its opening day exemplifies the issue (Box 8.26; see also Chapter 6, Box 6.6). For there to be a problem, there also has to be low structural damping and conversely, if there is a problem, one potential solution is to add damping.

Box 8.25 Broughton Suspension Bridge, Manchester, UK (1831)

This was a chain-supported suspension bridge reported to have collapsed as soldiers marched across it. Reports hint at the soldiers deliberately exciting the motion. The immediate failure cause was a single bolt snapping in chains close to a support. This may well not have been a true resonance problem but perhaps some extra loading caused by a full live load enhanced by some dynamic effects which precipitated a failure on a poor unit (the bolt was reportedly badly forged). Whatever the precise causes, the Army then ordered that all troops should break step when crossing a bridge. A sign to this effect can still be seen on Chelsea's Albert Bridge. The reality of co-ordinated pedestrian movement on a bridge can be seen from the performance of the Millennium Footbridge across the River Thames, in 2000 (Box 8.26).

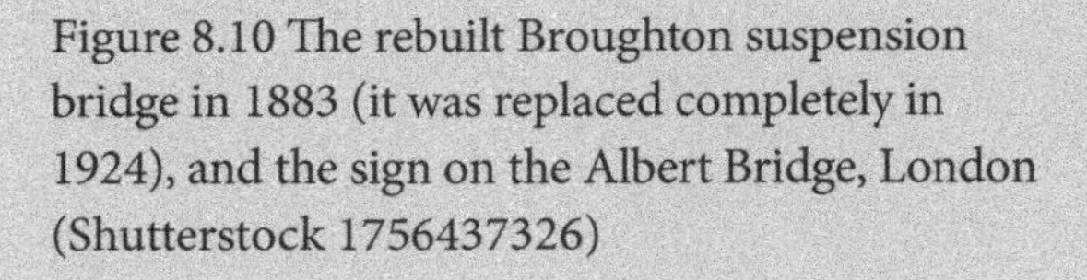

Figure 8.10 The rebuilt Broughton suspension bridge in 1883 (it was replaced completely in 1924), and the sign on the Albert Bridge, London (Shutterstock 1756437326)

Box 8.26 Millennium Footbridge, London, UK (2001)

This is a splendid footbridge spanning the River Thames opposite St Paul's. It is shallow and supported by draped cables. On the day of opening, large crowds crossed the bridge, but it began to vibrate and as it moved, the crowds began to walk in synchronisation with the motion, which only served to amplify the displacements. The bridge was then closed off and significant damping had to be added to make it serviceable. It was eventually reopened satisfactorily (Dallard et al., 2001).

Whilst vibratory effects can occur in any structures, the greatest concerns lie within stadia since these necessarily contain large numbers of people. There are video records of fans jumping up and down on cantilevers, creating alarming motion. The fear is that such excess motion might generate panic, with an additional factor being the knowledge that cantilevered structures can have a high susceptibility to collapse if local structural damage or failure occurs. In assessing risk, there are several uncertainties, and these occur on 'both sides of the equation', i.e. in the behaviour of people to cause structural response and in the mathematical prediction of that response. Some relatively modern structures in the UK are known to respond dynamically to rhythmic crowd movement. In 2010, a grandstand collapsed during a car race in Brazil (a video can be seen on the Web: Brazil grandstand collapse, 2011). A nasty stand failure also occurred at Indianapolis in 1960 and real-time footage of that incident can also be found on the Web (Indianapolis stage collapse, 2011).

8.8.11 Wind engineering

When the Tay Bridge collapsed in 1879, not much was known about wind loading, and its effects on the bridge had been severely underestimated. By the time Ronan Point fell down about 80 years later, knowledge was much more advanced; even so it was realised that the then current codes were quite weak at describing wind phenomena and changes were hurried through. These led to more complexity and not a little consternation. Putting aside the low pressures assumed for the Tay Bridge, a second lesson is to beware when wind loads are dominant. In many building structures, the governing load combination is dead + live + wind and there would have to be a severe error in the assessed wind load component to cause gross global failure. In other structures, the combination might be just dead + wind, say on a transmission tower where live load is non-existent. Wind loads then obviously become more significant and the risks of error within wind load computation become more relevant. If, in that combination, dead is low and wind is high, then even greater care should be taken in the accuracy of the wind loading prediction. It should be recalled that wind pressure is proportional to (wind speed)^2 potentially making adequacy in speed prediction that much more critical.

Problems that commonly recur are often due to lack of appreciation of wind suction, which leads to roof cladding being ripped off, damage at points of wind speed acceleration over ridges and around corners, and damage to 'non-load bearing' panels. There is a recurrent theme in CROSS Reports of hoardings and walls being blown down during the construction phase; such walls are often internal and at risk until buildings are enclosed CROSS Reports 12, 75, 82, 94, 99, 116, 134, 135, 144, 162, 282, 295). (The failure in Box 8.27 was not due to wind. Nevertheless. it was due to an unanticipated lateral load on a wall.) Assessing failure risk in temporary conditions can be uncertain when exposure time is short and when wind is the dominant loading. The example (Box 8.24) of a large TV screen blowing down explains why. Owners may inadvertently add to risk and some CROSS Reports (CROSS Report 159) refer to the danger of attaching large hoardings, sheeting or banners to temporary structures (the Milwaukee crane failure (Big Blue crane failure at Milwaukee, 1999) appears to have occurred because wind drag on the roof part being lifted was not accounted for). Many scaffolding towers are now enclosed to limit the risk of dropped objects but of course the enclosing sheeting adds to wind load. Several reported failures relate to clad scaffolding collapses.

It has been pointed out within CROSS that reductions in wind loading on temporary structures should only really apply if there is an expectation that at time of high wind the

Box 8.27 Fatal wall collapse at school due to 'wall climbing', Edinburgh, UK (2014)

There have been numerous cases of free-standing wall collapses causing fatalities, particularly to young persons. A SCOSS Alert, 'Preventing the collapse of free-standing masonry walls' was issued in September 2014 (SCOSS Alert, 2014). Cross Report 447, Newsletter No. 51, 2018. This collapse was of a free-standing changing room privacy wall 2.1 metres high, formed of a single layer of bricks laid in stretcher bond. Young pupils 'climbed' the wall by moving up the gap between an adjacent wall, setting their backs against one and pushing with their feet against the other. The privacy wall then toppled, fatally injuring an 11-year-old.

Horizontal loading applied by people exerting pressure has been observed previously. In the 1970s, tests were carried out at Liverpool FC on barriers in the Kop end, which was then used for standing only. It was found that the highest loads were applied when, to resist pressure from the crowd behind, supporters put their feet on the rail and pushed back. Had they not done so the pressure could have crushed their chests. This data was used when recommending barrier loading in the original Guide to Safety at Sports Grounds (Sports Grounds Safety Authority, 1973).

structure will not be in use, thus making the consequences of failure possibly tolerable, i.e. merely financial. Where this is not true, great caution should be exercised and the example of the Birmingham TV screen and the pop festival stage collapses (Section 8.8.9) illustrate well both the danger and need to be more conservative about wind load on temporary structures when people are at risk (Institution of Structural Engineers, Advisory Group on Temporary Structures, 2017).

8.8.12 Wind dynamics

If knowledge of structural dynamics is scarce, knowledge of wind aerodynamics is even scarcer. There are several mechanisms which might cause structures to vibrate in wind flow; one is vortex shedding; others are galloping and flutter. There are some more obscure forms related to bridge deck response and forms of vortex shedding that only arise in rain and snow conditions (see below). Generally, any slender or circular object exposed to wind flow is potentially at risk.

The reality of how bad dynamic response can be is exemplified by the well-known collapse of the Tacoma Narrows Bridge in 1940. This wasn't the first bridge to suffer problems, as Section 8.8.4 reports the collapse of the Union Bridge in 1820 and problems with the Menai Straits Bridge (1826). The failure of the Tacoma Narrows Bridge is, however, one of the most famous civil engineering disasters of all time, partly because it was so bad and partly because it was filmed. What we should recall is that in the 1940s, wind engineering remained in its infancy and very little at all was known about wind dynamics: the physical processes were not understood, the mathematics were not available, and neither were the analytical tools (technically the response at Tacoma was a form of 'flutter'). Avoidance of dynamic response of the deck overall probably wasn't solved until a deck aerodynamic shape was evolved for the Severn Bridge in 1966. Even so, later bridges have suffered. The Erasmus Bridge in Rotterdam (completed 1996) had to be damped, post construction, to prevent its deck lifting up and down.

The mechanism of vortex shedding is actually quite simple (Bolton, 1978, 1983). In any fluid flow, the frequency of vortices being shed off a circular object in that flow is regular. Vortices shed either side sequentially so as to create a pulsating cross-wind force, applied at regular intervals. If that frequency coincides with the structure's natural frequency, a possibility of strong oscillation is set up. Standard cases exist on lamp posts and chimneys, especially lightweight steel chimneys, and in all these the risk might be exacerbated since peak bending moment occurs at the base. which is also the location for maximum corrosion weakening. (Many examples of such shaking can be seen on the Web.) Ways to prevent shaking include installing tuned mass dampers or adding strakes at the top. Similar vibrational responses can be generated on cables, which often require dampers.

An allied case creating a dynamic response exists if circular objects are in line (the objects need not be the same diameter). In this configuration, vortices shedding off the front object pound the one in the wake at regular beats. A case is known of vortices coming off a cooling tower and destroying a smaller structure in its wake.

Sometimes the vibration danger only exists during construction, when members are exposed and very slender. A standard case exists when slipforming a chimney. The structure's natural frequency varies according to its height and the design objective is to ensure that the vortex shedding frequency (related to diameter) does not match the cantilever's frequency, certainly when at full height. During construction, the chimney will vary in height and, at some stage, the two frequencies may well coincide. A construction objective is to get through this height as quickly as possible to limit the time at risk.

Some years ago puzzling vibrations of suspension bridge hanger cables were noticed. Vibrations did not always occur but tended to be worse in rain or in snowy conditions. The cause was found to be that water (or snow) attached to the cable was

Box 8.28 B of the Bang Sculpture, Manchester, UK (2012)

B of the Bang was a sculpture consisting of a multitude of tapered spikes radiating from one central point. Early in its life, several spikes vibrated in light winds and fell off. Eventually the sculpture was dismantled. The cause was reported as a form of vortex shedding with press reports suggesting that the failure cause was wind-induced oscillation which generated a fatigue failure (CROSS Reports 136, 157).

Box 8.29 Emley Moor TV Mast, UK (1969)

This was Britain's tallest TV mast, at 385 metres high. It was a tubular guyed structure located in the Pennines. In winter, the cables iced up and chunks of ice used to fall off, leading to a warning system being deployed around the base. However, worse still, the structure vibrated (probably affected both by cable mass and iced-up diameter changes) and this eventually brought about total tower collapse.

In 1985, yet another large TV mast collapsed in Germany, in similar wind and ice conditions, though brittle fracture of the cable attachment lug might also have been a contributory factor. In Minnesota, USA, in 1991, the 259 metre WDIO TV tower became covered in ice and then collapsed in gusting wind. A feature of this failure was that the mast fell onto power lines and thereby caused significant consequential damage.

able to change its shape as far as wind flow was concerned and thus alter the degree of vortex shedding. The phenomenon is known as rain-induced vortex shedding. A similar phenomenon exists if cables ice up and when that happens both shape and cable mass will change, both of which affect dynamic response. The failure of Emley Moor TV mast (Box 8.29) was partly put down to these effects (which at the time were argued on the frontiers of technology).

Galloping is another form of vibration. This is caused by differences in pressures on opposite sides of a component/structure which might be affected by variations in the flow angle of incidence. Galloping is an interaction of the aerodynamics with the deflection of the structure in an unstable manner. In 1959 an unknown form arose when power conductor cables over the River Severn vibrated. The cause was found to be linked to the nature of the cable stranding. When the wind flow was not normal to the line, but at an angle to it in plan, flow on one cable face was in line with the lay of the strands (so smooth) whilst on the opposite face, the flow would then be normal to the strand lay (so rough). Those differences caused the cable to bounce up and down. The motion was eventually controlled by wrapping the cables in tape. In another case, a light latticed tower formed part of an amusement ride. The owner decided to clad it and thereby altered it to a light box section. This box then started to oscillate violently back and forth. Tower deflections had been enough to move the structure in the wind flow from a stable to an unstable condition.

8.8.13 Fatigue design

Fatigue is invidious form of failure (Chapter 7, Section 7.4.9). There are two phases: crack initiation and crack propagation. Fatigue life depends on the absolute value of stress range applied and on the number of cycles endured. The problem with the first demand is that life is governed by the real stresses in the structure rather than the notional stresses mostly used for structural strength assessment and there is a good deal of uncertainty about what the real stresses might be, partly for reasons discussed under Section 8.8.7 (Analysis). Moreover, life is inversely proportional to (stress range)^3 so relatively minor increases in the stress range can drastically reduce fatigue life. Any moving structure, or any structure carrying moving loads, is potentially at risk. Any structure subject to vibrations is at risk because a huge number of load cycles can be applied over relatively short periods. Once a crack starts, it will propagate, though the extension rate depends on the geometry of the structure and on the general stress field: the crack rate growth might rapidly escalate, or it might stop if the tip progresses into a compression field.

Fatigue design is required on all bridges and on offshore structures (where the source of stress change is the constant ebb and flow of the water). Chapter 7 describes some catastrophic offshore structure failures attributable to fatigue damage (Sea Gem, 1965 and Alexander Kielland, 1980). Any structure moving itself or carrying a moving load is at risk and structures for roller coasters are a good example. Fatigue design for them is often the factor governing ride life. In highway bridges, the dominant loads are dead and live, with live perhaps quite a low proportion of the total applied stress, so the fluctuating range is maybe only a relatively small proportion of the total applied stress. In contrast, on coaster track, the track self-weight is low and stress fluctuations are mostly governed by the car's weight (with passengers) and live load is thus dominant. Numbers of cycles will be large because of the multiple wheels on a typical car and the fact that trains might pass any point on the track every few minutes. On top of all this, downward forces will be much enhanced above gravity due to radial acceleration effects. On coasters, fatigue safety strategy is a combination of initial design followed by frequent checking. What is sometimes overlooked is the premise behind the propagation phase. Structures do not become immediately dangerous the moment a crack starts. They move into a danger zone as the crack length propagates. Hence, it is necessary in design to be reasonably certain that a crack can be detected before it gets too long and that also means making some judgements over the part at risk. For example, the crack length available around a small tube is not going to be very long and hence the potential time from initiation to danger is not long either.

Crane support structures are potentially at risk of fatigue damage dependent on their duty: cranes used for occasional maintenance are unlikely to be at risk. Some forms of fatigue failure that have occurred are useful in highlighting the reality of cracking being governed by real stress rather than nominal stress. Failures have been observed several times on the interface between crane rails and flanges of supporting girders (Chapter 7, Section 7.4.9, Examples).

Fatigue failures occur, exacerbated by lack of awareness of the issue and underestimation of the stresses. Some of the serious train derailments in recent years have been caused by track breakage, partly due to rolling contact fatigue, via a poorly understood mechanism. Mann (2008, 2011) describes how initial cracking can develop within the rail body generated by rapid application of contact stresses. Once a surface-breaking crack has formed moisture may penetrate, but high wheel stresses, superimposed later on from above, will seal the crack instantaneously yet at the same time compress it; this in turn creates a huge hydrostatic pressure at the crack tip sufficient for propagation until eventually full rail fracture occurs. Originally rails used to be connected at their ends by bolted fish plates and the discrete joints were responsible for the distinctive sound of rail travel. Repeated bolt loosening and joint fatigue failures in fish plates were one of the factors leading to the introduction of all-welded (continuous) track.

Chapter 7, Box 7.6, describes the Silver Bridge failure in the USA. Fatigue was a contributory factor.

8.9 Civil design stage

8.9.1 Introduction

Civil design, as opposed to structural engineering generally, concentrates on activities such as foundations, tunnelling, roads, dams, water supplies and drainage. As with other categories of construction, civil engineering failures have helped shape current safety and design thinking.

8.9.2 Foundations

There is perhaps no more famous ground failure in history than that of the tower in Pisa (Burland, 2006, 2008) and there are plenty of other building settlements to be observed which vary between the picturesque and financially ruinous. The ground on which our structures sit is inherently variable both regionally and locally (many of the tunnel failures that can be studied result from intersecting local pockets of water or fine-flowing material). Well into the 20th century, scientific understanding of ground behaviour, that is its strength and predicted displacement, was quite limited. Burland reports (2006) that consequently it was hardly surprising that there were many failures of slopes and retaining walls. The birth of soil mechanics, as we now know it, can be traced back to Karl Tezaghi, who started his career in the 1930s. He constantly emphasised that, in most situations, prediction precision is impossible because of the inherent ground uncertainties.

Foundation design is required to be reliable for obvious reasons and one that ought to be obvious is that, if foundations are wrong, there is little opportunity to correct them. Foundation design is generally predicated on an intention to control absolute and differential settlement and to that extent, foundation design should not be separated from superstructure design and the capability of the structure to absorb deformations to an acceptable extent. When wrong, as at Pisa, the consequences can be alarming. Longevity is no guarantee of survival, as many cities are suffering overall regional settlement due to excessive extraction of water from aquifers. In 2011, it was reported that the Taj Mahal was settling, manifested by cracks appearing in its superstructure. It was suspected that low water levels in the adjacent river were allowing drying out and degradation of the timber piles which form the foundations. Venice as a whole was sinking at about 4 mm/year due to excess aquifer extraction, though that rate has now been slowed. That same depletion of underground water is potentially an issue as well in the preservation of Venice's timber piles. Decay is only inhibited so long as piles remain submerged in mud.

Most homeowners fear the word subsidence, which will manifest in building cracks or floor tilt, with the very word blighting their asset's value. A distinction has to be made between settlement and subsidence. All foundations settle to some extent under imposed loads. Subsidence is excess movement due to external effects (Institution of Structural Engineers (2001) gives a full description). Subsidence might be due to ground shrinking; the lowering of water tables or from water extraction consequent on hot summers; by tree roots; by building on shrinkable material or on poorly compacted ground. The reverse of subsidence is heave; perhaps frost-heave or heave due to sulphate expansion of fill (Finnegan and Forde, 2014). Such expansion can be great enough to lift whole buildings off their foundations. The precautions against both subsidence and heave lie with adequate investigation in the first place and thence consideration of all the relevant risks. These might include mining subsidence in some areas. Many parts of the UK are underlain by old workings, or caverns from quarrying or from salt extraction. All over the world, dramatic sinkholes appear, sufficiently large to swallow up whole houses. Some are natural and some are caused by leaking drains washing away fine material; many are dramatic (SCOSS Alert, 2017).

Problems occur when extending structures, partly as a result of differential settlement between the two parts and partly because of undermining adjacent structures, causing their foundations to fail. In some cases, this is caused by lack of appreciation of foundation function. In one case, a floor slab within an industrial shed cracked after some extensions had been carried out. The frame was a standard portal and the extension was made by adding a new portal alongside, which required excavation for the foundation. Clearly the designers had not considered the role of the removed ground in providing lateral thrust resistance to the original frame's pad foundation. Hence on excavation, the old base moved sideways. More tragic instances have occurred when drains or similar have been laid in front of retaining walls, so unwittingly removing soil that provided horizontal stability. In 1998, two volunteers were killed in Scotland when a concrete slab fell on them whilst working in a hole below a retaining wall that collapsed (Chapter 7, Box 7.23). Trench and trial pit collapses have been responsible for many injuries and the HSE website warns:

> Every year people are killed or seriously injured by collapses and falling materials while working in excavations. They are at risk from:
>
> Excavations collapsing and burying or injuring people working in them;
>
> Material falling from the sides into any excavation; and
>
> People or plant falling into excavations.

CROSS Reports 175, 241, 297, 361, 385, 423 and 539 give other examples.

One example of severe failure caused by excavation collapse is that of the Nicoll Highway failure in Singapore (Box 8.30).

There cannot be many more spectacular failures than that which occurred in Shanghai in 2009 when a complete block of flats at the Lotus Riverside complex, which was supported on piles, overturned (Chapter 7, Box 7.24). The cause was excavation on one side with piled up material on the other. Care always needs taking in excavation in case that activity destabilises adjacent foundations. CROSS has a report of a domestic basement collapse caused by opening up the cellar area to remove stabilising cross walls, then excavating to below existing foundation level, so allowing the boundary walls to kick in (CROSS Reports 175, 241, 297, 361, 385, 423, 539).

Box 8.30 Nicoll Highway collapse in Singapore (2004)

This was a major ground collapse occurring during works for the metro line. The supporting structure for a deep excavation work failed, resulting in a deep cave-in that spread across six lanes of the highway. The collapse killed four people and injured three more. The tunnel was being constructed by cut and cover and temporary works supporting diaphragm walls failed. Predictions of forces on the wall were inadequate as was the design of the temporary works such that the strut/whaler system had only half the strength it needed.

According to the Committee of Inquiry, the cause was: 'Use of an inappropriate soil simulation model which over-estimated the soil strength at the accident site and (as a consequence) underestimated the forces on the retaining walls within the excavation.'

With regard to numerical modelling in geotechnical design, the Committee of Inquiry went on to recommend that:

> Generally, numerical analysis or modelling should not be over relied upon. It can only be used to supplement and not supplant sound engineering practice and judgement. It must be well undertaken by competent persons. Those who perform geotechnical numerical analysis must have a fundamental knowledge of soil mechanics principles and a clear understanding of numerical modelling and its limitations.

As with the Heathrow collapse, there also seems to have been inadequate monitoring instrumentation and disregard of different warnings (SCOSS Failure Data Sheet, 2004).

Scour affecting bridge foundations is a generic risk potentially leading to failure. The failure of the Malahide Viaduct by scour was one example (Chapter 7, Box 7.22), Workington's Northside Bridge in Cumbria is another and a CROSS Report (CROSS Report 138) describes a near miss when a historic brick bridge collapsed after its foundations were weakened by flood. A dramatic example of this occurred in huge storms in Newcastle in 2012 when the ground surrounding a block of flats supported on piles was washed away, exposing the piles: the flats had to be demolished.

CROSS has several reports of retaining wall failures. Most of these relate to boundary walls in domestic properties which often seem to have been constructed without any calculation and with little apparent strength (e.g. high blockwork walls). The effect of water often seems to be ignored. In 2012, a long length of riverside retaining wall suddenly shifted in Bridgwater, Somerset, and the retained roadway collapsed.

8.9.3 Tunnelling

Tunnelling is one of man's oldest activities and the story of how Brunel nearly lost his life when constructing a tunnel under the River Thames is well known. In 1828 his tunnel suffered inundations from the river above and six men died in the flood. A further 100 navvies died constructing Brunel's Box Tunnel, which opened in 1841, for his Great Western Railway. In more modern times, new techniques for tunnelling have been evolved but there have still been spectacular failures. By its very nature, ground make-up is uncertain and all sorts of perturbations might be encountered along the tunnel line, allowing cave-ins; flooding by water or fine sand; or collapse due to faults or pockets of weak ground. As with many failures, safety issues are related to the rate of failure. Clearly, rapid inundations or sudden roof falls are critical. Additionally, when there is a cave-in, the wider consequences of what might happen on the surface have to be reckoned with. In the Munich Metro collapse of 1994, a bus drove into the surface crater, killing three passengers.

Tunnelling is a specialised skill, but lessons to be learned from a review of numerous failures have generic implications for all engineers:

- Many failures are reported to be due to a lack of adequate ground investigation and a failure to identify weak ground, e.g. by probing ahead.
- There needs to be good communication between those working on the face (who can detect changes, for example, in drilling rates) and those managing the overall project strategy.
- Gas leakage might be a hazard followed by explosion.
- When tunnelling is aided, for example, by freezing, plant failure and its consequences are a factor. The generic issue is, as always

(Section 8.5.3 and Section 8.6.2), to consider the possibility of mechanical failure. Equally, as flood control generally relies on pumping, the need to consider loss of power and loss of individual pumps should be obvious.

Many aspects of tunnelling require keen observation of ground performance as the tunnel progresses; indeed, this is the essence of the New Austrian tunnelling method (NATM). A failure to monitor adequately and then respond to circumstances has led to many serious NATM failures (there have been about 40 over a 20-year period). So many occurred that in 1996 the HSE carried out a special investigation of the methodology's safety (Health and Safety Executive, 1996). One case can be highlighted as an example (Box 8.31):

Box 8.31 Heathrow Express tunnel collapse, London, UK (1994)

The tunnelling technique was NATM in London clay. A collapse took place directly under part of Heathrow Airport, generating a 10 metre crater on the surface. The consequences were severe: chaos at the airport, settlement of surface structures and huge costs. The recovery costs were three times the original contract price. Of particular interest here, the cause was not some unforeseen condition but largely attributed to a combination of design and management errors, combined with poor workmanship and poor quality control. These deficiencies were known about but not acted upon; there was too much pressure to build quickly. The HSE described the event as the worst civil engineering disaster of the last quarter-century (Health and Safety Executive, 1996; SCOSS Topic Paper, 1994; New Civil Engineer, 1994).

8.9.4 Dams

Several dam failures have been described in Chapter 11. One really bad disaster which occurred in China in 1975 was that of the Banqiao Dam. Design recommendations for the number of sluice gates were cut back and in a violent storm, the installed gates were unable to cope with the overflow, partly as a result of sedimentation blockage. The dam overtopped, and a huge flood resulted, wiping out large downstream areas; thousands died (this was the third-deadliest flood in history). The population warning system failed leading to a lack of evacuation, which increased the death toll.

8.10 Detailing stage

Steel structure performance is largely governed by connection capacity. In most such structures, connections are the weak link. Concrete component capacity depends on

Box 8.32 Collapse of shopping centre roof, Riga, Latvia (2013)

In 2013, the roof of the Zolitūde shopping centre collapsed in two stages, killing 54 people, and injuring another 41. Many others were trapped inside the wreckage. The roof was supported by steel trusses. According to the investigation, the cause of failure appeared to be connections that were too weak. A factor appears to be that the original scheme envisaged the support structure utilising one continuous truss but this had been split in two for reasons of transportation. It was the splice between these two parts that gave way. Reports suggest this change was not authorised or verified (see Management of Change in Chapter 6, Section 6.6). As with all post-mortems, several irregularities were revealed and it emerged that there had been plenty of warning signs of danger; but all these had been ignored.

successful detailing overall and in many continuous concrete structures, joint capacity is dependent on the arrangement of rebar. For example, in corner 'opening joints', the ability to develop 100% of target strength, i.e. joint efficiency, is governed by the manner of detailing. The crucial importance of detailing can be observed from the failure of Charles de Gaulle Airport (Chapter 6, Box 6.7) or the failure on the I-35W bridge (Box 8.33). Poor detailing was feature of the Ronan Point collapses (Box 8.9) and in the Silver Bridge collapse (Chapter 7, Box 7.6). It might even be said that the Ramsgate collapse (Box 8.11) was precipitated by failure of a connection detail. The failure at Camden School (Box 8.6) is illustrative of detailing importance, as was the failure of the tower crane base described in Chapter 6. Examples of gross failure consequent on steel connection failure are given in Boxes 8.32 and 8.33.

In terms of failure discussion, a worrying point is that catastrophic consequences can result from failure at a highly localised point (as in Box 8.33). In 2013, there was a horrific crash on an Italian motorway when a coach plunged off a bridge and fell 30 metres, killing 38 passengers. In terms of safety, no one can avoid crashes, but designs should seek to mitigate consequences as far as possible. A coach careering off a road into a field offers a chance of survival; falling 30 metres off a bridge renders the chances of survival slim. There was a safety parapet on the bridge, but there are suggestions that it was inadequately anchored. As in the crash barrier example of Chapter 3, it is essential to control modes of failure. And in this case, that depended on the detailing. As a generality, in any structure designed for hazard loading where dependency is placed on energy absorption, sufficiency of capacity is absolutely dependent on sufficiency of detailing.

Figure 8.11 shows the construction detail in the bottom flange of a bridge. The central plates cracked, undermining the whole tensile value of the bottom flange. The cause was lamellar tearing (Chapter 7) but neither the designers nor fabricators had

Box 8.33 The I-35W Mississippi River Bridge, USA (2007)

The I-35W Mississippi River Bridge was a wide bridge supported on trusses. During one evening rush hour it suddenly collapsed, killing 13 people and injuring 145 more, with the failure being caught live on CCTV. Failure was attributed to an original design flaw exacerbated by the concrete decking added progressively over time (increasing dead weight by some 20%) plus additional weight from stored construction materials. The investigation found that some of the truss gusset plates had been undersized and were inadequate to support loads. It had been clear from interim inspections that something was amiss with these plates since they were bowing (eventually they fractured). In their report, the National Transportation Board (National Transportation Board, 2008), criticised the designers for not undertaking proper gusset designs and criticised state officials for inadequate attention to the gussets during inspection. The investigation excluded pre-existing cracking as a cause but does not mention that a fracture of the form shown in the failure photographs is indicative of non-ductile material. The investigation identified safety issues as:

- insufficient quality control procedures for designing bridges, and insufficient federal and state procedures for reviewing and approving design plans and calculations;
- lack of guidance for bridge owners with regard to the placement of construction loads on bridges during repair or maintenance activities;
- exclusion of gusset plates in bridge load rating guidance;
- lack of inspection guidance for conditions of gusset plate distortion;
- inadequate use of technologies for accurately assessing the condition of gusset plates on deck truss bridges.

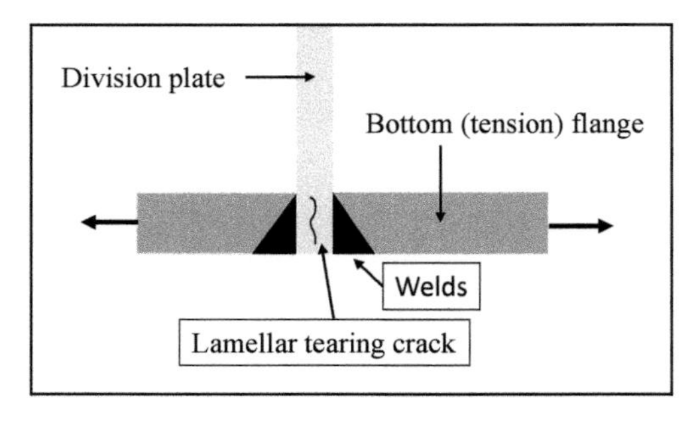

Figure 8.11 Lamellar tearing

contemplated this potential failure mode and thence the need to avoid it to maintain bridge capacity.

The failure of the Canadian de La Concorde Bridge (Chapter 7, Box 7.18) was a shocking event, highlighting the danger of incorporating the type of deck support half-joints (and their detailing) that were used (Wood, 2008).

8.11 Site construction stage

8.11.1 Introduction

In many ways, the construction phase of any project is the riskiest; certainly that view seems to be borne out by descriptions of classic failures occurring during construction and by the sheer number of reported cases. There are reasons for this. First, main designers tend to assess stresses and stability in the structures as finally constructed. Whilst understandable as a starting point, this approach neglects consideration that structures must be assembled piece by piece and during that assembly process the structures will be in varying states of stability and carrying loading not checked by structural conditions under final load. It is perfectly possible that any of these intermediate states might be governing. As an example, consider the process of erecting a column. When first lifted a column ought to be stable and in fact the main function of holding down bolts may well be to hold upright columns in position before their incorporation into a more substantial frame. There have been accidents and deaths when such items have toppled due to inadequate base fixity (CROSS Reports 20, 324, 1038). Once erected, the foundation is subject to almost no vertical load but perhaps substantial overturning from lateral wind, compounded by column lean, and these foundation design conditions are quite different from those under final design, which might simply be to resist high axial compression. In its final condition a column will be carrying axial load and be laterally restrained. In contrast, in its erected position, the column is a vertical cantilever with bending at the bottom. This bending can be substantial especially when, to save splices, double length columns are utilised.

This simple case illustrates three important points:

a. The loading and stability conditions during erection differ from those at completion and may be governing.
b. To assess safety in construction, it is essential to break the process down into discrete steps and then examine the strength and stability of components at each successive stage.
c. The demands on foundations might be quite different between construction and final state.

It will be seen from the examples below that these three characteristics are universally applicable to the challenge of assessing failure cause during construction.

8.11.2 Loading during construction

Figure 8.12 shows the lifting of the London Eye along with a diagrammatic representation of the forces (Berenbak et al., 2001; Mann, 2001). At the moment of initial lift, the uplift

Figure 8.12 Lifting the London Eye

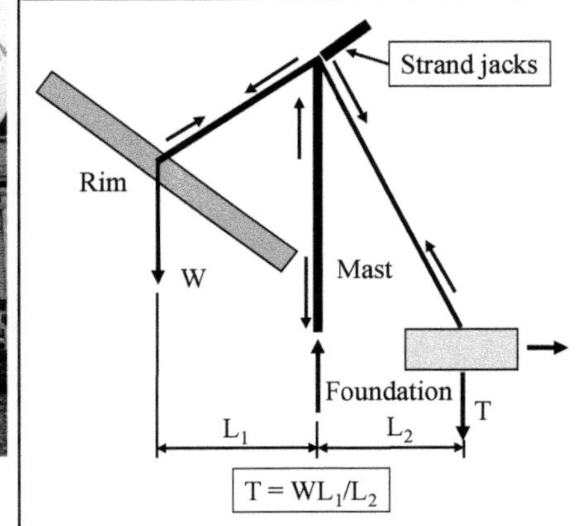

forces on the rear foundation peak, since at that stage the mass of the rim being lifted is most eccentric to the rear foundation and it was that condition that governed base sizing. This is a good example of the need to integrate a methodology of construction with final design. Although much larger than the introductory example of a column being erected, the principles are just the same as those set out in (a) to (c) above. It will also be appreciated from the diagram that during the lift, the Eye's rim is loaded out of plane by self-weight whereas in its final condition, the gravity load acts in the vertical plane.

Another illustration can be seen in Figure 8.13, which shows the lifting of a heavy truss.

- In stage (1) the part truss is lifted in such a way that there is reverse bending on the top boom. There are also potentially high local cross-bending stresses at pick-up points.
- In stage (2) the assembled truss is now in position but unrestrained along its top flange since incoming boom members/purlins have not been erected. At this stage, the loading on the truss will probably be less than existing in the permanent case but equally, the compression boom is less stable. Such conditions might also exist during a lift.

Figure 8.14 shows an arch bridge being lifted. In the permanent case the small bottom member is a tie. During the lift it might be in compression (and in the example illustrated here it actually buckled). Figure 8.15 shows a plate girder bridge launch. In the permanent condition, weld to flange web capacity is determined solely by horizontal shear = QAy/I. But during launch, a vertical reaction is applied which must be dispersed through the flange to the web (similarly to the issue causing fatigue explained in the Chapter 7, Section 7.4.9, 'Examples'). This reaction condition, which might govern weld capacity, applies all the way along the flange as the girder is progressively launched over its support roller.

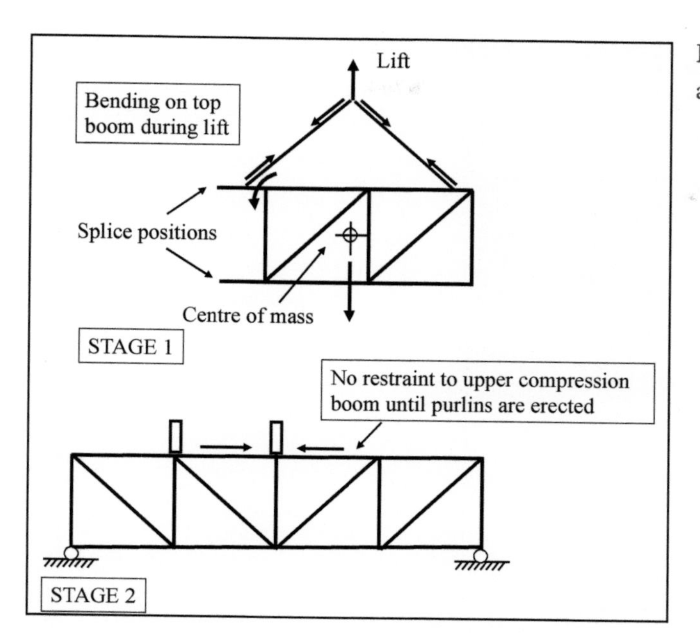

Figure 8.13 Lifting a heavy truss

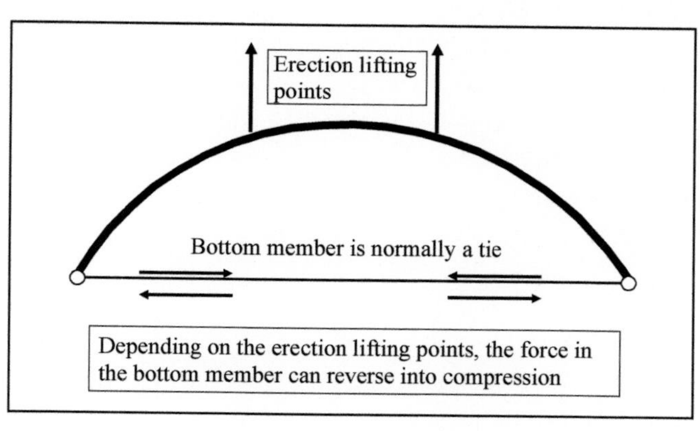

Figure 8.14 Lifting an arch bridge

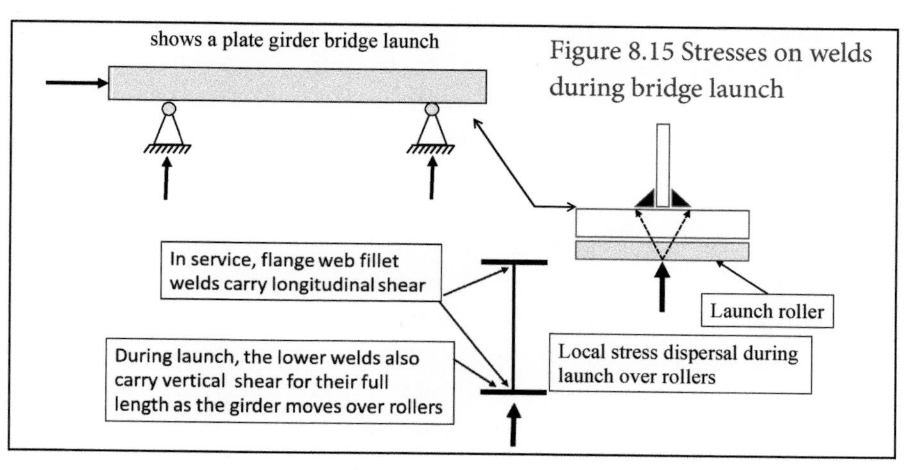

Figure 8.15 Stresses on welds during bridge launch

Figure 8.16 Uncertain loading from stacked material during construction

The loading existing during construction is very uncertain, as illustrated by Figure 8.16, albeit the stacked material might equally have been pallets of bricks. The relevance to scaffolding capacity should be obvious. The lesson is that all surfaces need to be robust and the site crews informed over support capacity.

Box 8.34 Gerrards Cross tunnel, UK, 2005

In this failure, a precast arch collapsed onto a railway line whilst being backfilled. Fortunately, a train driver about to enter the tunnel noted something wrong and halted his train, thus preventing what might have become a major disaster. The exact cause of the failure has not been made public. However, it seems obvious that the loading conditions on the arch in a state of partial backfill, plus the stability of the arch in that condition, differ substantially from those that apply in the final condition (Mylius, 2005).

8.11.3 Stability during construction

The manner of placing material induces stress states which are not readily predictable and Chapter 4, Section 4.2.2, 'Checking stability', referred to the Bragg Report, which was prepared following several falsework collapses (Section 8.11.4, below). Figure 3.5 in Chapter 3 shows a suspended floor being concreted from one side. Considering the mid span area, the worst beam shear occurs there under the partial loading state. Furthermore, the eccentric vertical loading associated with partial span concreting causes a lateral sway tendency that does not exist in the permanent case. Such sway, or the mere existence of a heavy mass up in the air, points to a need to consider lateral loads in falsework design just as much as vertical load capacity. In many cases, known falsework collapses could have been prevented simply by the addition of a small amount of diagonal bracing. In 2005, a disastrous failure took place at Almuñécar in Spain, during concreting or shuttering adjustment; a 60 metre long part failed and six workers died. Exact details of the failure cause have not been revealed. The Structural Engineer (2016–2017) cites several examples of falsework failures.

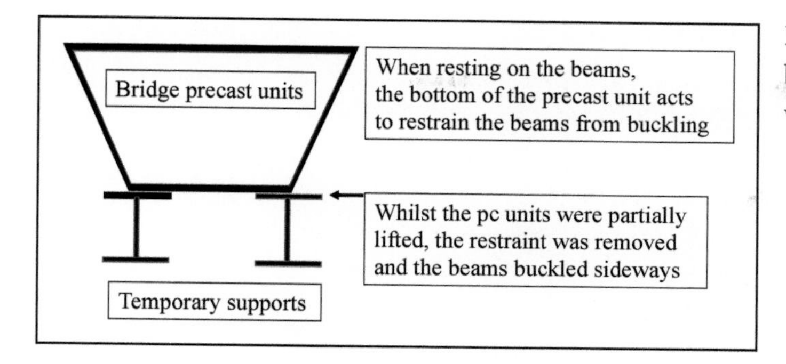

Figure 8.17 Lateral buckling of temporary works support beams

Box 8.35 M60 motorway bridge, Barton, Manchester, UK (1959)

The bridge form was one of long-spanning deep plate girders. In their final state, the top girder flanges would be embedded in deck concrete, which would act compositely with the steel and provide lateral restraint. In the erection state, the girders were supported on temporary supports. From this position the girders just toppled over sideways (many images can be seen on the web at 'images for Barton bridge collapse 1959'). This was the second accident during construction of the bridge; a tower falsework had collapsed just months earlier (altogether six workers died) (Short, 1962).

Figure 8.18 1959 failure of Barton motorway bridge, Manchester (courtesy *Manchester Evening News*)

In 2017 a new bridge was being built alongside the older motorway bridge referred to above. This new bridge had a liftable central section. During lifting trials, it appears the centre section was just dropped or perhaps the winches lifting its four corners were uncontrolled (the failure cause has not been revealed). The failure can be seen as it happened (M60 bridge collapse, 2016). This is another example highlighting the need for engineers to be conversant with concepts of mechanical failure and lack of control.

Figure 8.17 shows a bridge being erected. Initially the lower flange of the precast unit acted as a horizontal diaphragm, restraining the top booms of the temporary support girder out of plane via friction between the deck and the steel compression flange. To complete installation, the concrete unit was then jacked up at discrete points. At that instant, the continuous lateral restraint to the compression flange was removed and the supporting girders buckled and toppled sideways into the river below, killing

Box 8.36 Roof collapse during extension work at the stadium of FC Twente in Entschede, the Netherlands (2011)

This roof collapse was extensive, and workers who were on the roof at the time fell a great distance; two were killed and others were injured. An investigation was carried out by the Dutch Safety Board who concluded that although the failure was primarily caused by instability of the roof structure, there were actually several causes:

- Work was progressing in parallel, not sequentially, and the roof structure was overloaded and not stable enough in its temporary condition.
- The roof was loaded up with construction materials and because of geometrical problems, certain members had been force-fitted, so adding to member stress.
- The construction work was inadequately co-ordinated and not checked.
- There was confusion over responsibilities and transfer of information in that the steel subcontractor was still working when the main contractor started to use the structure and instructed other subcontractors to start work.
- The contract did not assign proper division of responsibility.

All in all, there was a mixture of predictable technical reasons compounded by human failings (SCOSS Topic Paper, 2012; Brady, 2012).

four men. The basic erection concept was flawed and rendered the temporary works unstable.

Some of the most dramatic and tragic failures were those that occurred during construction of box girder bridges. There were three infamous ones: the bridge at Milford Haven (Cleddau Bridge, 1970), a bridge being built over the River Rhine at Koblenz (1971), and the Westgate Bridge in Melbourne, Australia (1970). The most complete information is available on the Westgate Bridge, since this was a very significant civil engineering disaster in Australia, and prompted a Royal Commission of Enquiry (Royal Commission of Inquiry into the Failure of West Gate Bridge, 1971). This bridge was a long box type and to erect the boxes, they were split down their middle and intended to be rejoined on site at the longitudinal seam. Thus, during erection, the strength and stability of the sub-units differed substantially from the state that would apply when fully assembled. During construction, it was found that the two sides of the boxes could not marry up along their length due to some plate buckling. Collapse was then initiated by erectors removing bolts from the upper flange splice to try and correct the buckle and achieve a fit. Although the precise technical cause could be pinpointed, the inquiry afterwards heavily criticised the designers for having failed to give proper consideration to the erection proposals, which returns to the topic of communication between parties.

There are similarities within the series of box girder bridge collapses, where lack of understanding of stress and strength conditions at various stages of erection led to the failures; though it might be argued these were also precipitated by being on the frontier of technology.

8.11.4 Scaffolding

Scaffolding is required to provide safe access to a structure during construction. It must be adequately strong (to support workers and their materials) and stable in both horizontal directions. Stability away from a building is often achieved by anchoring back to some solid structure. In recent times it has become customary to sheathe scaffolding to provide weather protection and to limit the risk of objects falling off onto persons below. Such sheeting increases the wind forces to be carried. Many scaffolds are quite complicated, having to span over protrusions and carry loads down to ground via less than straightforward routes. All scaffolding should be designed by a competent person and the HSE website (HSE website, undated) provides a whole schedule of specific types that require bespoke design. Global collapses are not unknown, and CROSS has multiple reports (CROSS Reports 79, 111, 152, 230, 407, 511, 571, 702); see also HSE Reports (HSE website, undated). The one below is 'typical' (Box 8.37).

In 2006, a very significant collapse occurred at Milton Keynes (CROSS Reports 333, 337). The scaffold was 40 metres high when it failed. One end collapsed, causing all supported levels to fall with it and according to eyewitnesses, the rate of failure was rapid. The HSE prosecuted and it was said that a combination of factors led to the failure: the scaffolding was neither strong nor stable enough for the work being carried out. Furthermore, inspection of the scaffold had been inadequate. One man died and two others were seriously injured. Large fines were levied.

Scaffolding systems can be designed to carry relatively large loads and are often used within falsework, for example, to support the loads of concrete pours. CROSS reports another failure in this mode (CROSS Topic Paper, 2002). The incident occurred

Box 8.37 Scaffolding collapses

Photographs sent in by a reporter (CROSS Report 152) show a scaffold collapsed onto a parking area with the pictures showing the scaffold peeled away from the building. Luckily no one was injured, and only 11 vehicles were damaged. The report noted that front to back diagonals and horizontals were offset by half a bay and that the horizontal members were only held with putlog clips. The diagonals and some verticals were wrapped in polythene before the clips were put in place; the clips appear to have slipped.

Other CROSS Reports related to deficiencies in anchors holding scaffolding back to buildings.

during an 800 m^3 concrete pour. Shortly after the pour began, buckled standards were spotted in the falsework. An inspection revealed numerous deviations from the scaffold arrangements shown on the certified design drawings, primarily that transverse diagonal bracing was fitted to every fourth bay and not to every bay. Collapse was fortunately prevented by various secondary systems. This was a near miss, but another wasn't, as also reported in the CROSS Topic Paper (2002). In this event, there was a partial collapse of falsework across an area of approximately 300 m^2. A forensic investigation led to the discovery of a multitude of errors with regard to falsework design and construction, any one of which could have led to the collapse. These included: no falsework design for this area of complicated geometry; large floor to floor heights; a mixture of different falsework systems used; overly large gaps between table forms; incorrect arrangement of table forms, missing bracing and a dangerous overuse of timber packing underneath scaffolds.

CROSS has reported (CROSS Topic Paper, 2002) that the 1970s saw a series of falsework collapses of major significance within both the building and civil engineering sectors of the industry. Other collapses had preceded these, for example, the falsework collapse on Barton Bridge in Manchester in 1959 (Short, 1962). To allay growing concern over an area of work not well regulated at the time, an advisory committee was established under the chairmanship of S. L. Bragg, with a wide remit to investigate. His committee produced the report known by his name – the Bragg Report (Bragg Report, 1975 and Institution of Structural Engineers, 1976). Subsequent to the Bragg Report (and in compliance with one of its recommendations) industry produced the first edition of British Standard BS 5975 in 1982, 'The code of practice for falsework' (BS 5975). Although problems diminished in the UK, failures continued elsewhere: in 1997 travelling falsework on the mighty Tagus Bridge in Lisbon, Portugal, collapsed, killing six workers.

8.11.5. Dropped loads

Most site activities involve lifting and that activity is linked to the hazard of failure during a lift or just dropping something (see Box 8.35). There are innumerable cases of cranes toppling over (The Structural Engineer, 2016–2017) along with the consequential damage that toppling might cause; the hazard is ever-present as heavy masses are lifted high in the air at an eccentricity. Lack of attention to simple stability, or to the added lateral loads from the dynamics of swing, or from wind gust, are all factors to be accounted for. The risks are much higher where proper attention is not given to ground support under wheels or stabilisers. The higher and the heavier the lift, the more care is required. Spectacular examples of failure can be found on the Web (e.g. Big Blue crane failure at Milwaukee (1999) shows the lifting of a roof in Milwaukee, USA, and its side sway failure apparently due to lifting in windy conditions. This is also an example of chain events: there was much collateral damage, including deaths. Once a load starts to move sideways, the process is essentially destabilising since P-Δ effects contribute to the overturning force always present from wind. A hazard associated with cranes is one of dropping the counterweight (CROSS Report 47). In nuclear plants a standard hazard is that of 'dropped load' and coping with that eventuality can dominate a design (Jordan and Mann, 1990).

Any object dropped from height, even a nut, is potentially lethal. Hence great care is required to minimise the 'dropped item' risk and mitigation measures include the fixing of kickboards on scaffolds and avoiding working in lower areas when other work is being carried out overhead. Any item being lifted needs to have a proper attachment for lifting, a check on its capacity during the lift and a 'loose items' check before the lift. There have been instances of large objects, such as precast units, just being dropped. Particularly tragic cases occurred at Manchester (Box 8.35) when the toppling girders fell on site huts clustered below and West Gate Bridge (Melbourne) (Chapter 6, Section 6.7.2) when the collapsing bridge parts also fell onto site huts. Box 8.35 also reports the dropping of a long bridge section.

The danger of parts falling off buildings came to the fore in Scotland when a waitress in Edinburgh city centre was killed by falling masonry (2000). In the aftermath, the Scottish Building Standards Agency conducted a survey and produced a report examining the risks (Construction Industry Council for Scotland, 2003; SCOSS, 2008). Throughout the UK, many incidents involving masonry and other materials falling onto public footpaths and other pedestrian spaces have been reported to local authorities during each year of the study. Since 1988, at least one person has been killed or injured each year in Scotland as a result of masonry or roofing materials falling from buildings. When discussing such incidents with those responsible for public safety in local authorities, comment is frequently made regarding the number of near misses that also occur.

8.12 Commissioning stage

Commissioning is a process of checking and adjustment of systems post construction, before handing over to clients for use. Primarily the terminology refers to mechanical items and services, but some structures are subject to testing (including load testing) before handing over and in some special moving structures, commissioning involves both load testing and assessment of fault conditions on the control system before handover. Water-retaining structures must be filled for testing and perhaps leakage points repaired before being put into service. Likewise, drains are checked for leakage.

Commissioning is a risk period, since it is the first time many structures will see their full design loading, so it needs to proceed with caution. Almost comic cases are known about when trains did not fit around bends (Chapter 6, Box 6.5) and in one case, a moveable structure simply ran off its end rails whilst being trialled for the first time.

The Sleipner collapse (Chapter 6, Section 6.4) occurred during installation; another North Sea platform just sank. The Chernobyl disaster occurred during a test. Section 8.8.7 describes the failure of the Seneffe water tower, which occurred on filling for the first time. The 2016 Barton Bridge failure (Box 8.35 and M60 bridge collapse, 2016) seems to have occurred during testing.

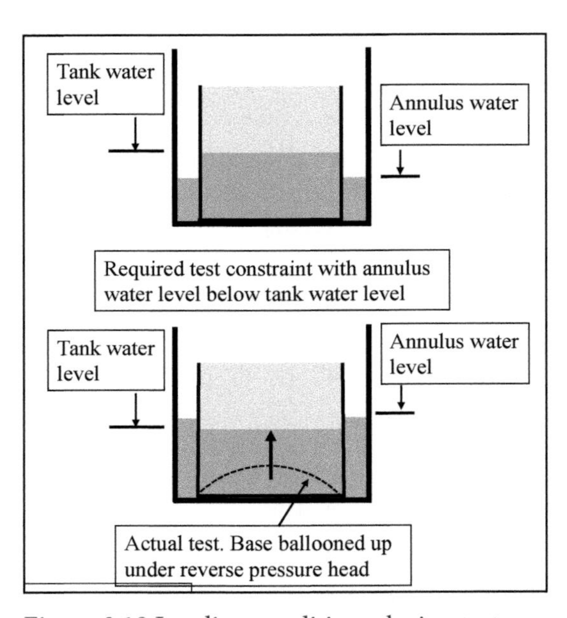

Figure 8.19 Loading conditions during test

Figure 8.19 shows a high-integrity tank that was protected via a containment annulus. To ensure watertightness, the tank had to be tested and this involved filling the tank and annulus and then draining in such a way that the annulus water head lay below that of the tank. Unfortunately, draining of the inside occurred at a rate allowing the annulus head to exceed the inner tank height. Consequently, the thin base of the main tank was back pressurised and ballooned up. This failure illustrates the wider principle of a need to consider the case of pressure reversal or out of balance loading in any structure where hydrostatic pressure is a factor. Groundwater may cause upwards base bending (when tanks are temporarily empty) or structure flotation in any dry dock or underground reservoir when operating empty. Such failures are most likely during construction when structures might float on groundwater before counterbalancing weight is applied.

8.13 In use stage

8.13.1 Introduction

Nothing lasts forever, and Chapter 7 describes a variety of degradation mechanisms that might apply to structures. Just because a structure is still standing does not mean it remains safe or even serviceable and structures require managing throughout life. The safest structures give some indication of impending failure well before collapse (Chapter 5) but it will be noted that some of the failures scheduled in this chapter occurred instantly. The fact is that the world has a legacy of old structures that just cannot all be replaced. Some of these were erected when knowledge and control weren't as good as they are now, and some are ancient structures that society would not want to change anyway. Retaining

Box 8.38 Collapse of Viale Giotto 120 building, Foggia, Italy (1999)

This was a six-storey apartment building which collapsed without warning in just 19 seconds. The final cause was determined to be the poor workmanship and materials of the original build. This failure was not unique and a number of similar collapses have taken place in Italian dwelling blocks that are suggestive of inherent weakness, all linked to hasty building following post-war demand, allied with using poor materials, poor workmanship and weak regulation. In this failure, the attempted repair of poor concrete involved stripping column cover off back to the vertical rebar and links and this seems to have been enough to permit bar buckling and thence general column collapse.

ancient structures is a challenge, and conservation of Britain's stock of ancient cathedrals is a never-ending project: the stones decay, the glass gets brittle and the foundations settle. In the 1970s the piers under York Minster's (72 metre) main tower (dating from the 14th century) were known to be settling, highlighted by stonework distortion. By any modern standards, the foundation bearing pressures were just far too high. Consequently, the mediaeval bases all had to be underpinned to prevent a gradual degradation that would otherwise have culminated in collapse (Dowrick and Beckmann, 1971). One example of a complete failure is that of the ancient Venice campanile, located in St Mark's Square. In 1902, the brick tower just disintegrated (the current campanile is an exact copy). The campanile failure was preceded by signs of cracking which progressed rapidly.

8.13.2 Examples

One of the worst 'instant' building disasters of modern times occurred in 2013 in Dhaka, Bangladesh, when a whole multi-storey factory disintegrated, killing more than 1,000 workers in the process. The first pictures showed utter collapse of the factory's concrete frame. The causes are described in Chapter 11. The Dhaka collapse is similar to one that occurred at the Sampoong department store (Korea) in 1995 (see Chapter 11).

Judging by the large number of reported failures, bridges too are at risk of collapse in service. Threats come from durability (such as the Morandi Bridge (Figure 8.19) and Chapter 7, overloading, impact, from fatigue and foundation failure and from bearing failure (see Thelwall Viaduct in Chapter 7, Box 7.13). Even fire can cause failure (see Chapter 9). Authorities accept the need for in-service inspection to detect the onset of degradation, but some failures have occurred when parts have been un-inspectable (see Chapter 7, Boxes 7.6. 7.19 and 7.20). It is especially important to look after bridges, obviously for safety, but also because they are frequently a costly asset owned by the communities and because failure results in significant on-cost from lack of availability. Because bridges are costly, the aspiration is for a life of about 100 years and that puts a greater demand on the durability aspects of their design.

Figure 8.19. Collapsed Morandi Bridge (Shutterstock 1185155212)

Chapter 7, Box 7.22 (Malahide Bridge) and Section 7.7 (Workington Bridge) described the effects of foundation scour, and the potential for sudden failure.

8.14 Demolition stage

Demolition is the reverse of construction but historically has been associated with greater danger, and the HSE website contains plenty of cautionary tales. By law, all demolition, dismantling and structural alteration should be carefully planned and carried out by competent practitioners. Discounting here the risks to personal injury (which are about twice those in construction), key issues are:

- injury from falling materials;
- risks from connected services;
- hazardous materials (such as legacy asbestos);
- uncontrolled collapse.

Thoday (2004) and Clarke (2010) give general guidance, and there is more on the HSE website (HSE website, undated).

Uncontrolled collapse might occur either through bad practice, or because of defects in the construction that are not evident, or because there are no records to describe the stability system, or through degradation. There are considerable uncertainties, because buildings might be weaker than assumed or they might be a lot stronger, which carries dangers in affecting the mode of any planned collapse. New forms of construction, such as large

Box 8.39 Warsaw radio mast, Poland, 1991

The Warsaw mast was extremely tall: 646 metres. In 1991 it was undergoing repair. Guy wires were being exchanged but some sort of error occurred leading to a catastrophic failure. The mast first bent than snapped off at about half height. Blame was assigned to the company which built and maintained the mast. A very similar failure occurred in Canada in 1998 when replacing guys on the WLBT 609 metre high mast.

panel buildings, carry their own risks. In the past there has been considerable concern over how to release the energy in pre-tensioned structures during demolition. Accidents have occurred with heavy equipment punching through into basements or being supported on slabs which then collapsed. A recent terrible failure was the complete collapse of Didcot power station (UK) in 2017. Four men were killed. As yet no cause has been published.

Failure can take place during alteration as in the incident of the Warsaw radio mast 9 (Box 8.39). The generic issue, as with erection and demolition, is stability in intermediate phases.

CROSS has relevant reports; for example, Reports 96 and 186 report on problems with certain large panel structures during demolition. The buildings were 13-storey-high tower blocks. During progressive reduction, the panels and floor slabs forming the bottom eight floors collapsed progressively and unexpectedly. New methodologies had to be evolved for continuing demolition but even with these, some surprise failures took place. Of considerable concern was a very clean shear failure across the slab/cross wall panel junction, indicating a significant lack of connection (mechanical tying) between floor panels and the cross walls forming the structural cells. (There was a significant absence of ties in the Ronan Point flats.) Harvey (1994) describes a progressive collapse occurring during demolition of a masonry arch in 1992. Two men were killed. The structure collapsed after the arch crown of one span had been partially cut away.

8.15 Chapter summary

- There have been innumerable failures during the design and construction of buildings and infrastructure. Some examples have been given in this chapter. It is not hard to find others.
- The examples used herein are to show that failure can occur in any stage of a project's life cycle. Moreover, they can occur in major structures; nothing is immune, so it is never wise to presume the worst cannot happen. Active scrutiny of safety is always required.

- The purpose of this book is to show that lessons can be learned from disasters, which will equip engineers with a skill set to reduce the risk of failure recurrence. With hindsight, the causes of failures cited in this chapter have not been mysterious. Most of the failures could have been avoided and many are not fresh incidents but repeats of previous failures. Hence it is certainly possible to learn from them.

References

ACEC (American Council of Engineering Companies) (2003), 'A guideline addressing coordination and completeness of structural construction documents', CASE Document 962 D.

Alexander, S. J. (2014), 'Design for Movement in Buildings', CIRIA C734.

Beeby, A. (1999), 'Safety of structures, and a new approach to robustness', *The Structural Engineer*, **77**(4).

Berenbak, J., A. Lancer and A. P. Mann (2001), 'The British Airways London Eye. Part 2, Structure', *The Structural Engineer*, **79**(2).

Big Blue crane failure at Milwaukee (1999), www.youtube.com/watch?v=ZXr1IeWbP10.

Board of Inquiry (1955), 'Collisions, sinkings, etc.: enquiry into failure of no 3 dock caisson and accident involving HMS *Talent* and death of four dockyard workers on 15 December 1954'.

Bolton, A. (1978), 'Natural frequencies of structures for designers', *The Structural Engineer*, **56A**(9).

Bolton, A. (1983), 'Design against wind-excited vibration', *The Structural Engineer*, **61**(8).

Bosela, P. and P. Bosela (2018), 'Tropicana parking garage collapse', https://ascelibrary.org/doi/10.1061/9780784482018.108.

Brady, S. (2012), 'De Grolsch Veste stadium failure', *The Structural Engineer*, **90**(11).

Brady, S. (2013), 'Interstate 90 connector tunnel ceiling collapse', *The Structural Engineer*, **91**(4).

Brady, S. (2015), 'Hartford stadium collapse: why software should never be more than a tool to be used wisely', *The Structural Engineer*, **93**(8).

Bragg Report (1975), *HSE, Final Report of the Advisory Committee on Falsework (Bragg Report)*, HMSO.

Brazil grandstand collapse (2011), www.youtube.com/watch?v=aXXnPrbHbtE.

British Constructional Steelwork Association (2010), *National Structural Steelwork Specification*, 5th edn, BSCA.

BS 5975:2008+A1:2011, Code of practice for temporary works procedures and the permissible stress design of falsework.

Building Research Advisory Service (1973), 'Report on the collapse of the roof of the assembly hall of the Camden School for Girls', Department of the Environment, Special Investigation Report No. 2761.

Building Research Establishment (BRE) (1988), 'The structural adequacy and durability of large panel system dwellings'.

Burgoyne, C. and R. Scantlebury (2006), 'Why did Palau Bridge collapse?', *The Structural Engineer*, **84**(11).

Burland, J. (2006), 'Interaction between structural and geotechnical engineers', *The Structural Engineer*, **84**(8).

Burland, J. B. (2008), 'Foundation engineering', *The Structural Engineer*, **86**(14).

Bussell, M. N. and A. E. K. Jones (2010), 'Robustness and the relevance of Ronan Point today', *The Structural Engineer*, **86** (23/24).

Carpenter, J. (2008), 'Partial collapse of Ronan Point: what can we learn from the past?', *The Structural Engineer*, **86** (5).

Carpenter, J., J. Wicks, B. Clarke, A. Borthwick and R. Falconer (2013), 'The importance of understanding computer analysis in civil engineering, *Proceedings of the ICE – Civil Engineering*, **166**(CE3).

Chapman, J. C. (1998), 'Collapse of Ramsgate Walkway', The *Structural Engineer*, **76**(1).

Clarke, Rob (2010), 'Role of the structural engineer in demolition', *The Structural Engineer*, **88**(11).

Construction Fixings Association (2015), *Best Practice Guide: Selection and Installation of Top Fixings for Suspended Ceilings*, FIS (Finishes and Interiors Sector).

Construction Industry Council for Scotland (2003), 'Risks to Public Safety from Falling Masonry and other Materials'.

CROSS Report 6, Ceiling collapse.
CROSS Report 12, Masonry walls.
CROSS Report 20, Steel Portal Frame Erection.
CROSS Report 47, Counterweight failure – Plank Lane Bridge (news).
CROSS Report 75, Some press reports on wall collapses in January 2006.
CROSS Report 79, Scaffold collapse.
CROSS Report 82, Wind on internal masonry walls during construction.
CROSS Report 94, Serious injury from free standing wall collapse.
CROSS Report 96, Demolition risks.
CROSS Report 99, Collapse of a wall during construction.
CROSS Report 100, Ceiling collapse in an educational building.
CROSS Report 101, Ceiling collapse in cinema No. 1.
CROSS Report 102, Ceiling collapse in cinema No. 2.
CROSS Report 103, Ceiling collapse in cinema No. 3.
CROSS Report 111, Unauthorised scaffolding.
CROSS Report 116, Death from wall collapse (news).
CROSS Report 124, Office ceiling collapse.
CROSS Report 130, School ceiling collapse.
CROSS Report 134, Deadly retaining walls.
CROSS Report 135, Critical wall failure.
CROSS Report 140, Underpass ceiling collapse.
CROSS Report 136, Public art structures.
CROSS Report 138, Brick bridge collapse.
CROSS Report 144, A failure survey of free-standing walls.
CROSS Report 148, Suspended granite ceiling collapse.
CROSS Report 152, Scaffold collapse and slipping clips.
CROSS Report 157, More about public art.
CROSS Report 159, Decorative lights and banners can damage buildings.
CROSS Report 162, Collapse of blockwork retaining wall.
CROSS Report 175, Unsuitable underpinning.
CROSS Report 186, Collapse of large panel system (LPS) buildings during demolition.
CROSS Report 203, Collapse of recently installed suspended ceiling.
CROSS Report 230, Defective imported scaffold ties.
CROSS Report 241, Lack of control on site when underpinning.
CROSS Report 251, Swimming pool ceiling collapse (news).
CROSS Report 264, Snow loading in Scotland – summary of reports.
CROSS Report 282, Hoarding blown onto child (news).
CROSS Report 295, Masonry fall due to high wind.
CROSS Report 297, New basements beneath existing properties.
CROSS Report 304, Partial collapse of suspended ceiling.
CROSS Report 324, Lack of experience on steel column erection.
CROSS Report 333, Falsework support to a bridge – a near miss.
CROSS Report 337, Falsework collapse during slab pour in SE Asia.
CROSS Report 361, Basement party walls.
CROSS Report 385, Failure of existing basement wall.
CROSS Report 407, Unsafe timber scaffolding.
CROSS Report 423, Temporary works design for basements.
CROSS Report 426, Swimming pool ceiling partial collapse.
CROSS Report 442, Apollo Theatre London ceiling collapse.
CROSS Report 511, Collapse of infill wall panel.
CROSS Report 539, House collapses during basement conversion (news).

CROSS Report 571, Injuries from falling scaffold tube.

CROSS Report 702, Scaffold lifting beams incorrectly installed.

CROSS Report 1038, Column stability during erection.

CROSS Topic Paper (2002), 'Falsework: full circle', SC/T/02/01 (updated 2010).

Dallard, P., T. Fitzpatrick, A. Flint, A. Low, and R. Ridsill Smith (2001), 'The Millennium Bridge, London: problems and solutions', *The Structural Engineer*, **79**(8).

Dowrick, D. J. and P. Beckmann (1971), 'Summary of Paper 7415: York Minster structural restoration', *Proceedings of the ICE*, **49**(2).

European Steel Design Education Programme (undated), ESDEP Lecture Notes. Lecture 1B.8: Learning from Failures, http://fgg-web.fgg.uni-lj.si/~/pmoze/esdep/master/wg01b/l0800.htm.

Finnegan, P. and P. M. Forde (2014), 'Sulphate-generated heave – the effect on structures in Dublin', *The Structural Engineer*, **92**(5).

Godfrey, E. (1923), 'The failure of the Knickerbocker Theatre', *Journal of the Institution of Structural Engineers*, **1**(2).

Harvey, W. J. (1994), 'Demolition of arch bridges', *The Structural Engineer*, **82**(5/1).

Health and Safety Executive (1996), *Safety of New Austrian Tunnelling Method (NATM) Tunnels. A Review of Sprayed Concrete Lined Tunnels with Particular Reference to London Clay*, HSE Books.

Health and Safety Executive (2003), Control Systems Out of Control: *Why Control Systems Go Wrong and How to Prevent Failure*, HSE Books.

HSE website (undated), www.hse.gov.uk/. Examples include: Scaffold checklist; Milton Keynes scaffolding collapse 11 April 2006; Avoiding structural collapses in refurbishment. A decision support system; A technical guide to the selection and use of fall prevention and arrest equipment.

Indianapolis stage collapse (2011), www.youtube.com/watch?v=Fe5HNtfTdGE.

Institution of Structural Engineers (1969), 'The implications of the Report of the Inquiry into the collapse of flats at Ronan Point, Canning Town: Report of the open discussion meeting held by the Institution of Structural Engineers at The City University, London ECI', *The Structural Engineer*, **47**(7).

Institution of Structural Engineers (1976), 'Institution Comments on Bragg Committee Report', *The Structural Engineer*, **54**(6).

Institution of Structural Engineers (2001), *Subsidence of Low-Rise Buildings: A Guide for Professionals and Property Owners*, 2nd edn, IStructE Ltd.

Institution of Structural Engineers (2002), 'Guidelines for the use of computers for engineering calculations'.

Institution of Structural Engineers, Advisory Group on *Temporary Structures (2017), Temporary Demountable Structures: Guidance on Procurement, Design and Use*, 4th edn, IStructE Ltd.

Institution of Structural Engineers' Health and Safety Panel (2017), 'Managing Health & Safety Risks (No. 61): Unexploded ordnance', T*he Structural Engineer*, **95**(4).

Jordan, G. W. and A. P. Mann (1990), 'THORP Receipt & Storage: Design and Construction', *The Structural Engineer*, **68**(1).

M60 bridge collapse (2016), www.manchestereveningnews.co.uk/news/greater-manchester-news/m60-bridge-collapse-video-barton-11355496.

MacLeod, I. A. (2005), *Modern Structural Analysis: Modelling Process and Guidance*, ICE Publishing.

MacLeod, I. A. (2008), 'The ascent of structural mechanics', *The Structural Engineer*, **86**(14).

Mann, A. P. (2001), 'Building the British Airways London Eye', *Proceedings of the ICE*, **144**(2).

Mann, A. P. (2002), 'Safety of hanging systems: lessons from CROSS reports', *The Structural Engineer*, **97**(9).

Mann, A. P. (2008), 'Learning from failures at the interface', *Proceedings of the ICE – Civil Engineering*, **161**(6).

Mann, A. P. (2011), 'Cracks in steel structures', *Proceedings of the Institution of Civil Engineers – Forensic Engineering*, **164**(1).

Menzies, J. B. and G. D. Grainger (1976), 'Report on the collapse of the sports hall at Rock Ferry Comprehensive School, Birkenhead', BRE Current Paper 69/76.

Metrodome sports stadium collapse (2010), www.youtube.com/watch?v=Y8eV96EulJc.

Mylius, A. (2005), 'Backfill operation probed in Gerrards Cross tunnel collapse', *New Civil Engineer*, www.newcivilengineer.com/backfill-operation-probed-in-gerrards-cross-tunnel-collapse/532257.article.

National Cyber Security Centre (undated), www.ncsc.gov.uk.

National Transportation Board (2008), 'Highway Accident Report: Collapse of I-35W Highway Bridge, Minneapolis, Minnesota, August 1, 2008', NTSB Number HAR-08/03, NTIS Number PB2008-916203.

New Civil Engineer (1994), Heathrow collapse special, 28 October, 3–8.

Rolton, D. (2008), 'Construction methods and techniques', *The Structural Engineer*, **86**(4).

Royal Commission of Inquiry into the Failure of West Gate Bridge (1971), *Report of Royal Commission into the Failure of West Gate Bridge*, www.parliament.vic.gov.au/papers/govpub/VPARL1971-72No2.pdf.

SCOSS (2005), *15th Biennial Report from SCOSS.*

SCOSS (2008), 'Confidential Reporting on Structural Safety for Scottish Buildings', Scottish Buildings Standard Agency.

SCOSS Alert (2010), 'Temporary event structures: "saddle span" type tents', www.cross-safety.org/sites/default/files/2010-10/temporary-event-structures-saddle-span-type-tents.pdf.

SCOSS Alert (2012), 'Temporary stage structures', www.cross-safety.org/sites/default/files/2012-01/temporary-stage-structures.pdf.

SCOSS Alert (2014), 'Preventing the collapse of free-standing masonry walls', www.cross-safety.org/sites/default/files/2014-09/preventing-collapse-free-standing-masonry-walls.pdf.

SCOSS Alert (2017), 'Sudden loss of ground support', www.cross-safety.org/sites/default/files/2017-07/sudden-loss-ground-support.pdf.

SCOSS Alert (2018), 'Effects of scale', www.cross-safety.org/sites/default/files/2018-11/effects-scale.pdf.

SCOSS Failure Data Sheet (2004), 'The Collapse of the Nicoll Highway on 20 April 2004' (A summary is derived from the official report entitled 'Report on the incident at the MRT circle line worksite that led to the collapse of the Nicoll Highway').

SCOSS Topic Paper (1994), 'Collapse of the New Austrian tunnelling method (NATM) tunnels at Heathrow Airport', www.cross-safety.org/srms/safety-information/cross-topic-paper/collapse-new-austrian-tunnelling-method-natm-tunnels-heathrow.

SCOSS Topic Paper (2008), 'Risk issues associated with large TV/video screens at public events', SC/08/008, Issue 01.

SCOSS Topic Paper (2012), 'FC Twente stadium roof collapse – learning from the fatal consequences', www.cross-safety.org/sites/default/files/2014-07/fc-twente-stadium-roof-collapse.pdf, and www.onderzoeksraad.nl/en/onderzoek/1824/collapsed-roof-grolsch-veste-7-july-2011.

Short, W. D. (1962), 'Accidents on construction work with special reference to failures during erection or demolition', *The Structural Engineer*, **40**(2).

Sports Grounds Safety Authority (1973), *Guide to Safety at Sports Grounds*, HMSO, https://sgsa.org.uk/green-guide/.

Sunshine Bridge (1980), www.youtube.com/watch?v=U2FxkgvyQPM.

Sutherland, J. M. (2008), 'The impact of materials', *The Structural Engineer*, **86**(14).

Taylor, R. (2010), 'On the rack – structural failure of pallet racking systems', *The Structural Engineer*, **88**(22).

The Structural Engineer (2016–2017), 'Temporary Works Tool Kit', Parts 1 to 15.

Thoday, J. (2004), 'Demolition – assessing the risks and planning for safety', *The Structural Engineer*, **82**(15).

Thomas, D. (2016), 'What is the Temporary Works Forum?', *The Structural Engineer*, **94**(9); see also www.twforum.org.uk.

Wood, J. G. M. (2008), 'Implications of the collapse of the de la Concorde underpass', *The Structural Engineer*, **86**(1).

9 Fires

9.1 Introduction

Self-evidently fires are potentially devastating. Everyone is aware of what can happen. It might be thought that dangers have receded in modern times, but periodic catastrophic events and constant domestic incidents offer no support for such a view. Chilling observation of the World Trade Center (2001); Buncefield (2005); Deepwater Horizon (2010) and Grenfell Tower (2017) (Box 9.1) show there is absolutely no room for complacency. Fire remains a generic hazard for us all to consider.

In terms of design, traditional protection rules have evolved from observation. Relatively recently, more analytical approaches have been developed to model fire events and there has been a significant growth in understanding fire effects on complete structures with the specialisation of structural fire engineering emerging as a distinct discipline. Within the UK, it is possible to take courses on this, but even for those designers who do not specialise, a wide understanding of fire hazard, risk and consequence remains a fundamental skill to acquire.

Some reflection suggests there are sound reasons for continuing study. We no longer have a stock of timber houses with thatched roofs at risk from open fires. Instead we have a whole new generation of materials which burn in different ways. We no longer warm our houses by burning coal but instead have a society of cigarette smokers or perhaps buildings with faulty electrics to act as different sources of ignition. New risks emerge, such as in 2013 when a Birmingham plastics factory (Section 9.4.5) ignited quickly from a stray floating Chinese lantern (Fire Brigade websites discourage such lantern use). As a society, we are forced into higher rise and denser accommodation, where perhaps older empirical rules for fire protection and rescue no longer suffice. We have new forms of construction for which active study is required to assess behaviour; the fire at Grenfell Tower in 2017 was a turning point. We have a complex infrastructure of bridges, tunnels, power supplies and factories, all of which may be vulnerable.

9 Fires

The government produces annual fire statistics (Home Office, 2012, as updated). Statistics have been broadly consistent in recent years. Notable findings in the reports from 2020 to 2021 are:

- There were 151,000 fires in Britain (broadly consistent since 2012).
- There were 240 fire-related deaths, lower than in any year in the last 50 years (the highest number of fatalities recorded was 967 in 1995–96). The majority (186) occurred in dwellings.
- Being overcome by gas, smoke or toxic fumes was the most common cause of death.
- Of the 186 fatalities in dwellings, most were from accidental causes. The main cause was careless handling of fire or hot substances (e.g. careless disposal of cigarettes).

The UK statistics follow the same pattern year on year, save that 2017 will be characterised by the 72 deaths attributed solely to the Grenfell Tower fire. Comparable annual reports are available from other countries and ones from the USA paint a similar picture: most deaths occur in homes; cooking is a leading cause of fire; and fires originating in smoking cause most deaths. In countries with poor regulation, death rates

Figure 9.1 The Grenfell Tower fire (Shutterstock 667008772)

Box 9.1 Grenfell Tower, London, UK (2017)

Grenfell Tower was a residential tower block (public housing), 24 storey high. It had been externally clad to improve its insulation and appearance. On the evening of 14 June 2017, the cladding caught fire, with the fire apparently ignited by a faulty electrical kitchen device, and the internal flames spread externally via a route around the windows. In a very few minutes, flames dramatically spread up the outside, engulfing the whole tower (Figure 9.1). Fire brigades were called, but the flame spread was so swift very little could be done. As in many other fires, molten debris could be seen falling on rescuers. Residents were trapped and 72 died whilst 70 more were injured. The whole event could be seen on TV.

As of late 2021, the inquiry remains underway. However, there appears little doubt that a prime fault was the overall application of flammable cladding. Smoke accumulation in the single staircase also seems to have been a factor in the tragedy.

The fire was so dramatic and so horrific that it traumatised the nation and led to widespread calls for better regulation and enforcement. This fire, like the Ronan Point collapse and the World Trade Center disaster, is likely to be one of those seminal events that change the course of building design. A sobering lesson is that it took this tragedy to stimulate change, although, as in so many other cases, it had been preceded by comparable fires whose clear warnings had passed unheeded.

are significant multiples of those for the UK and USA. (In the US, fire death numbers in 2018 will be magnified by those occurring in the autumn wild fire event in California.)

Although such statistics are useful, they paint an incomplete picture. They do not inform about huge annual commercial losses. They do not inform of what might have happened. No one was injured at Buncefield (2005), which was the UK's biggest fire since the Second World War. And as for many safety issues, raw numbers do not reflect the occasional horror of multiple deaths in periodic incidents like the Bradford stadium fire (1995) (Section 9.3.6) or the King's Cross Underground fire (1997) (Section 9.3.8) or the absolute horror of Grenfell Tower in 2017 (Box 9.1).

There is a downward trend in dwelling fires and deaths, presumably due to better design, better protection and better awareness (though some figures show an increase in firefighter deaths). However, financial losses rise year on year. Insurance reports in the UK suggest that annual losses exceed £1 billion, yet such costs may not include allowances for business disruption or the inability to replace precious items. The UK's largest fire losses have been in warehouses. In the USA in 1991, the cost of the Oakland firestorm alone was put at $1.5 billion and the California firestorm of 2007 cost $1.9 billion (the fires of late 2018 and 2020 may cost even more). Most structural failures are partial, but fire has the potential to cause total and widespread destruction and thus disproportionate cost.

In the wider context of safety, fire safety is critical after earthquakes. Fire was a major cause of destruction after the great San Francisco earthquake of 1906: several fires around the city burned for days and destroyed nearly 500 city blocks. In more recent times, fire was a major contributor to death and destruction after the terrible Kobe earthquake of 1995 and was a feature of several incidents in the Fukushima earthquake of 2012. Frequent causes are upsetting domestic fires or cooking activity or the rupturing of gas mains and so on. Precautions include the installation of automatic cut-off valves to close off fuel supplies when pressure loss is detected.

9.2 Aims of fire protection

The chief aim must be to protect human life and this objective has implications for building layout in terms of escape and becomes a highly significant issue in very tall structures. Protection of life also involves making sure that structures remain standing for a certain amount of time for the dual purpose of allowing occupants opportunity to escape and for protecting firefighters on the scene. Excess heat from fire is a clear risk, fumes and smoke are just as dangerous and so the propensity for building materials to produce smoke (and for that smoke to spread) is a matter of significant design importance. Apart from smoke being an irritant, it may contain deadly carbon monoxide or other toxins.

More generally, an objective is to prevent fires from starting and thence spreading. To avoid starting, sources of ignition need to be considered. To avoid spreading, there is a requirement to eliminate materials that could promote rapid spread plus a task of arranging the building such that, if a fire does start, the damage is at least contained.

Detection and suppression are other objectives, with detection much aided by various services and monitors along with systems that allow alarms to be sent quickly. Many structures require installation of dedicated services to aid firefighters in their task of extinguishing a blaze. These might range from simple fire extinguishers to firefighting water mains and dedicated means of safe access.

In summary an overall strategy might be:

- Minimise the risk of ignition.
- If fires ignite, make sure flames cannot spread (especially not quickly).
- If fires spread, make sure they are volume-contained.
- Assume fires will develop and make sure people can escape (or be adequately protected).
- Provide a means of fire detection and suppression.
- Protect the firefighters.

The Design Principles of Fire Safety (Department of the Environment, 1996), may be consulted.

Although the chief aim of protection must be for human life, avoiding commercial loss is a factor not to be overlooked and one not reflected merely by application of the Building Regulations or code requirements. An overall basic strategy to minimise commercial damage (complementing the list above) might be:

- Contain spread of fire and smoke to a single compartment to minimise financial loss.
- Prevent premature collapse and, in high-rise structures, prevent progressive collapse.
- Identify and provide a level of fire resistance appropriate to occupancy/usage.
- Consider where active fire protection measures may complement passive protection measures.

The HSE's view is that most fires are preventable and their website spells out legal responsibilities on employers (and/or building owners or occupiers) to carry out fire safety risk assessments and keep them up to date. Based on such assessments, employers must ensure that adequate and appropriate fire safety measures are in place to minimise the risk of injury or loss of life. The basics are to identify: what could cause a fire; who might be at risk, and how such fires might be controlled. Note again, the legal obligation is only set out to control the risk to occupants; commercial matters are excluded, although it may well be prudent to consider them. An understanding of all these matters can be gained from a study of past incidents and the CROSS database includes reports on matters related to fire safety.

9.3 Some historical fires

9.3.1 Introduction

Everyone is aware of events such as the burning of Ancient Rome or the Great Fire of London (1666). The latter event lasted three days and destroyed much of the medieval city, some 95% of its dwellings and 99 churches. It is thought there were few deaths; rather the event is recalled for its scale and consequences. The fire is traditionally thought to have started from a bakery so then, just as now, cooking was a common initiation scenario. The devastation was made worse by winds fanning the flames and by a lack of firebreaks, which were the period's main measures for fire containment. In the Moscow fire of 1812, following Napoleon's invasion, 75% of the old city was destroyed.

An indication of what might happen in modern cities can be gleaned from the Great Chicago Fire of 1871, which destroyed about 9 km^2 of the city and made 100,000 residents homeless, or even more recently from the effects of bombing in the Second World War. Firestorms developed in London during the Blitz, and also in Hamburg, Dresden and Tokyo. In those events, victims numbered tens of thousands. A firestorm is a conflagration attaining such intensity that it creates and sustains its own wind system. Modern examples occur in regional fires such as those described in Chapter 10 (Natural disasters) and the devastation they cause graphically illustrates the need for alertness and control (see Oakland firestorm, Section 9.1). An extreme example of modern conflagration might be that in Kuwait after the first Gulf War of 1991. The retreating Iraqi army deliberately ignited 700 oil wells and it took nearly ten months to bring the fires under control. The devastation was horrifying, with the pollution visible from space. Losses were about $1.5 billion.

Another huge event in London was the 1834 burning of the old Houses of Parliament, famously witnessed and painted by Turner. The blaze started in chimney flues overheated by burning heaps of scrap tally sticks. The spread of the fire was rapid, developing into the city's biggest conflagration since the Great Fire. The collection of buildings was known to be a fire risk, having a large fire load (lots of combustible material) plus features such as wooden panelling and furnishings which would promote spread. The sequence is thought to have been from the overheated flue brickwork igniting supported joists and so on, a chain reaction. Firefighting services existed but were incapable of tackling so large a blaze. During the event, priceless historical records, and buildings, were lost. In terms of general safety principles, if the consequences of a catastrophe are that high, then prevention measures need to be pitched accordingly (see also Section 9.4.2 on fires in historic buildings).

The 1834 London fire was nothing new. Fire problems had been a city and building plague from time immemorial and efforts to reduce the risk had been the spur behind some of the earliest developments in structural engineering during the Industrial Revolution. In 1803, the original cotton mill at Belper in Derbyshire was destroyed by fire. To reduce the risk in the new building, columns and beams were made of cast iron whilst the floors between were brick arches interlinked with wrought iron ties. The whole concept was promoted as a 'fire-proof building'. Likewise, the Albert Dock in Liverpool (designed by Jesse Hartly, and opened in 1846) was built in a similar manner and claims to be the first non-combustible warehouse system in the world. Safety was really needed since such buildings constituted a potentially huge fire risk with their large stored volume of cotton, molasses, rum and the like. Even today, warehouses are classified as a 'special case' in potential insured losses.

9.3.2 Theatres

Old theatres were a fire risk: stages included flammable scenery yet had to be illuminated by naked flame. Interiors were often lit by gas. Stage limelight was first used in the Covent Garden Theatre in 1837 and thereafter was widely adopted around the world and obviously acted as a potential ignition source (limelight is an intense illumination created by playing an oxyhydrogen flame onto a cylinder of quick lime). Additional risks existed because many theatres had mainly wooden interiors and clearly there were potential problems in audience evacuation. In 1794, the Drury Lane Theatre, London introduced the first fire safety curtain between stage and audience, a precaution which would eventually become a statutory demand. Drury Lane also had a large water tank on its roof to extinguish any fires in the stage area.

In 1887, the Exeter Theatre Royal was gutted by a fire that occurred during a performance. The source was backstage, when gas lighting ignited gauze. Panic ensued, and with an inadequate number of exits, there were 186 deaths, with only 68 bodies ever being recovered. Another bad fire occurred in 1903 at the Iroquois Theatre, Chicago. An afternoon matinee was crowded out with an estimated 2,000 people in the audience. Standing room areas were so packed that some patrons sat in the aisles, blocking exits. During the performance, an arc light shorted and its sparks ignited a muslin curtain; flames spread quickly over the several thousand square feet of highly flammable scenery. A separating fire curtain was present but snagged on lowering; later tests suggested it would have been of little value even if lowered. Escape was hindered partly by exits being blocked and partly because people could not open the doors; many were crushed and trampled in the chaos. Dancers and performers backstage escaped to the rear but as they did so, air rushed in, feeding the fire and making it even bigger. This effect was compounded when the large doors provided to accommodate scenery entry were opened, thence providing a through route for yet more air, which created a fireball.

The Iroquois Theatre had been described as 'fireproof' but post-incident investigations found many deficiencies:

- inadequate provision of exits; blocked-off stairways and confusing exit routes;
- skylights above the stage roof, which were intended to open automatically during a fire to vent the heat and smoke, were fastened closed;
- lack of testing of the safety curtain such that its snagging in use ('on demand') was not detected;
- no extinguishers, sprinklers, alarms, telephones, or water connections were present;
- dismissive and uninterested management.

After this disaster, three major changes were introduced:

- implementation of panic bars; first invented in the UK after the Victoria Hall disaster of 1883 (Section 9.9.2);
- ensuring that all doors in public buildings had to open outwards in the direction of emergency egress;
- raising and lowering the fireproof asbestos safety curtain before each performance as a test and lowering it afterwards to separate the audience from any fire on the stage.

Theatre fires seem to cause continued problems. In 1774, Venice's Opera House burnt down and had to be rebuilt. It opened again in 1792, renamed 'La Fenice' (the Phoenix) in reference to its survival. In December 1936, it was again destroyed by fire, rebuilt and reopened. In January 1996, it was completely destroyed by fire. Arson was suspected, and two electricians were found guilty of starting the blaze.

The theatre blazes described above illustrate how risks and preventative rules have evolved by learning lessons. Some more modern fire disasters can be described to highlight further lessons.

9.3.3 Summerland, Isle of Man, UK, 1973

Summerland was a leisure centre capable of accommodating 10,000 tourists. Its street frontage and part of the roof were fully clad in 'Oroglas' (a transparent acrylic glass sheeting). One day in 1973, a fire started, ignited by boys smoking in a disused kiosk alongside. The kiosk was set alight and fell against the leisure centre wall, which had a bitumen-coated surface, and the fire then spread to soundproofing material, which ignited the highly flammable Oroglas. Flames spread very rapidly across the whole surface (the development can be seen on the Summerland fire video (2013)). As the sheeting melted, it allowed more air to be sucked in and the building's open plan allowed the air to flow freely, so feeding the fires. Melting material fell, causing yet more fires and injuring those trying to escape.

Emergency management appears to have been inadequate as there was no call to the fire services and no orderly evacuation. Subsequently about 3,000 occupants started to panic, a situation made worse by some escape doors being locked. All this led to congestion at the free exits and some deaths by crushing. During the fire, power supplies and emergency generators failed (again a flaw in design called 'failure on demand'). Altogether, 50 people died and the centre's structure was seriously damaged. In the ensuing public inquiry, deaths were attributed to misadventure, whilst the use of flammable building materials was condemned.

9.3.4 Woolworths' fire, Manchester, UK, 1979

The Woolworths' fire in Manchester city centre exemplifies the dangers of smoke.

Woolworths was a multi-storey department store and the fire started in its furnishing department on the second floor. Initiation is believed to have been from a faulty electrical cable which had furniture stacked in front. Unfortunately, this furniture incorporated flammable polyurethane foam, which gave off highly toxic smoke when alight. Some of the dead were found close to exits where they had presumably been overcome by fumes or were unable to locate escape routes in the smoky conditions.

During the fire, the second floor was gutted while the third floor above suffered severe smoke damage. The ground, first and second floors all suffered extensive damage from extinguishing water. No calls to the fire brigade were made by the store itself, so clearly there were weaknesses in emergency procedures. Ten shoppers and one staff member died. Polyurethane foam has since been banned from household furnishings.

9.3.5 MGM Grand Hotel, Las Vegas, USA, 1980

Another fire highlighting that smoke effects can be worse than flame is an incident which occurred in the MGM Grand Hotel. The fire killed 85, mostly through smoke inhalation and carbon monoxide poisoning, while a further 650 were injured. The fire started in one of the hotel's restaurants and spread to the upper part of the casino. Smoke dispersed upwards through vertical shafts (lifts and stairwells) and through seismic movement joints towards the top floor. Many escapees had to be rescued off the upper floors by helicopter.

9.3.6 Valley Parade stadium fire, Bradford, UK, 1985

This event was the worst fire disaster in the history of English football: recordings can be seen on YouTube. During a game a small fire started below a crowded stand. Ignition was probably due to a discarded match, with the fuel being a gross accumulation of rubbish that management had failed to clear. Fire progress was rapid and in less than 4 minutes (promoted by wind) the whole stand was ablaze. Its wooden roof (impregnated with asphalt and bitumen) also ignited. Fans were trapped and suffered from burning and molten roof material falling onto them. There was panic and some trying to escape to the rear had to break down locked doors to get out; no stewards were on hand to unlock the exits. There was free access to the front onto the pitch and this offered a successful escape route, which was lucky since other stadiums had barriered their pitches off to deny access to hooligan fans. Nevertheless, 56 fans died and 265 were injured.

9.3.7 British Airtours flight 29M, Manchester, UK, 1985

This was a bad fire that occurred on a holiday flight whilst the plane was preparing for take-off. The fire started due to an engine failure and take-off was aborted. Although evacuation started, only 82 people managed to escape, leaving 54 inside who died mostly due to smoke inhalation. Investigations highlighted problems with the seats and plane surfaces (as sources of fuel and fumes) and a need for clearer evacuation rules. Other effects exacerbated the loss of life: the crew thought a tyre had burst (rather than realising there was a fire) so the pilot braked slowly, but this time lapse then allowed flames to spread. The plane also stopped in a direction whereby wind fanned the fire which then spread rapidly through the fuselage, whilst bottlenecks in the layout hindered escape. Subsequently, detailed studies were made of routes using escape dynamics, leading to an introduction of over-wing exits. Recommendations were also made for fire-resistant seat covers, fire-resistant wall and ceiling panels, floor lighting and more fire extinguishers.

9.3.8 King's Cross, London, UK, 1987

The King's Cross underground fire of 1987 was a turning point in fire history – a traumatic event. Below the streets a warren of intersecting tube lines were reached by escalators. In the early evening, a fire was reported by passengers; staff investigated, then called the fire brigade. But the fire was ablaze below a wooden escalator, difficult to reach with extinguishers, so a quick decision was made to evacuate the station. At this stage, the fire was still quite small, but within about 15 minutes flashover[1] occurred and flames reached up out of the escalator shaft due to the chimney effect,[2] filling the ticket hall above with intense heat and thick smoke so rapidly that most people in the hall were killed or injured. With the fire in this upper location, passengers lower down became trapped, though many managed to escape via other tube lines still running. For those remaining, once again escape in some directions was prevented by locked gates. Altogether 31 people died (including one firefighter) and 100 were hospitalised.

After the King's Cross fire a public inquiry found that the cause was most likely a discarded match dropped down the side of the escalator. Arson was ruled out. Before this, smoking had already been banned but the ban was weakly enforced and widely ignored. This time smoking was firmly prohibited, with the prohibition enforced and publicly accepted. Investigators found a build-up of grease under the escalators, which was a known risk but judged difficult to ignite and slow to burn. However, the grease under the particular escalator that caught fire was heavily impregnated with fibrous materials and when tested with dropped matches, it did ignite, and the fire did spread. Thereafter, it was shown that the metal sides of the escalator contained the flames and funnelled hot air along and upwards to the extent that wooden treads several metres ahead were heated, promoting flashover. The inquiry also found evidence of poor emergency response management and found that firefighters' work had been

1 Flashover: a flashover is the near-simultaneous ignition of combustible material in an enclosed volume. When confined, fires initially grow at a rate limited only by fuel availability. Thereafter, in a confined space, development is checked by the rate air can reach the flames. Nevertheless, as materials get hotter, they reach a point where they will auto-ignite (including ignition of fumes given off by the heated solids); then there is a sudden intense fire growth.

2 Chimney effect: this effect is significant in many fires; buoyancy causes hot gases to rise and this sucks air in from below, which serves to intensify the fire.

hampered by their lack of familiarity with the complex underground layout. Finding their way was a lot harder because there were no dedicated maps showing the layout, location of key exits and key hazards. After the fire, improvements were made including installation of better detection equipment and removal of all wooden escalators across the underground system.

9.3.9 Piper Alpha, UK, 1988

Piper Alpha was a North Sea oil production platform. In July 1988 it endured an enormous explosion and fire with the images forming some of the most iconic 'disaster' pictures of all time. The catastrophe was described in Chapter 6 (Section 6.6) with a focus on causation by poor control of change (amongst other factors). The catastrophe is also covered in Chapter 11.

9.3.10 Ostankino Tower, Moscow, Russia, 2000

This tower is 540.1 metres tall, making it one of the highest freestanding structures in the world. In August 2000, a fire broke out in its electrical equipment, started by a short circuit (located about 459 metres above ground level). The fire necessitated evacuation but three firefighters died in attempts to extinguish the blaze, and one lift operator died when his cabin crashed to ground due to the fire. Losses were substantial since failure of the fire suppression system permitted a great deal of destruction. Given the height of the fire, tackling it was difficult. Heavy equipment, including specialist fire extinguishers, had to be hauled up by hand. The tower top also tilted, raising fears that it might collapse.

9.3.11 World Trade Center, New York, USA, 2001

The terrorist attack on the twin towers of the World Trade Center in New York must be one of most well-known disasters of all time; there is no shortage of images of the hijacked planes colliding with the towers and the aftermath. For engineers, the performance of the towers and their (delayed) failure mode is of immense practical importance. The towers responded remarkably to initial impact but were eventually brought down by the fire, which was intensified by the load of burning jet fuel from the planes. The South Tower collapsed after burning for about an hour whilst the North Tower collapsed after burning for about 1.75 hours. In both towers, intense heat weakened steel frames, already damaged by the planes' impact. Ultimately, 2,753 deaths were verified, these included 343 firefighters. In terms of fire the issues are:

- firefighting techniques in high-rise structures;
- difficulties of escape from high-rise structures;
- deaths from heat and smoke;

- firefighter deaths;
- the effects of heat on structural form/members;
- the performance of fire protection;
- mode of collapse.

9.3.12 Buncefield, UK, 2005

The Buncefield fire was caused by the igniting of the Hertfordshire Oil Storage Depot, at the time the fifth largest depot in the UK, containing 60 million gallons of fuel. The depot had 20 large tanks and the fire and explosion started in one whilst it was being filled. The tank overfilled and this resulted in a fuel–air explosion of such intensity that it progressed to all the other tanks, destroying the entire site and burning all the fuel. The fire was the largest of its kind in peacetime Europe, the biggest UK explosion since Flixborough (see Chapter 11), and it generated an enormous smoke plume. Part of the cause seems to have been a faulty level gauge and switch system which did not alarm as the tank overflowed. This is a good example of the relationship between security of a control system and 'safety' (see Chapter 8).

Following the fire, problems were encountered with water pollution linked to firefighting foam chemicals seeping into the ground.

9.3.13 Summary of historical events

The above events might seem randomly distributed over time and over many types of structure. Yet there are common themes:

- the nature of fire initiation from many causes which might be accidental, mechanical failure or deliberate;
- the need to always accept that a fire might start and to have a strategy to contain it;
- the role of different types of material in promoting spread of flame;
- the role of materials in smoke generation and the role of the spatial geometry in smoke dispersal;
- the role of external air flow in promoting or intensifying the fire;
- the need for an escape strategy taking account of exits, the masking of routes in darkness and smoke and the possibility of crowd panic;
- the need for an overall strategy and hazard management, especially in complex plant;

- the need to understand exactly what is happening in an emergency;
- the need to protect firefighters and to consider how a fire might be fought;
- the need to understand structural performance in fires;
- the need to anticipate the worst and have an escape strategy.

What should also be learned is that many of the events that took place during the fires described above would be hard to predict without background. Yet with background, the lessons appear obvious in hindsight. It is especially important to observe the disastrous consequences of chains of events, one problem leading to another.

9.4 Broad groups of fires

9.4.1 Domestic fires

There are always domestic fires caused by cigarettes, faulty electrics, cooking, and sadly, arson. Things have improved since the 1950–1970 era, with a reduction in the use of coal fires, paraffin heaters and flammable upholstery. Tighter building control has also helped, along with the introduction of smoke alarms, but there are still too few of these in use. There is also a steady number of domestic gas explosions. Altogether, community fire safety remains an important role for firefighters. More stringent laws regulating factories, offices and hotels have been introduced as result of lessons learned and these should be enforced.

Because the risks of a fire starting can never be eliminated, it remains important to enforce building standards and to continue appraising lessons. A recent cause of concern has been the fire resistance of timber-framed dwellings. A few major fires in these have happened during construction: the cause often being arson. Such fires can develop and spread at extraordinary rates, rapidly bringing the structure to collapse. Moreover, the heat generated has been such as to spread the damage to adjoining properties where firebreak distances have been too small. The biggest risks seem to be during construction, or where houses have had external timber cladding.

A severe illustration of what can happen in high-rise blocks is given by the London Grenfell Tower fire of 2017 (Box 9.1). The form of the Grenfell fire was not unique, not even within UK experience, and many other structures are similarly clad with flammable material. Another fire in Barking (London) in 2019 highlighted the risks from wooden balconies which offer a fire route up the outside. At Barking, the whole six-storey facade was engulfed in flames. These scenarios could have been foreseen, as illustrated by precedent. In 2010 in Shanghai, there was bad fire in a 29-storey apartment block, which was undergoing refurbishment and so was scaffolded out over its exterior. Sparks from welding ignited the scaffolding and the fire spread upwards over the whole of the building height, progressing into its interior. Due to the height it was impossible to hose water to the top. Helicopters sent in to aid rescue could not operate because of the smoke. It took 90 fire engines and several hours to contain the blaze during which 53, mainly elderly, people died.

9.4.2 Historic buildings

Introduction

We have a huge legacy of historic structures and these pose particular risks, partly because such buildings obviously will not meet modern concepts of fire resistance either in layout or materials. Moreover, there is conflict in that it is hardly sensible to strip a room of its historical timber panel lining, even if this makes for sensible fire protection. It is also the case that the fabric and contents of these buildings represent an incalculable loss if they are destroyed. Unfortunately, there have been major fires in historic buildings; protecting them poses challenges. The fires at York Minster, Hampton Court and Windsor Castle are examples of what can happen, and there have been many others.

York Minster fire, UK, 1984

There have been at least five fires in the Minster, with two being very serious. During one night in 1829, a deranged man set fire to a heap of cushions and prayer books. The resulting blaze melted roof lead and cracked some of the stonework. Tragically, the medieval choir stalls, organ and pulpit were all destroyed. In 1984, the whole timber roof of the south transept was destroyed by a spectacular fire, ignited by lightning. The fire brigade was automatically alerted and soon arrived and despite molten lead and debris crashing down, there was an opportunity to salvage priceless artefacts whilst the fire was still spreading. Though the fire brigade was soon at the scene, firefighting was severely hampered both by the difficulties of getting up to the roof and then by the roof's configuration, which rendered it very hard to spray with water, making hosing largely ineffectual. With the fear that the fire would spread, perhaps causing a chimney effect up the central tower, a decision was made to use water jets to bring about a collapse of the burning roof timbers onto the transept floor. The entire roof was lost but the fire was contained and the stonework (and East Window) remained intact.

The Minster roof did have a lightning protection system, with conductors connected to the roof's lead sheathing. But there were electrical circuits in conduits and metal control boxes running beneath the lead, all earthed, but fixed to wooden roof members. There is the suspicion that in a lightning strike, a heavy spark occurred between the roof surface and nearby electrical services sufficient to start the fire in the wooden parts.

Hampton Court Palace fire, UK, 1986

Hampton Court Palace, on the outskirts of London, is one of Britain's top historic attractions. But it has many uses and part of it was given over to 'grace and favour apartments'. The 1986 fire broke out in one of these apartment bedrooms, but wasn't discovered for about four hours. Voids behind old panelling and through floors provided routes for the fire to travel undetected by the alarm system. As a consequence, the apartments (designed by Christopher Wren) were gutted, and the Cartoon Gallery roof left crumbling. Damage costs were about £5 million. A need to rescue artwork was a feature of the incident response.

Uppark House fire, UK, 1989

Uppark House is a 17th century 'stately home' owned by the National Trust. It was undergoing refurbishment when a fire was started by workmen repairing the roof's lead flashing. Because the fire developed during opening hours, it was spotted rapidly, so staff managed to carry out works of art and pieces of furniture from lower floors. Meanwhile, the garret and first floors burned and collapsed onto the lower floors and unfortunately their contents were lost completely.

Windsor Castle fire, UK, 1992

During this fire, some of the castle's most historic parts suffered severe damage. The fire began when a spotlight ignited a curtain. Alarms did detect the fire, but it spread rapidly and a good part of the state apartments were soon ablaze. Within an hour St George's Hall was on fire. The castle has its own firefighting team, but they soon called for reinforcements; at its peak, the blaze was being tackled by 39 appliances and 225 firefighters. During the event, floors and the roof of St George's Hall collapsed. As well as fighting the fire, huge efforts had to be made to rescue irreplaceable works of art and the logistics were such that vast numbers of removal vans were required. Major losses were to the castle's fabric and restoration costs reached £36.5 million.

Summary of fires in historic buildings

There is no shortage of examples of fires in historic buildings. Causes can be arson, or carelessness, but many seem to start during refurbishment and thence spread rapidly. So, the lesson is obvious for any site work: prepare for fires starting and for getting out of hand. Early detection is vital. A common theme seems to be the need to remove precious items at short notice: it is not for nothing that in historical times, key documents and money were stored in portable chests. Another theme seems to be the difficulties of firefighting, even if response is rapid, so detailed contingency planning is suggested. Nothing is immune: in 2019 the much-loved Notre-Dame cathedral in Paris caught fire during refurbishment and almost collapsed. There were strong parallels with the York Minster incident.

9.4.3 Tunnel fires

Introduction

Tunnel fires create special problems of escape, of smoke and of presenting conditions where fire problems are exacerbated. Many tunnels are used for transport of some kind and thus on road tunnels, all manner of crashes can take place that involve vehicles with fuel loads, and which may be made worse by the cargo carried. Several incidents show the need to think through carefully the entire fire safety strategy to minimise overall consequences.

Channel Tunnel fires, UK/France

The Channel Tunnel is 31.4 miles (50.5 km) long and consists of dual traffic tunnels with a smaller service tunnel running in between. This service tunnel is pressurised to prevent smoke entering from any fire in the running tunnels. There have been four fires in the tunnel since it opened. In 2008, there was a freight train fire about seven miles from Calais. The 32 people on board were led to safety, some suffering minor injuries after inhaling smoke. The blaze lasted 16 hours. An earlier fire in 1996 had been more damaging. It occurred on a train carrying trucks (as with all the other fires). The affected train stopped about 12 miles into the tunnel and then both it, and its passenger coaches, became covered in thick smoke. Heat from the fire was intense, burning out overhead power lines and thence immobilising the train. The subsequent localised fire caused very severe damage to the concrete tunnel lining over a length of about 500 metres, with the thinnest area being left just 200 mm thick. No one died, but passengers did suffer from smoke inhalation. The tunnel was not fully reopened for about six months and financial losses have been put at about £200 million.

The 1996 incident highlighted several failures in hardware and procedures.

Physical responses

- The intense smoke prevented crew members from seeing the service tunnel doors.
- The supplementary ventilation system was turned on but ran for several minutes with its blades set incorrectly, so did not clear smoke. When the blades were set correctly, the smoke cleared sufficiently for the passengers and crew to evacuate.
- Close to the fire, services were destroyed, including power cables, communications and lighting.
- The French crossover doors and one of the piston relief duct doors failed to close properly allowing smoke to enter the other running tunnel.

Technically, for structural engineers, the performance of the high-strength tunnel lining was of considerable interest. Fire-induced temperatures are thought to have exceeded 1,100 °C, producing severe thermal shock within the concrete. Explosive spalling took place where the concrete strength was over 100 MPa, a response affected by the material, its moisture content and the thermal gradients. Originally the tunnel lining design did not account for fire resistance and similar spalling had taken place in the Great Belt Tunnel fire (Denmark 1994) and in the Tauern Tunnel fire (Austria 1999).

Firefighting

- Initially there was confusion as to the train's location.
- The firefighting water supply was restricted by leaking pipework and the number of available jets was reduced until an engineer reconfigured the valves.
- Because of the conditions, each shift of firefighters could only work for short periods before returning to the service tunnel.

Emergency Response

- There was a concept of an 'unconfirmed alarm' meaning that that the incident was not at first treated seriously.
- Later, control centre staff were overwhelmed. They had not been sufficiently trained for an emergency and were using procedures and systems that were too complex. One consequence was the late switching on of the supplementary ventilation system and of running it incorrectly for several minutes.
- The original strategy in the event of fire was for trains to continue out of the tunnel. But in this incident, the fire had damaged the train, forcing it to stop and then consequently the overhead power line failed just four seconds later, meaning the train could not then be moved at all (cf. 'Chains of events' as discussed in Chapter 5, Section 5.5).
- In the aftermath, liaison between Eurotunnel and the emergency services was improved. Joint exercises and exchanges of personnel between the British and French fire brigades were held to ensure experience of the other's operational procedures. Communications were also improved.

Great Belt Tunnel fire, Denmark, 1994

The Great Belt fixed link includes twin rail tunnels about five miles (3.1 km) long. The tunnel lining was made of precast (high performance) segments. During construction, a fire broke out on one of the tunnel boring machines, probably due to an oil spillage. The consequence was severe damage to the lining, with up to 270 mm being removed by layered spalling.

Mont Blanc Tunnel fire, France, 1999

The Mont Blanc tunnel is 7.2 miles (11.6 km) long. In 1999, 39 people died when a truck carrying flour and margarine caught fire. Dense smoke soon filled the tunnel and there was huge heat from the burning cargo (over 1,000 °C). The smoke was poisonous, containing both carbon monoxide and cyanide, and it caused several problems. First, the tunnel ventilation system drove toxic fumes down the tunnel faster than anyone could escape. Second, and worse, on the day of the incident, the natural air flow was from the Italian to the French side, yet the Italian authorities pumped more air in so providing more oxygen for the fire and pushing the poisonous fumes throughout the tunnel length. Moreover, these fumes depleted oxygen downstream and caused vehicles there to stall, including fire engines, which had to be abandoned. All the deaths were on the French side. The intense heat melted the wiring and plunged the tunnel into darkness. Burning fuel escaped and spread down the roadway surface causing tyres to burst and fuel tanks to explode.

The tunnel was provided with refuges, but these were only rated for four hours and the fire persisted for 53. Drivers reaching refuges were overcome. Fifteen firemen were also forced into refuges; of these escapees, one died and the other 14 were left in a serious condition. It was over five days before the tunnel cooled down.

After the event, there was serious criticism that a controllable incident had evolved into a disaster. The closest smoke detector was out of order and there was poor planning of responses because the two states did not take a coordinated approach. French emergency services did not use the same radio frequency as those inside the tunnel.

The tunnel underwent major changes in the three years it remained closed after the fire. Better detection equipment was added, extra security bays were provided, and a parallel escape shaft constructed with clean air flowing through. A mid length fire station was added complete with double-cabbed fire trucks. People in the security bays have been provided with control centre video contact for better communication. There is better management via a cargo safety inspection system created on each side. Financial losses to the economy at large have been put at about 2.5 billion euros.

Kaprun funicular railway fire, Austria, 2000

This incident took place within an inclined tunnel. Skiers travelled up the mountain on a funicular railway running through the tunnel, which was about 3 km long. An electrical fire started below the floor of the rearmost carriage and its heat melted plastic pipes carrying flammable hydraulic fluid. Apart from acting as fuel, loss of the fluid caused loss of braking and an automatic stop. However, because the hydraulics had been lost, the driver was unable to operate the doors, so the passengers were trapped. Furthermore, the fire then burnt out the power cable causing a loss in communication and a blackout. The driver did eventually manage to open the cabin door manually, but by that time many passengers had lost consciousness in the toxic fumes. Many of those who did get out made the mistake of fleeing upwards away from the fire but became asphyxiated by rising smoke.

The whole fire was made worse by the chimney effect linked to the tunnel's inclination as air was sucked in at the bottom, fuelling the flames and pushing poisonous black smoke upwards. As the first smoke reached the top, escaping workers accidentally left exit doors open which only exacerbated the problem. The train was destroyed and about 150 people died.

Apart from the being a fire tragedy on a large scale, the Kaprun incident is another classic example of the benefits of conducting a complete HAZOPs analysis to assess cause and effect within a system (see Chapter 13).

St Gotthard Tunnel fire, Switzerland, 2001

The St Gotthard Tunnel is very long tunnel, at about 17 km in length. In 2001, two trucks collided inside it, leading to the deaths of 11 people through smoke inhalation. Intense heat made firefighting difficult and parts of the tunnel roof collapsed. Temperatures are thought to have reached over 1,000 °C.

Summary of fires in tunnels

Fires in tunnels are a specialism within the topic of fire engineering as whole. Across the reported catastrophes, there is commonality in the way smoke disperses, influenced by confinement and by the ventilation system; commonality on problems of refuge and commonality on difficulties of firefighting systems failing on demand. The overall strategy of fire protection and response should have a profound influence on the overall tunnel design e.g. by provision of separation and escape tunnels, ventilation and emergency procedures. Although tunnels are a specialism, there is similarity with the difficulties encountered at King's Cross and the generic problem of all buildings which have deep basements, with their potential need to deal with smoke extraction. Above all, the complex interaction of cause and effect (chains of events), notably in Kaprun, the Channel Tunnel and Mont Blanc fires, are textbook examples of the value that HAZOP-type analyses can bring to understanding a problem and so shape the overall design (Chapter 13). HAZOPS are especially valuable when guided by observations of what has gone wrong in similar systems. The topic of tunnel lining performance in fires is of direct interest to structural engineers (Beard and Carvel, 2004).

9.4.4 Fires on construction sites

Introduction

Every year there are lots of fires on construction sites, raising significant commercial and safety issues. Estimates for the UK are of many events each month with corresponding huge costs (many millions of pounds). With such losses, obtaining insurance cover requires a demonstration of very good fire safety standards. Site fires can be particularly hazardous for reasons such as:

Ignition: there are potentially many sources of ignition associated with building activities, such as hot works including welding or the hot laying of asphalt. Arson is a risk that must be seriously considered (see also comments on fires occurring during refurbishment of historical buildings).

Fire load: during construction, lots of materials are collected together, including rubbish, any of which may act as a fire load. Flammable materials such as timber, insulating materials and soft furnishings may be abundant and are frequently present. LPG has been involved in many serious site fires and explosions, particularly where there have been gas leaks in confined areas. Strict precautions are required where LPG is stored and used.

Fire spread: buildings will naturally be unprotected at various stages; compartmentalisation will be incomplete because doors and windows will be absent and fire stopping may not have been applied.

Means of escape: access is an issue during construction. Likewise, facilities for rapid egress are also going to be problematical. Stairways may be incomplete; lighting and signage will be absent and escape routes confused. On high-rise structures, even if there is adequate warning of fire, the time needed to escape from upper levels to a place of safety may be considerable.

Firefighting: standard provisions for firefighting, including water supplies, may be incomplete. The 2018 fire which destroyed the famous Glasgow School of Art during its refurbishment (ironically after a previous fire) broke out just two weeks before sprinklers were due to be installed. On high-rise structures over 30 metres tall, fire ladders will not reach to the top. So, fire rescue becomes much more problematical and because compartmentation may be imperfect, fire spread might be rapid.

As a generality, any work activity generating heat, sparks or flame should be very carefully controlled, which includes keeping the area clear of combustible materials, ensuring that fire extinguishers are present and maintaining a watch. It is beneficial if there is a 'permit to work' system operating (which includes activating a fire watch).

Two examples are of fires occurring during construction/refurbishment are given in Boxes 9.2 and 9.3.

Other significant construction site fires have been those at Broadgate (Box 9.13) and in the Great Belt tunnel (Section 9.4.3)

Summary of fires on construction sites

As the introduction states, there have been many very serious fires on sites during construction or refurbishment projects, and it highlights a few reasons. Causes of ignition can be many: a blow torch at Uppark House; a lamp on a curtain at Windsor Castle; welding sparks at Düsseldorf; and arson can never be ruled out. Another lesson

Box 9.2 Minster Court, London, UK (1991)

This fire occurred during fitting out of the new London underwriting centre. The fire started from some ignited rubbish in an atrium on the upper ground floor, then rapidly spread through the atrium to the upper building. Although quickly discovered, portable fire extinguishers were inadequate to control the blaze. The fire was probably made worse by the atrium acting as a chimney and the rising heat ignited the central escalator plus a heap of scaffolding planks. Fire then reached the upper floors and blew the atrium roof off. Costs were in the region of £105 million.

Box 9.3 Düsseldorf Airport, Germany (1996)

A severe fire broke out in the passenger terminal of the airport, requiring the attention of about 1,000 firefighters. Initiation was by welding sparks igniting large amounts of flammable polystyrene insulation within the ceiling and the fire spread widely before detection. Because of this, smoke began to emerge from ceiling vents and this was followed by the fire running out of control. A flashover took place involving about 100 m^2 of ceiling, and within seconds the burning produced a huge build-up of black smoke. The heat also melted PVC cables, releasing toxic vapours. The airport was fitted with neither sprinklers nor fire doors. Firefighting was inadequate since many within the airport fire brigade had only been trained to address aviation accidents. Other firefighters had no experience of polystyrene fires and necessary supplies were missing. Fire and smoke rendered Terminals A and B unusable, and the damage costs were estimated at 0.5 billion Euro). A total of 17 people were killed and many were injured.

is that once a construction fire starts, its spread may be rapid and the consequences disproportionate partly because firefighting systems may be incomplete. The only safe position is to take the risk of fire seriously, to follow careful procedures for all hot working and to follow the guidance of the Fire Protection Association (Fire Protection Association, 2015). The HSE website provides considerable resources on site fire prevention advice (HSE, undated).

9.4.5 Fires in industrial plant

Introduction

There are clear dangers from fire in any industrial plant. The fire loads might be huge, the released products and fumes might be poisonous and released over areas far outside the plant boundary. Access to the plant might be hazardous and specific firefighting skills might be required related to the plant type. Obvious examples are the major fires

Box 9.4 San Bruno pipeline explosion and fire, California, USA (2010)

This fire involved a 76 cm diameter natural gas pipeline, which exploded in flames within a residential neighbourhood. Eye witnesses reported a huge wall of fire which quickly engulfed nearby houses. Nine people died.

There have been several similar incidents, often caused by illegal attempts to tap off pipeline oil.

at Piper Alpha (Section 9.3.9), and Buncefield (Section 9.3.12). Chapter 11 describes the Windscale plant fire of 1957, which was Britain's worst nuclear accident. It might be recalled that a huge impetus for fire safety in the UK in the 19th century stemmed from the need to control losses in industrial premises, including mills and warehouses.

Certain fires in factories in other countries have resulted in major loss of life. The Ali Enterprise garment factory in Karachi caught fire in 2012 when a boiler exploded, igniting stored chemicals. There were between 300 and 400 workers inside, but many exit doors were locked, and many windows were barred. A total of 257 people died from suffocation and more than 600 were seriously injured.

Pipelines carrying flammable liquids or gases can also be fire hazards. Box 9.4 describes a fire originating from a gas pipeline.

In October 2013, there was a severe fire at the Walker's crisp factory in Leicester. It appears to have started in a fryer, and it created thick, black smoke, making for challenging firefighting conditions. Earlier that year, in July, in Smethwick, Birmingham, another factory was consumed by fire. It contained 100,000 tonnes of plastic recycling material and the blaze was the largest ever in the West Midlands. It produced a plume of smoke 2 km high and caused about £6 million of damage. This fire seems to have been ignited by a stray Chinese lantern (firework) that landed on the roof; thereafter the fire had taken just nine minutes to take hold.

Summary of fires in industrial plant

Fires within industrial plant can be some of the most catastrophic, posing dangers to the public well beyond site boundaries. Full risk assessments can only be made with the knowledge of hazards that are particular to the plant's purpose. A proper fire prevention strategy therefore requires full co-operation across the entire design team.

9.5 Ignition

It should be clear from any of the preceding paragraphs that a myriad of ignition sources is possible. Chapter 6 describes how fire problems on the new Boeing Dreamliner were traced back to a faulty lithium battery. Government studies and research by the Fire

Box 9.5 Torre Windsor office building, Madrid, Spain (2005) (Montalva et al., 2009)

This office building, 106 metres high, was built in 1979; in 2005 it was so badly gutted by fire that it partially collapsed. It is thus a good example illustrating how a relatively modern RC building might perform in a fire. The building was 30 storeys high and had a central RC core surrounded by six reinforced concrete columns and perimeter steel load-bearing mullions. Floors were concrete waffle slabs while strong transfer floors were included at the 3rd and 17th levels. The mullions were not fire protected and there were no sprinklers.

On the day of the incident, a fire was detected on the 21st floor, and it spread quickly throughout the entire building, leading to the collapse of the outermost steel parts of the upper floors (the mullions weakened in the heat). Initially the fire spread downwards to the 2nd floor and then it moved upwards to the top of the building, burning for about 19 hours altogether. Although local progressive collapses occurred, a total collapse was narrowly averted, probably aided by the intermediate strong floors; the structure did have an inherent robustness. The new staircase performed well. Nevertheless, the building was eventually demolished. It was later concluded that the fire spread through a combination of the lack of operational fire doors and fire stopping, and through radiant heat. Cladding failure worsened the situation by allowing the fire to spread between floors. Several floors burnt simultaneously, giving rise to very high temperatures as the size of conflagration was intensified by a lack of cooling from external air sources. The fire cause might have been an electrical fault or arson.

Protection Association (FPA) suggest that accidental fires in commercial buildings are mainly caused by faulty appliances or by equipment misuse. Yet about half of large fires at business premises were thought to be deliberate. Likewise, in domestic properties and schools, many fires are ignited by arson (in schools, arson is the most likely cause). The Grenfell fire is thought to have started from a faulty electrical device. That same cause was cited for the disastrous fire in the Torre Windsor office block (Box 9.5).

In some cases fire cause seems incredible. In 2009, the iconic Beijing Television Cultural Centre was consumed by a fire linked to firework displays for Chinese New Year. Some of the terrible nightclub fires that have occurred (Section 9.9.1) were ignited by pyrotechnic displays.

In other circumstances, the source might be thought of as unfortunate. The York Minster fire was started by lightning even though a conductor system was in place. But surely lessons can be learned from the subsequent investigation which suggested that wood might be ignited by sparks between metal pieces. The overarching lesson is that no matter what precautions are taken, an ignition is always possible.

9.6 Fire load

A fire needs fuel; hence the extent of damage, or the intensity of burn, is going to be linked to the amount of fuel present. The risks posed by an oil depot (as at Buncefield) are obvious as is the potential damage during a plane crash (Chapter 11). In houses, cooking fats and furniture are combustibles. At the Bradford stadium fire, it was burning rubbish that fuelled the fire and a flammable roof that allowed it to spread. In tunnels there will be motor fuel plus the risk of flammables in transported cargo. In other buildings, there might literally be nothing to burn. Any risk assessment must therefore consider the potential fuel load plus how that fuel load might react on burning; it also has to consider allied risks such as gas cylinders exploding.

Risk assessments need to be kept up to date with changes in society. Many bad historical fires have been in warehouses, and those in warehouses containing whisky or sugar have been spectacular (see Cheapside Street whisky bond fire, Box 9.6). This example is included mainly to illustrate threats to firefighters. Modern warehouses have increased in height and plan area and the logistics of retrieval demand large uncompartmented spaces; densely packed goods; high-bay storage and largely automated systems, with the types of goods stored changing daily. Potentially there are also huge volumes of flammable packaging. Collectively, all these factors increase the risk and/or consequences of fire.

9.7 Compartmentation and spread of flame and smoke

9.7.1 Compartmentation

The principle is simple: if there is going to be a fire, efforts should at least be deployed to contain that fire within bounds and stop it spreading elsewhere. The general idea has been around for a very long time with early planning laws controlling building separation (street width) or the floor area of warehouses and legislating on the

Box 9.6 Cheapside Street whisky bond fire, Glasgow, UK (1960)

This incident was at the time Britain's worst peacetime fire services disaster, and so illustrates the dangers firefighters face. A fire breakout was reported, causing a number of fire engines to race to the scene. The warehouse 'fire load' was huge, consisting of over a million gallons of whisky and rum under one roof, all stored in wooden casks. As the temperature of the fire increased, some of these casks ruptured, causing an explosion and the ensuing pressure wave pushed the front and rear warehouse walls out, causing large quantities of masonry to collapse into the street. This wall collapse killed 19 firefighters.

Box 9.7 Tooley Street fire, London, UK (1861)

This was major warehouse fire resulting in significant property loss. It led to the London Fire Brigade proposing that a cube of 60 ft (~19 metres) was the maximum size of compartment that could be safely fought (metricated to 7,000 m^3) and this has remained as the basic compartment volume size ever since. The limit of 60 ft was based on the height accessible by ladders and the horizontal dimension to a distance that could be fought with hoses.

boundaries between buildings to ensure some sort of barrier. In the late 19th century, the notion of a compartment (i.e. a defined volume) bounded by fire resisting surfaces emerged as a fundamental control strategy.

Compartment sizing stems from the Tooley Street fire in London (Box 9.7). Compartment boundaries need to prevent transmission of fire, radiated heat, and smoke or hot gases, at least for a finite amount of time, to allow any fire to be tackled and to prevent spread. But there are natural inherent weaknesses. All walls and floors are penetrated by doors, windows or services passing through. Hence these need to be controlled by, for example, using fire doors (with smoke seals), by fire stopping all surface penetrations, or by adding dampers inside any large ventilation ducts crossing the boundary.

One problem is the potential for hot gases or smoke to percolate through concealed spaces, which might be ducts, cavities, voids above ceilings and so on (see Hampton Court fire (Section 9.4.2) or the fire at Düsseldorf Airport (Box 9.3) or at MGM Grand Hotel (Section 9.3.5)). An added danger is that, as well as these passageways providing a ready route for fire spread, the spreading smoke can proceed undetected until emerging some distance away from the origin thus leaving occupants unaware of the danger for too long. To prevent such spread, cavity barriers are required. Unfortunately, barrier effectiveness can be undermined by poor workmanship. In other incidents, alterations have been made and unwittingly removed defences, or fire doors have been left wedged open.

Special care is required in residential blocks where individuals will be at risk from the activities of neighbours and where residents might be sleeping or unable to escape through infirmity (Box 9.8) (some bad fires have also occurred in prisons). As buildings increase in height, the problems obviously get worse (see fires in Shanghai (Section 9.4.1) and Torre Windsor (Box 9.5)). No fire can be more exemplary of the potential consequences than the Grenfell Tower fire of 2017 (Box 9.1). Much of the debate since that fire has centred on the fact that the supposed compartmentation of individual flats was inadequate. During the fire, occupants were advised to stay in their homes and await rescue by the firemen, the presumption being that compartmentation would ensure their safety. This transpired to be incorrect.

Box 9.8 Fairfield Old People's Home, Nottingham, UK (1974)

This fire was in a home for elderly persons and resulted in 19 deaths. There was extensive fire in cavities above suspended ceilings and this spread then affected parts of the building remote from the fire's origin. Together with Summerland, this fire prompted the introduction of controls on hidden cavities.

9.7.2 Spread of flame

Surface spread of flame is a key issue as shown at Summerland (Section 9.3.3) and Grenfell Tower (Box 9.1) and is controlled by selective use of resistant material. The key issue is the rate of spread, and dramatic rates can be readily appreciated from YouTube videos of Summerland or Grenfell. Figure 9.2 shows fire spread up the outside of a building. At a deeper level, air feed has to be considered with regard to rate of spread. Several instances are known of the effects of wind and its direction on the spread of regional fires but in the whole evolution of a building fire, features which might control the general air supply are of concern. Escape doors have to be opened, or they might be

Figure 9.2 Spread of fire up the outside of a building (Shutterstock 379205551)

opened to let firefighters in, yet this may have a downside, the adverse effect being to intensify fires, as in the Iroquois Theatre fire (Section 9.3.2). In tunnel fires such as the Mont Blanc incident (Section 9.4.3), there were adverse effects consequent on pumping in more air. And in the Kaprun tunnel and King's Cross incidents, the up-rush of air created chimney effects intensifying the blaze. Interestingly, one way of putting out an oil well gusher fire is to blast it: the explosion temporarily eliminates the oxygen feed and snuffs out the flames.

9.7.3 Spread of smoke

Shopping malls and atria present special problems. Malls can form giant compartments (violating the basic principles of separation) and if multilevel, fire and smoke can spread rapidly upwards from floor to floor both internally and externally. Persons on upper floors can thus be at immediate risk from a fire below, so raising the number of occupants put at risk. Recommendations intended to promote safe aspects of design and management in buildings containing atria are given in BS 5588-7:1997 and Hansell and Morgan (1994). The key issue is smoke control and, allied to that, ensuring evacuation safety of potentially large number of shoppers.

Basements present problems in terms of smoke passage and means of firefighting (see King's Cross, Section 9.3.8). Smoke will rise from any basement fire and make firefighting access downwards difficult. To reduce these problems, basement fires may be vented to allow escape of heat and smoke and so improve visibility, reduce temperatures and make search and rescue easier. Venting can be by natural means (via removable covers) or by using mechanical extraction, but clearly the extraction power supply needs protection (i.e. a common cause failure must be prevented).

Hospital fires, prison fires and hotel fires pose obvious risks in flame or smoke spread either because the occupants cannot escape or, as in hotels, they may all be asleep and unaware of the dangers.

9.7.4 Summary of compartmentation and spread of flame and smoke

Section 9.2 summarised some aims of fire strategy. The policies of compartmentation and controlling flame and smoke spread were amongst them, but were violated in the examples of fires given above. In turn, that prevented occupant escape and so led to deaths. Chapter 5 set out a key safety attribute of structural safety as the need to give warning before collapse and so provide time to save lives. The speed of flame spread in the Grenfell Tower tragedy was a classic example of the downside of failures taking place so rapidly that escape was prevented. Grenfell was also a reminder yet again of the need to learn lessons. There had been similar fires in high-rise blocks before, yet the lessons had been ignored.

9.8 Alarms, detection and suppression

9.8.1 Detection

In earlier times, the only method of fire detection was observation and fire watchers were routinely deployed. In many cities, high towers were purposely erected as district observation centres. Nowadays, modern technology offers all kinds of detection systems for both heat and smoke, plus automatic or activated alarms. Fire brigades throughout the UK promote the use of smoke alarms in houses yet current indications are that about 90 deaths each year occur because smoke alarms are not working.

9.8.2 Alarms

The object of any alarm is first, to permit and promote timely evacuation and second, to alert those who might be able to put the fire out. Domestically, working smoke alarms tend to shorten discovery time and generally the shorter the interval between ignition and fire discovery, the lower the death rate. Timely fire brigade response is also desirable and promoted by modern technology, which automatically links detection to alarm activation in the local fire station. This is not always successful; crews attended the York Minster fire within minutes, but there were difficulties nevertheless (Section 9.4.2). Likewise, crews appeared at the Grenfell Tower not long after the alarm was raised.

On construction sites, audible alarms are required, implying that the alarm system needs to be extended as the construction front progresses. Advice is given in Fire Prevention on Construction Sites: Joint Code of Practice (Fire Protection Association, 2015), which points out experience that suggests shortfalls in the provision of audible alarms on site. Modern technology offers the possibility of wireless controlled systems that can be easily extended and remain functional over a wide area.

Alarms are even more important when those at risk may not be able to help themselves. Examples might be in hospitals, old people's homes, prisons and hotels.

9.8.3 Suppression

Sprinkler systems have been demonstrated as being very effective in controlling fires and in limiting the scale of commercial loss; so much so that insurance companies offer premium discounts when sprinklers are installed. Fire brigades also promote wider usage. Obviously, sprinklers remain functional whilst buildings are unoccupied. After many major fires there have been calls for more extensive installation, as, for example, after the Lakanal House tower block fire of 2009 (Box 9.9), and they can be mandatory in certain higher-risk structures. The world's first system was the one installed in the Theatre Royal, Drury Lane (Section 9.3.2). Later, throughout the 19th century, systems began to be used in textile mills, albeit these were not automated; an automated system

Box 9.9 Lakanal House fire, Camberwell, London, UK (2009)

Lakanal House was a 14 storey tower block of flats dating from the 1950s. During the blaze (reckoned to be Britain's worst tower block fire at that time) six people died, including three children. The fire swept up the building after a television caught fire on the ninth floor. Those who died were on the 11th floor. The front door of the flat with the burning television collapsed, allowing smoke and fire to spread along the corridor when containment failed. There had been significant lack of fire stopping around pipework that had been added during renovations.

There were no fire seals on the door of the 11th floor flat. Moreover, panels beneath the flat's bedroom windows did not have the requisite resistance against spread of flame and fire and smoke penetrated through them and through heat-distorted aluminium window frames.

The coroner referred to the absence of sprinklers and urged wider adoption within high-rise multi-occupancy residential buildings. Alas, that advice was not heeded; the Grenfell Tower also had no sprinklers.

was patented in 1872 but then improved upon. The system used today effectively dates from around 1890 (the glass disc sprinkler).

Statistically, the odds of sprinklers going off accidentally are very low. And when they do activate, it is only sprinklers local to the heat source that activate. Consequently, the damage caused either by fire or water is reduced as the fire is extinguished more quickly using less water. Fire hoses, if used as an alternative, cause much more damage.

9.8.4 Summary of detection, alarms and suppression

The sooner a fire is detected the better. Early detection increases the possibility of escape for those in danger, it increases the possibilities of containing the fire and it lessens the risk to firefighters. Thus overall, alarms, detection and suppression are key components in any overall fire protection strategy. Modern technology offers much in support.

9.9 Emergency exits

9.9.1 Introduction

It took several disasters to formalise demands on emergency exits; one of the worst being the Triangle Shirtwaist Factory fire of 1911 in New York. This factory occupied the upper three floors in a ten-storey building. The fire is supposed to have started in a waste bin, probably via a discarded cigarette, but a contributory cause was the bin overflowing, not having been emptied for a long time. The fire spread within minutes and there was no alarm system. Worst still, managers had locked stairwell doors so many workers were trapped. One of the exterior staircases was so flimsy it just collapsed under the weight of those fleeing. Some workers jumped from the upper floors to the streets below. Altogether 146 workers died from burns, asphyxiation, impact injuries, or a combination of all three.

An inability to escape because of blocked exits features in several historical incidents described in this chapter and blocked exits still regularly feature in contemporary incident reports. Blocking is often in locations where large groups gather and where proprietors are interested in keeping gatecrashers out (night clubs seem to be special offenders). It cannot be said that the need to maintain emergency exits and keep them free is novel. That need was firmly established in the Triangle Shirtwaist Factory, with a further bad example occurring in Boston (USA) in 1942, when 492 people died in the Cocoanut Grove fire, during which there was a stampede to the club's single exit, a revolving door that became blocked. Current regulations prescribe a set number of exits (opening outwards) with routes to them being maintained clear of obstacles. However, those lessons have not been uniformly applied; it is not hard to find examples. Crowd panic and crushing can occur even without fire as a cause, and an appalling UK landmark incident occurred at the Victoria Hall, Sunderland, in 1883, when about 1,000 children stampeded down the stairs towards a door which had been opened inwards and bolted in such a way as to leave a gap only wide enough for one child to pass at a time. Order was not maintained; children at the front became trapped and crushed by the weight of those following and 183 children died. The Sunderland incident led to the invention of 'push bar' (panic bar) emergency doors.

In the 2001 World Trade Center disaster, some of the emergency exits inside the building were inaccessible, while others were locked. In the 1981 Stardust nightclub fire disaster in Dublin, 49 died and 214 were injured. The fire probably started from an electrical fault and featured the generation of thick smoke, with the ceiling melting and dripping onto clubbers. Mass panic resulted, but escape was hindered by locked fire exits (some with chains and padlocks) whilst others had chains around their panic bars. Failure of the lighting exacerbated the stampede, creating confusion as to what was an exit: many mistakenly escaped into the men's toilets thinking they were exits but the windows were barred. A similar event occurred in Brazil (the Kiss nightclub fire, 2013) when a fire was started by pyrotechnics igniting foam insulation. Flames and smoke filled the club and there was mass panic, with 230 people dying, many in the stampede.

In a 2006 Moscow hospital fire, emergency exits were locked, and most windows barred shut. This was a hospital for drug addicts, and 45 women died from smoke asphyxiation and nine more suffered carbon monoxide poisoning. Those who died were trapped between fire at one end of their second-floor corridor and a locked metal grille that barred stairs at the other end.

The Station nightclub fire occurred in Rhode Island, USA in 2003. It was started by the band letting off pyrotechnics which ignited flammable sound insulation foam on the walls. Flames spread over the surfaces, engulfing the club in just a few minutes. Billowing smoke and blocked exits hindered evacuation, whilst the exit routes were poorly designed, including an exit door that opened inwards not outwards. A total of 100 people died and 230 were injured.

In the British Airtours Flight 29M fire at Manchester Airport (Section 9.3.7), escape from the plane was hindered by smoke obscuring routes. Consequently, all planes are now fitted with floor-level illuminations to show exit routes. Similar requirements exist in buildings. In the UK, the Health and Safety (Safety Signs and Signals) Regulations 1996 define a fire safety sign as an illuminated sign or acoustic signal that provides information on escape routes and emergency exits. Well-designed signage is necessary for emergency exits to be effective. In any hotel, wall plans are required to show occupants the way out.

9.9.2 Summary of emergency exits

Time and again tragedy has been caused by identical design and management failings. Nobody can prevent fires occurring with certainty, so a key part of overall safety strategy is to ensure there are adequate means of escape and ensure occupants know how to escape. Regular drills are advisable. A common failing is not being prepared for mass panic and stampede, both made worse when fire spread is rapid. Other failings are not anticipating the needs of escape in dark, smoke-filled conditions and finally, allowing exit routes to be closed off.

9.10 Firefighting

9.10.1 Firefighting development

The first formal firefighting forces are thought to have been formed in Rome. In Britain, formality lapsed during the Dark Ages, although William the Conqueror decreed that all household fires should be put out at night. In later centuries, Church or town authorities occasionally provided very simple fire equipment, such as thatch hooks or buckets, and after the Great Fire of London, simple fire engines were built, and the first organised fire brigades began to appear.

At about the same time, companies were created to offer fire insurance (from 1667). These early insurers soon realised the advantages of protection and so they provided teams of men and fire appliances. Insured buildings had a metal plate fixed to an outside wall as a fire mark. For a century, these insurance company brigades provided the main firefighting capability (but only insured building blazes would be tackled!).

As the number of fire losses increased in the 19th century, local authorities began to provide fire services. Reliable steam-powered fire engines were introduced about the 1850s and these allowed a far greater amount of extinguishing water to be deployed. Fire extinguishers were invented from the mid 19th century onwards. Breathing apparatus was introduced in 1890. The very first 'municipal' fire service was formed in Edinburgh in 1824, and two years later Manchester had its own service. Insurance brigades were not finally disbanded until 1929. In the 20th century, fire services became much more professional as brigades adapted to new problems, such as fires from petroleum, celluloid and explosives. Much was learned as a result of dealing with munitions and chemical fires during the First World War. In 1939, a Fire Brigades Act was passed which, for the first time, gave local authorities the responsibility for providing an efficient, free, fire service.

In the later part of the 20th century, the significant rise in road transport meant that fire services accepted a wider role in emergency response to deal with crashes. The Fire and Rescue Services Act of 2004 recognised this wider role and today's fire and rescue service must deal with major environmental problems, large-scale chemical incidents and terrorist attacks (fire brigade action in Manchester after the IRA attack of 1992 and London Tube bombings of 2005 and in Northern Ireland was significant). Technical developments have assisted in communication, better firefighting equipment, breathing apparatus, heat-seeking cameras and advanced hydraulic cutting systems to aid victim extraction.

9.10.2 Building design and interaction with firefighting

One aim within an overall fire strategy is to ensure access for firefighting and to protect firefighters, recognising that firefighters may well be at risk entering burning buildings for rescue or fire suppression duty. In the UK several firefighters are killed every year and in the World Trade Center disaster, 340 New York firefighters died. The threats to firefighters are well highlighted in Boxes 9.6 and 9.10.

Building regulations provide for building-in protected routes and lobbies for firefighter access plus provision of pipes for water supplies and for providing firefighting equipment at higher floor levels. Further regulations cover the extraction of smoke from basements. Ensuring building survival for a certain time is an aim of structural fire design (see Section 9.11).

Box 9.10 Atherstone Industrial Estate, Warwickshire, UK (2007)

This fire was started by arson in a vegetable warehouse. Three firefighters were inside when the roof collapsed, killing all three, whilst a fourth man later died in hospital. The men had entered the building wearing breathing apparatus, but became trapped. The event is controversial since manslaughter charges were laid against senior officers in charge at the scene on the grounds that firefighters were put at risk when no other lives were actually at risk in the blaze. The jury found the senior officers not guilty. Whatever the verdict, the case is an illustration of the managerial dilemmas faced in tackling serious blazes.

The primary objective in any fire is to rescue people. First, that means providing effective escape routes, heat-protected, smoke-free and illuminated to allow occupants to get out. But beyond that, firefighters might have to enter zones which are hazardous, dark and smoke-filled. A lesson from certain fires is that this activity has been severely hampered by lack of plans. In the King's Cross Fire (Section 9.3.8) there was significant confusion in the labyrinth of underground tunnels and a key inquiry recommendation was for the provision of fire plans to assist firefighters. Similarly, after the Düsseldorf Airport fire (Box 9.3), the absence of operational procedures for terminal building fires was an omission; firefighters were missing floor plans and access keys.

9.11 Structural fire protection

9.11.1 Introduction

Providing adequate fire resistance is a key part of the design process for any building and that activity includes compliance with Building Regulations for means of escape; for firefighting; and for avoidance of materials that have adverse rates of flame spread. All civil and structural engineers require a general appreciation of these factors and the incidents described in this chapter illustrate why. Most of the issues require resolution in conjunction with the architect, services engineers and fire specialists. However, a key direct interest for civil and structural engineers is the topic of structural fire resistance. That is, design to ensure that the structure itself survives for a sufficient duration to ensure occupant escape time and firefighter protection. A basis of design will require that the structure has a certain notional resistance defined in terms of hours. For reinforced concrete structures this can have a direct influence on the cover thickness and for steel structures it can have an influence on the degree of fire protection provided for the steel. Thus, for all structures, the demands of fire resistance have a direct influence on detail design and cost.

The rules for standard buildings have evolved over many years and these prescriptive rules are not too onerous to apply. But new challenges emerge, particularly with regard to high-rise residential blocks. To cope with such demands, a performance-based design approach has emerged, and the techniques are reflected in the Eurocodes (Eurocode 1, 1996; Eurocode 2, 1996; Eurocode 3, 1996). As Section 9.4.3 shows, for example, the link between fire resistance and tunnel design is very strong. Fire resistance also has an influence on bridge design, though the reasons may be less obvious. For bridges, the normal source of fire and excessive fire load may be, as for tunnels, vehicle crash and vehicle burn. Two examples serve to show the potential importance (Boxes 9.11 and 9.12).

9.11.2 Fire resistance period

For all buildings, fire resistance is defined in terms of time, normally somewhere between zero (none required) and four hours. The amount of fire resistance time is

Box 9.11 M1 bridge, North London, UK (2011)

In 2011, a fire started in a scrapyard located under an M1 motorway bridge north of London. The bridge was a concrete one and it suffered very severe spalling to a degree that left its rebar exposed. The weakening was so bad that the motorway had to be closed for several days whilst deck propping was installed. It was later alleged that the fire had been started deliberately. Whilst the damage costs alone were obviously significant, the added community costs from delay and disruption were huge.

Box 9.12 New Little Belt Bridge, Denmark (2013)

There was lorry fire on the New Little Belt Bridge in Denmark. This is a suspension bridge (1,700 metre span) and the fire occurred right next to one of main suspension cables at its lowest point causing damage. If that cable had been severely weakened by the heat, the potential consequences (and costs) could obviously have been huge.

defined in various regulations, or might be determined by reference to BS 9999 which sets out a risk-based approach (BS 9999: 2009). A demonstration of component survivability is given via standard fire tests (as in BS 476-20 or BS EN 1363-1); such tests have their origins at the end of the 19th century when tests were carried out by insurance companies. However, these tests do not necessarily reflect performance in real fires (where temperatures rise and fall), nor reflect performance of systems overall since component performance as part of a wider system may differ from component performance in isolation. Tests are best thought of as giving ranking comparisons between different elements.

9.11.3 Steel structures

The load-bearing capacity of steel members falls off as temperatures rise and given steel's high thermal conductivity, the rates of temperature rise can be rapid. Much strength is lost by the 500–600 °C range whilst virtually all strength is lost when temperatures reach 1,000–1,200 °C, values easily reachable in real fires. However, the word 'failure' needs definition. At high temperature, beams may sag into a deep catenary shape which will be a 'failure' in conventional terms but nevertheless tolerable for 'survivability' in a large fire. Research and experience have also shown the importance of differential heating, for example, on beams supporting slabs where the slab provides enough protection to the beam top to allow its bottom section to survive greater temperatures than would otherwise be the case. Effects of differential expansion and contraction are important.

Box 9.13 Broadgate fire, London, UK (1990)

This fire developed in the partly completed 14-storey building which had concrete floors composite with steel framing. The fire began out of hours in a contractor's hut and spread undetected because fire detection was not yet installed (and neither was the sprinkler system). There was great deal of smoke, which damaged the facade. During the fire, temperatures reached about 1,000 °C, yet despite that heat, the structure was considered to have behaved well, even though it was only partially protected. There were some large deflections, but no overall collapse, nor was there any collapse of the exposed columns, beams or floors. These observations were an impetus to study the behaviour of complete frames in more detail.

A significant demonstration of real performance in one construction occurred in the Broadgate fire (Box 9.13) with verification of performance later carried out in a series of celebrated full-scale fire tests at Cardington (BRE Centre for Fire Safety Engineering, 2000). There is also experience of the way standard portal frames behave where, if the columns are protected, they can remain vertical and just allow the pitched roof to sag. TATA Steel and BCSA (2013) provides overall guidance.

In the aftermath of Broadgate, the UK steel industry carried out a great deal of research into fire performance. This had the aims of improving understanding, providing safer structures and economising on fire protection, which is normally provided by board or intumescent paint. Observations from Broadgate and Cardington (between 1994 and 2003) showed the great importance of concrete floor slabs in protecting beams and so contributing to overall fire resistance. The tests showed the ability of beams to survive in catenary action and the capability of concrete slabs themselves to survive, relying on membrane action with the notional slab reinforcement playing a key role. The work was also a spur towards a wider understanding the performance of steel structures in fires, not just of the building type tested at Cardington.

9.11.4 Concrete structures

Concrete is an inherently useful material, ensuring a fire-resistant structure. Concrete does not burn or melt or emit fumes. Its high thermal mass and low conductivity results in slow rates of heat gain away from heated faces. Thus, whilst concrete itself, and reinforced concrete, do lose strength with heat, the rate of strength loss is relatively low and low enough to provide adequacy of escape time.

Concrete itself starts to lose strength at about 250 °C but to a degree related to the exact mix details, including aggregate type and moisture content. Other structural parameters such as modulus and stress/strain characteristics are affected in a complex manner. Spalling can occur in different forms, with explosive spalling being perhaps of most concern. The resistance of reinforced concrete is linked to the amount of bar cover provided.

Spalling of standard concrete structures may not have immediate implications for structural capacity, though it might be an issue in special circumstances. One of these is the explosive spalling of high-strength concrete which has been observed in tunnel linings under fire (linked to moisture content) (see Section 9.4.3, Channel Tunnel fires). Control methods are available through the addition of fibres to the mix. Bailey and Khoury (2011) provide overall guidance.

A fire which highlighted how even concrete structures may be destroyed is the one that occurred in the car park of the Liverpool Echo Arena in December 2017. The fire was unprecedented in scale and it gutted the seven-storey structure, destroying up to 1,400 cars. Although the car park's concrete frame 'survived' (albeit badly damaged), several of the thin concrete floors were so damaged as to have virtually disintegrated. The fire seems to have started in one car (presumably through a fault) but then rapidly spread to many other cars, which collectively formed the fire load. Fires in cars and in car parks are not unusual but are usually contained and limited to one or two vehicles. This fire is an illustration of what can happen if fires are not contained (SCOSS Alert, 2018).

9.11.5 Timber structures

Timber structures might be thought of as excessively vulnerable in fires and it is true that timber burns. But when sections are large enough, initial surface charring provides insulation that can delay ignition and preserve strength for a period. Members can also be protected and treated with fire retardants. Serious fires have occurred in the construction phase and through arson, where the structural frame elements had not been protected by the in-service fire protection finishes (SCOSS Advice Note, 2010; the TRADA website, www.trada.co.uk, provides considerable advice). Moreover, new forms of material (such as cross-laminated timber, CLT) are being introduced and timber designs are being extrapolated to new heights: collectively these raise new issues of fire protection.

9.11.6 Masonry structures

Brick is inherently resistant to fire hence its use in chimneys. Nonetheless, in intense fires, walls can buckle and if hot enough, transmission of radiant heat might be a problem. The issue of most interest is more likely to be (lack of) fire stopping around penetrations.

9.11.7 Robustness

Fire is an accidental load case and coping with it essentially requires structures to be robust (see Chapter 4 and Cormie, 2013). The fire at the Torre Windsor complex in

Madrid (Box 9.5 and Section 9.5) highlighted the benefits of having intermediate strong floors and alternative load paths. The fire on the New Little Belt Bridge (Box 9.12) is an indication of what might happen when structural capacity is dependent on a single component. The ability of a fire to cause a progressive collapse can be graphically seen in the failure of a timber bridge in Texas, USA (Fire collapses wooden bridge, 2015).

9.11.8 Modern capabilities

The topic of fire resistance is so important and complex that several chairs of fire engineering exist in the UK. The output of their research has led to improved understanding, economy and higher safety and to new and evolving analytical methods. It is now mathematically possible to model fires, heat profiles in members, smoke discharge through buildings and even crowd movement in panic. But observation and feedback from real fire incidents remain important to validate theoretical prediction. For example, fires have highlighted unexpected structural behaviour of composite sandwich panels used as ceilings and internal walls exemplified by the sudden delamination and collapse which resulted in the death of firefighters in a fire in a food processing plant (Box 9.10).

9.12 Chapter summary

Fire is a generic hazard which is always going to be with us. But unlike certain natural hazards, it is a universal hazard affecting every structure in the world. Additionally, it poses an evolving risk linked to changes in building format and to the nature of certain industrial plant and processes. Precautions require attention to overall design, provision of firefighting services and attention to detailing. Real risks are also strongly influenced by management (or lack of it) and preparedness for the event: the human factor is, as always, present.

Appropriate design risk-mitigation rules have a long history and as will be seen from the events described in this chapter, many have evolved from observations of real fire events. The importance of this was brought to the fore in the aftermath of the Grenfell tragedy; the fire service's standard 'stay put' policy for those at risk in high-rise dwellings is now under active scrutiny. The fact that disasters recur can often be attributed to lessons not being learned. How else could the tragedy of Grenfell have occurred? How was it possible that so obvious a defect was not spotted by any of the professionals involved? Many of the incidents described herein reflect common and repetitive themes. There is much to learn by being informed over exactly what has happened before.

Fires can be complex phenomena and it can be insufficient to rely on arbitrary rules as a precaution. Events such as King's Cross, Piper Alpha, various tunnel fires and the Grenfell disaster demonstrate the need for comprehensive overall safety assessments requiring specialist knowledge. Such assessments are unlikely to be adequate unless supported by a study of what has gone wrong previously. All buildings and much infrastructure require fire protection. This can be a complex area where input from fire specialists is essential. Notwithstanding that, civil and structural designers need to be aware of the overall strategy justifying a structure's fire safety and thereafter understand that actual performance is strongly affected by the material selected and the structural detailing.

References

Bailey, C. G. and G. A. Khoury (2011), *Performance of Concrete Structures in Fire*, The Concrete Centre.

Beard, A. N. and R. Carvel (2004), *The Handbook of Tunnel Fire Safety*, Thomas Telford.

BRE Centre for Fire Safety Engineering (2000), 'Behaviour of steel framed structures under fire conditions: Main report', www.eng.ed.ac.uk/sites/eng.ed.ac.uk/files/attachments/research-projects/20150115/cardington-main-report.pdf.

BS 5588-7:1997, Fire precautions in the design, construction and use of buildings. Code of practice for the incorporation of atria in buildings (AMD 10546) (AMD 14991).

BS 9999: 2009, Code of practice for fire safety in the design, management and use of building.

Cormie, D. (2013), *Manual for the Systematic Risk Assessment of High-Risk Structures against Disproportionate Collapse*, IStructE Ltd.

Department of the Environment (1996), *Design Principles of Fire Safety*, HMSO.

Eurocode 1 (1996), *DD ENV 1991-2-2:1996: Eurocode 1: Basis of design and actions on structures: Actions on structures exposed to fire*, BSI.

Eurocode 2 (1996), *ENV 1991-1-2: Eurocode 2: Design of concrete structures: Structural fire design*, BSI.

Eurocode 3 (1996), *ENV 1993-1-2: Eurocode 3: Design of steel structures: Structural fire design*, BSI.

Fire collapses wooden bridge (2015), www.youtube.com/watch?v=bxyuVSvtn1Y.

Fire Protection Association (2015), *Fire Prevention on Construction Sites: Joint Code of Practice*, 9th edn, Construction Industry Publications Limited.

Home Office (2012), 'Detailed analysis of fires attended by fire and rescue services in England', www.gov.uk/government/collections/fire-statistics-great-britain.

Hansell, G. O. and H. P. Morgan (1994), *Design Approaches for Smoke Control in Atrium Buildings*, BRE Report, IHS BRE Press.

HSE (undated), www.hse.gov.uk/construction/safetytopics/generalfire.htm.

Montalva, A. et al. (2009), 'A catastrophic collapse: Windsor building fire', in B. S. Neale, *Forensic Engineering: From Failure to Understanding*, Thomas Telford.

SCOSS Advice Note (2010), SC/10/095, 'Timber framed buildings in fire situations: the role of the designer', www.cross-safety.org/sites/default/files/2010-11/timber-framed-buildings-fire-designers-role.pdf.

SCOSS Alert (2018), 'Fire in multi-storey car parks', www.cross-safety.org/sites/default/files/2018-02/fire-multi-storey-car-parks.pdf.

Summerland fire video (2013), www.youtube.com/watch?v=ORFFWAUJ8Po.

TATA Steel and BCSA (2013), 'Steel Construction: Fire Protection'.

10 Natural disasters

10.1 Introduction

Human society has always had to cope with natural disasters. Creation myths from early civilisations record catastrophic events, deluges being common. The Biblical flood is just one of them, that described in the Epic of Gilgamesh (Babylon, second millennium BC) is another (The Epic of Gilgamesh, 1964), with both stories probably stemming from periodic flooding in the Tigris/Euphrates plain. The Bible alluded to many other events: drought, pestilence and earthquake. We certainly know of terrible earthquakes in ancient history, of cities destroyed and of others now sunk below the waves. Whilst Egyptians depended on the recurrence of Nile floods for their survival, comparable Yangtze flooding in ancient China caused terrible problems, with annual events bringing misery to millions. As late as the 20th century, Yangtze floods caused thousands of deaths and billions of dollars in economic damage. In the 1930s, reports suggest three to four million people died. Millions have also died in earthquakes, and, like flooding, these are not just events in the distant past. Whilst the Shaanxi earthquake of 1556 is said to have killed upwards of three-quarters of a million people, the Tangshan earthquake in 1976 also killed somewhere between 250,000 and 750,000. Even in 2008, the Sichuan earthquake killed around 70,000. The Indian Ocean tsunami on Boxing Day 2004 killed maybe a quarter of a million and affected more than one million people in some way, and in Haiti, around a quarter of a million people died in the 2010 earthquake. In modern times these events are well publicised on TV, so pictures of drought, floods, severe heat and cold, hurricanes and forest fires are all commonplace.

There is credible evidence that the rate of natural disasters is increasing, though that might be due to more interest and better reporting. Either way, there is no doubting that disasters affect a lot of people across the globe with phenomenal annual costs. What is also certainly true is the world's population is still exploding; that the number of mega-cities is still rising and there is an ever-increasing shift from rural to urban dwelling. Thus, there are increasing concentrations of vast numbers of people in areas fundamentally at risk of natural disaster. Moreover, many mega-cities in the developing world lie close to oceans, which threaten a special set of disaster hazards. All this is putting aside the raising of risk linked to climate change, which is predicted to bring ever-more volatile and extreme weather with all the complications that entails.

The UK's position on climate change was set out in the 'UK Climate Change Risk Assessment 2017' (Committee on Climate Change, 2017) produced by a committee which reports to government. In their 'Key Messages' the committee spelt out:

> The greatest direct climate change-related threats for the UK are large increases in flood risk and exposure to high temperatures and heat waves, shortages in water, substantial risks to UK wildlife and natural ecosystems, risks to domestic and international food production and trade, and from new and emerging pests and diseases.

The Report schedules out the risks with implications and suggested actions. Overall, scientists have concluded that average annual UK temperatures over land and the surrounding seas have increased in line with global observations, with a trend towards milder winters and hotter summers (the weather over summer 2019 matched the predictions with a massive heatwave over Europe and torrential rainfall intensities. Summer 2022 was even hotter).

News reports from the US suggest that 2017 had been the costliest on record there for natural disasters, with total losses amounting to $306 billion (this being $100 billion more than the previous worst year, 2005). Over the 2017 season, there were sixteen $1 billion events including hurricanes Harvey and Ima. At the turn of the year leading into 2018, the East Coast then froze in severe low pressure systems. Over the same year, many records for rain and flooding were broken. As well as hurricanes, there were devastating fires in western states, particularly in California, requiring mass evacuations. The fires stripped Californian hillsides of vegetation and then, to compound the misery, heavy rains caused terrible mudslides. Generally, the US is annually afflicted with natural disasters: hurricanes, tornadoes, earthquakes, floods and droughts all costing a great deal and many resulting in significant fatalities. In 2012, the southern drought covering 48 states was reported to be the worst since the Dust Bowl of the 1930s (see Box 10.1).

Box 10.1 Dust Bowl, USA (1930s)

The 'Dust Bowl' is a term for the widespread devastation that occurred across the North American prairies in the 1930s. The problem was caused by severe drought, allied with inappropriate farming methods, which led to a spate of severe dust storms and soil erosion. Extreme drought and crop loss are also major problems in sub-Saharan Africa.

The US National Oceanic and Atmospheric Administration (NOAA, undated) also stated that 2017 was the third warmest on record for the US and the world's hottest since records began after correction for the El Niño effect, which gives an additional temperature boost when active. As El Niño was active in 2015–2016 (but not 2016–2017) the temperature then was at a peak. Overall, the years 2015 to 2017 have been the hottest ever recorded and 17 of the 18 hottest years since 1850 have occurred since 2000. In their annual report, the scientists observed that rising temperatures were linked to more weather extremes, from hurricanes in the US and Caribbean to heat waves in Australia and floods in Asia.

Early in 2018, the World Economic Forum (WEF) scheduled its predictions of the biggest risks to the global economy. The top three were: (i) extreme weather events; (ii) natural disasters; and (iii) failure to halt climate change (World Economic Forum, 2018).

The British Isles also suffered badly in 2012 (as an example). The year started off with fears of a drought, but the opposite occurred. The country endured its wettest April and June in 100 years, bringing extensive flooding. Thereafter, storms and flooding continued regularly over summer and into winter. As the year wore on, many flooding events were exacerbated by heavy rain falling on ground that was already saturated. There were numerous flash floods such as the ones in July which flooded some small Pennine towns. Some events were spectacular; none more so than the Newcastle evening rush hour storm of June which sent cascades of water down the city streets and a dramatic lightning strike on the Tyne Bridge. Shopping malls were evacuated, and the main station and metro were closed. Thousands were left without power. Later in the year, another Newcastle storm produced such a torrent that the ground surrounding the piled foundations of a block of flats was swept away, requiring the flats to be demolished. Seven years later, the 2018 UK summer was the hottest for nearly half a century but that record was broken in 2022 when temperatures exceeded 40°C.

The 2012 UK storms and flooding required many house evacuations and caused intermittent loss of power over the year across the country. There were significant landslides cutting roads and railway lines. There was at least one tornado and a storm that resulted in huge hailstones. There were deaths in the storms and floods and one couple were killed when their car was buried in a landslide. The bad weather of 2012 was followed by a long, cold winter with unseasonal and excessive amounts of snow in early spring 2013. The overall damage cost for the year has been estimated as approaching £1 billion. The heatwaves of 2019 and 2022 were so intense that rail services were badly disrupted. Following the 2019 heat, torrential rain was so bad that overtopping occurred at the Toddbrook Reservoir dam above Whaley Bridge, damaging its spillway. This raised alarm that the dam itself might fail and 1,500 residents were evacuated from the town below.

All this is relevant since a key role for civil/structural engineers is to create a built environment that protects populations from natural events. The challenges are enormous, but the rewards for taking on the challenges are also huge. To see this, it is simply necessary to contrast the loss of life from events such as earthquakes as between advanced countries like the USA and poor countries such as Haiti. The role of engineers in coping with disasters is well set out in Da Silva (2008).

10.2 Social and political effects

In 2011, a major earthquake and tsunami destroyed the nuclear plant at Fukushima in Japan (Chapter 11) creating partial fuel meltdown and radioactive release. Whilst the loss of life (about 2,000 deaths) was bad enough, it was not that great in comparison with other earthquakes. Of more widespread significance in terms of disaster implications have been the political and economic effects around the globe consequent on loss of confidence in the safety of nuclear plant. In some countries, such as Germany, the use of nuclear energy for power generation has become unacceptable and that in turn will have significant repercussions, certainly within a world struggling to reduce carbon emissions and lessen its dependency on fossil fuels.[1] In the aftermath of Fukushima, many more gas-powered stations were built and renewable energy technologies were much boosted.

Such political events consequent on natural disaster have often been significant. It has been not so much the disaster incident itself but the consequences on the built environment and society's handling, or not, of the subsequent crisis that mattered. In ancient times, natural disasters were often thought of as portents of Heaven's dissatisfaction. A most significant event in European history was the great Lisbon earthquake of 1775, which destroyed much of the city and devastated Portugal's economy (and incidentally caused a huge wave to travel as far as Cornwall). However, whilst many in contemporary society ascribed the devastation to divine displeasure, others were more enlightened. Thus the earthquake was widely discussed, with proper enquiries as to its potential cause and this was perhaps the basis for promotion of rationalism and the need to theorise on natural rather than religious causes. Post Fukushima, there was severe domestic criticism within Japan of the plant operators, the regulators and the government response, and Japan is still coming to terms with the fallout.[2] It will not have been the first natural catastrophe to have had such widespread effects.

The political effects after the Chinese Tangshan earthquake (1976) were equally significant: Mao Tse Tung died shortly afterwards and the repercussions, the secrecy over the Tangshan event and government mishandling all contributed to the end of the Cultural Revolution, initiating deep changes in Chinese society. In 2005, Hurricane Katrina devastated New Orleans. This was followed by severe criticism of the government at all levels, up to and including the White House, for lack of preparedness, seeming

1 Germany's avoidance of nuclear power and over dependence on gas (from Russia) have had severe implications following the sanctions between Russia and Europe in 2022.

2 In 2022, Fukushima executives were fined £80 billion for failing to prevent the disaster.

indifference and inept handling. Lessons were learned and in Hurricane Sandy (aka Superstorm Sandy, 2012), both local and national government were widely regarded as handling the crisis well. Certainly, the press images did no harm to Barack Obama's second term re-election campaign. By contrast, Donald Trump's response to Puerto Rico when it was devastated by Hurricane Maria (2017), virtually destroying the island's electrical grid, did nothing to enhance his image.

10.3 Tasks for civil and structural engineers

Civil engineering is about providing the fundamentals of civilised life; the most basic needs of shelter, provision of lifeline systems and the means for populations to earn their living. By and large, people cannot choose where to live; they must make do with their country of birth and cope with its inherent natural hazards as best they can. Engineers must understand what those hazards are, plus their nature, and provide protection or at least resilience, that being the quality of recovering after disaster has struck. Our fundamental concern should be the safety of those at risk.

This engineering task has become more urgent. Population growth has put many more people at direct risk (even in the UK). But changes in society have also played a part. The world is now much more interconnected: severe droughts or flooding have noticeable effects on worldwide commodity prices, so the effects of a catastrophe in one continent may well be felt in others. Hence, we all have a vested interest in coping with disasters wherever they occur. And society is no longer fatalistic; natural disasters are not now seen as acts of God or burdens that must be borne. The technology exists to mitigate their worst effects and there is an expectation of deployment.

Overall, the study of natural disasters has great significance. Engineers must first map out what the disasters might be; second, try to understand their cause and likelihood of occurrence; then third, learn how to design infrastructure against them or at least understand how to mitigate their consequences. Finally, engineers must contribute to preparedness and recovery. All these tasks present formidable challenges both technically and in terms of implementation. The costs of first build and longer-term maintenance can be formidable, but so too can the costs of recovery and reconstruction. Neither can be dodged.

Not all hazards affect the British Isles, but many do; moreover UK engineers work internationally, so an awareness of what can happen elsewhere across the globe is useful. Coping with disasters requires a different way of thinking, since some of the events have a low probability of occurrence yet very high consequences. In practice, that means making hard choices about what to protect and what not, or at least what level of protection to provide and how much scarce resource to allocate. The normal objective is to try and protect life but perhaps accept commercial loss to a degree linked to the likelihood of the event and palatability of its consequences. It means that engineers should look beyond the failure of the immediate structure to assess wider implications. Thus, if a river overtops and floods some farm fields that is one thing, but if whole communities might be swept away, that is another. If the failure of pylon structures leads to widespread power outages (in winter) that might indicate a need to look at more severe loading cases than normal for the transmission system design. In areas of the world more at risk from natural disasters such as earthquake, the margins of safety demanded on dams are raised, since failure of a dam and consequent loss of water reserves may have significance far beyond the immediate one of the earthquake itself. In the UK, such thinking has practical implications for the design of safety-related structures such as nuclear power plant and for high-cost infrastructure development (see discussion on the probability of extreme events in Chapter 4).

10.4 Natural disasters

A list of natural disasters of interest might be:

- earthquakes
- tsunami / tidal wave / storm surge
- extreme wind
- extreme flood
- extreme rain
- extreme heat
- extreme cold
- extreme snow
- extreme ice
- freezing rain
- hailstorm
- lightning
- fire
- pollution
- underground hazards
- landslide/mudslide
- other hazards.

10.4.1 Earthquakes

Earthquakes are one of the commonest natural disasters and regularly feature in the news (Figure 10.1). Large parts of the populated world are at risk, including many

Figure 10.1 Typical earthquake damage (Dreamstime 26258578)

mega-cities. Countries such as Japan, parts of China, northern India and the near East, southern Europe and the south-western USA are particularly prone. The risk is not totally random but linked to a locality's proximity to the edges of tectonic plates in the Earth's crust or nearness to particularly active faults. It is the motion of those plates (or fault movement) against each other that is the cause of ground vibrations which are so destructive to surface-based structures.

Earthquake action is frequently the cause of secondary disasters. An earthquake might, for example, cause a tsunami or landslip. Earthquake destruction might also initiate fire devastation; although this is not an immediate association, much of the destruction in, for example, San Francisco (1906) or Kobe (1995), or even in Fukushima (2011), was caused by fire consequent on disruption from seismic action.

Earthquakes do occur in the UK. The Colchester event of 1884 (also known as the Great English Earthquake), caused considerable damage in the town and its region. There was a destructive Dover Straits earthquake in the 16th century and many minor ones since. The direct relevance to the UK is that in statistical terms there is a chance of a rare event of significant magnitude, thus such earthquakes are a major design consideration for nuclear plant. Seismic consequences are also worth considering on very-high-value infrastructure projects. In very many other parts of the world, assessing response to an earthquake loading condition is a more or less standard demand.

10.4.2 Tsunami / tidal wave / storm surge

Before 2004 many people, even engineers, would not have known what a tsunami was. The term was common only in Japan, hence the word's etymology. Then on Boxing Day 2004, an earthquake generated an enormous tsunami in the Indian Ocean. Captured on film, the dramatic events were witnessed all around the world. The earthquake's epicentre was off the Indonesian coast and the incoming wave affected more than 11 countries, inundating coastal communities to a depth of about 30 metres. About 230,000 people were killed. Equally dramatic pictures were taken of the Fukushima tsunami incident, where the earthquake generated a 6 metre high wave, flooding about six miles inland, and leaving huge devastation and death in its wake (Figure 10.2). The power plant's tsunami-protective wall was overtopped which contributed to the severity of the nuclear incident. Once generated, tsunamis travel huge distances over water; many tens of thousands of miles of travel are possible with the wave motion travelling at great speed. The fact that the wave travels and takes some time to make landfall, at least gives the possibility of a warning system. The Pacific Ocean has such a system, based on floating sensors co-ordinated via a manned control centre based in Hawaii.

'Tidal wave' is a more general expression, used to describe any abnormally high wave, perhaps a bore or storm surge. The UK is not normally at risk from tsunamis, but they have occurred, as have storm surges. A 3 metre high wave reached Cornwall after the Lisbon earthquake of 1755 (and there may have been one behind extensive flooding in Bristol in 1607). A seiche is an oscillating sloshing action (with large wave height) generated by earthquake/wind action on enclosed water. In reservoirs and the like, the maximum wave height and pressure against boundary walls may be determined

Figure 10.2 Tsunami aftermath, Japan (Dreamstime 87965427)

by seiche action. In rarer cases, landslips terminating in water have created locally destructive tidal waves (cf. 1964, Valduz, Alaska).

A more common storm surge is the piling up of seawater on shore during a storm. The surge amplitude depends on wind speed and direction and local coastline topography. An exceptional amplitude (also known as a storm tide) can occur if a storm surge coincides with an astronomical high tide and a severe depression. It was such an event that created the North Sea catastrophe in the Netherlands and East Anglia in 1953 (see Sections 10.4.4 and 10.5).

10.4.3 Extreme wind

Typhoons/cyclones/hurricanes are all regional names for intense local weather systems that build up over warm oceans and then cross adjacent land masses bringing with them strong winds, torrential rain and storm surges. Areas at risk can normally expect a number of such storms over the season, with climate change suggesting that the number and intensity of such storms will increase. Wind forces are so high they will dominate the strength demands on buildings and infrastructure. Likewise, rainfall intensities are so high that drainage system designs will be dominated by precipitation rates. In the UK, extreme wind might be just associated with very rare storms, such as the infamous 'hurricane' of 1987, which caused significant devastation (the highest UK gust was measured as 190 km/hr = 53 metres/sec). In the wider world, much of southern Asia is at annual risk of typhoon or cyclone and the southern parts of the United States and the Caribbean are at risk during the annual hurricane season. Giant storm systems can develop and move up the US eastern seaboard, for example, Hurricane Sandy in 2012 (Box 10.2) and Hurricane Katrina in 2005 (Box 10.3).

Tornadoes are violent, funnel-shaped vortices where the rotating winds can reach phenomenal speeds. Typically, tornadoes might only be a few metres across and be short-lived. Nevertheless, those that persist and track will cause enormous damage to anything in their path. Tornadoes are mostly associated with the southern United

Box 10.2 Hurricane Sandy, USA (2012)

The worst US storm of 2012 was Hurricane Sandy, aka Superstorm Sandy. This formed in the south then moved up the eastern seaboard causing damage from the Caribbean to Canada. The effects in New York were severe: the storm surge flooded streets, tunnels and subway lines and cut power all around the city. Total damage was estimated to be many billions of dollars. Altogether, Sandy claimed 253 lives and has been rated as the second deadliest hurricane since 1900. Peak wind speeds were about 188 km/hour (52 metres/sec) (NB: the method of gust speed measurement may differ between countries so this speed does not necessarily compare with Britain's experience in 1987).

> **Box 10.3 Hurricane Katrina, USA (2005)**
>
> A storm that may be even better remembered is Hurricane Katrina in 2005. The devastation in New Orleans was intense as the levees were breached by the storm surge and 80% of the city was flooded. The event has been rated as the worst civil engineering disaster in US history. Thousands were displaced and over 1,800 died. The political repercussions were significant.

States (about 1,000 were reported in 2012) and eastern India, but they do occur in the UK, albeit at much smaller scale. There was significant destruction in Birmingham in 2005 when trees were uprooted and roofs sucked off, with reported damage costs of about £40 million. The most severe UK tornado recorded appears to have been one in Portsmouth in 1810, where winds are alleged to have reached 240 mph.

10.4.4 Extreme flood

Extreme flood might be associated with storms and frequently coincides with typhoons or cyclones or excess monsoon rain (Figure 10.3). It might, as in the UK, be linked to rare weather patterns which bring torrential rain. There are also possible coincidences when, for example, annual snow melt in the Alps sends large volumes of waters down the Rhine to be deposited in the North Sea. This can put the Netherlands at risk if the dykes are overtopped, especially if the flood coincides with adverse high tides. Floods might also come about through sea surges generated by storms. Inland flooding risk is a function of rain pattern but, as we have seen in the UK, that risk is intensified if the rain occurs after long periods of wet weather, when the ground is too saturated to absorb more water, or conversely, after long dry periods when the ground is hard and flash run-off occurs in advance of absorption. Run-off rates are affected by human activity such as deforestation, and by intense city development where large areas are hard-landscaped. Flooding risk intensifies as pressure on scarce building land requires development on low-lying areas which are inherently at risk.

In 2012, the UK suffered one of its wettest summers on record and there were many flooding incidents. However, the year wasn't as bad as 2007, during which 55,000 properties were flooded. The risk has been getting greater over the years, just due to the increasing area covered by houses and by the demands for housing land, which have meant building on areas previously considered unsuitable. The Met Office reports that six out of the UK's ten wettest years have occurred since 1998 (Met Office, undated). Historically there have been many catastrophic incidents. One occurred in London in 1928 when the Thames overtopped the embankment, but this was just one in a long line of periodic floods (Pepys recorded another in his diary). Nonetheless, the 1928 event was so significant that it figured in the decision to build the Thames Barrier, which

Figure 10.3 Extreme flood on the River Rhine (Dreamstime 107612699)

is one of the world's largest moveable flood barriers. The barrier protects 125 square kilometres of central London from tidal surges. Currently the UK government allocates around £300 million annually to flood protection works.

10.4.5 Extreme rain

Extreme rain, producing floods, is normally associated with regional effects. Characteristics of significance to engineers are the duration of rainfall, plus fall intensity. On very large roofs, the rate of fall might be so intense that overall drainage is significantly delayed leading to a surface covering of water and allied loading or the risk of ponding and roof collapse. Similarly, delayed run-off and local build-up can put basements at risk of flooding (especially if drainage systems back up) or put tunnels at risk of inundation. In Superstorm Sandy, New York underpasses flooded. Flooding of underground railway systems is a factor to be thought about, and before Sandy struck, the New York system was closed. In 2009, a bridge in Cumbria was destroyed (Box 10.4), brought down by scour undermining its foundations. The risk will have been increased by the temporarily much faster rate of storm-water discharging down the river. The Boscastle event of 2004 (Box 10.5) also illustrates what can happen in extreme rainfall events.

The consequences of heavy rain are not always obvious to outsiders. In 2013 there was very heavy rain and flooding in South Africa. As a consequence, a tourist attraction was forced to open some crocodile enclosure gates due to a storm surge from a nearby river. In taking this action 15,000 crocodiles escaped.

Box 10.4 Workington, Cumbria, UK (2010)

The Northside Bridge in Workington was swept away in a storm whilst a policeman was on it, diverting traffic away. He lost his life. Reports suggest the cause was probably scour and they suggest that losses of other bridges were due to the 'extremity and sheer force of the flood' (House of Commons Transport Committee, 2010).

Box 10.5 Boscastle, Cornwall, UK (2004)

About 75 mm of rain fell in two hours above the village, an amount that normally falls in the whole of August. This created a huge amount of water flowing down the twin rivers which join just above Boscastle. The result was a destructive torrent through the middle of the village. The church was filled with 2 metres of water and mud. Overall costs were huge.

10.4.6 Extreme heat

In President Obama's 2013 State of Union address, he raised the topic of climate change, pointing out that the US had suffered its highest summer temperatures ever, the link being to increased volatility in weather patterns. At that time, the UK's five hottest summers since 1914 had all occurred between 2002 and 2007 (Britain's hottest day was in 2003 at Faversham, Kent at 38.5 °C). Yet by 2018, the statistics changed such that eight of the ten warmest years overall had all occurred since 2002 (and all ten since 1990). The Committee on Climate Change report (2017) then predicted that heatwaves such as the one that occurred in 2013 would become the norm by the 2040s. True to the prediction, the UK summer of 2018 was the hottest for nearly half a century and summer 2022 was even hotter.

Excess heat affects death rates significantly and the European heatwave of 2003 is said to have caused about 35,000 extra deaths. The Committee on Climate Change report (2017) gives a UK estimate of 2,000 persons p.a. dying prematurely from heat-related conditions, with that trend likely to increase. Collectively, those rates suggest a demand for better climate control in buildings.

High temperatures themselves are of significance to engineers but so are prolonged high temperatures, as these give more chance for high thermal mass structures to heat up. Absolute temperatures are most obviously a design factor for long systems such as bridges (which might jam) and for railway track, where the risk of buckling is raised if expansion becomes too high.

Heat also raises the demand for water: for human consumption, for cooling and for irrigation. It significantly affects the risk of forest fire. Drought is often associated with areas at risk of both excess and prolonged heat. Civil engineering protection is by building and maintaining an adequate water infrastructure.

10.4.7 Extreme cold

Extreme cold is a feature of many climates. The need to protect populations from cold through the use of heating and insulation is obvious. In some countries with exceptionally harsh climates there is even an annual threat of death simply from large icicles falling. Low temperatures increase the risk of damage to many building materials especially by the freeze/thaw cycle and in steelwork structures by raising the risk of brittle fracture, which is a key design consideration on exposed structures such as steel bridges. But consequential risks exist as well. As a generality, cold weather might damage the infrastructure. Domestically this might be a flood from burst pipes (common when we had lead pipes) or more widely it might be a severe flood from water supply or drainage systems or it might be some crash consequent on railway track point failure due to freezing. The general issue (Chapters 2 and 5) is that when conducting a hazard analysis of any system, a cause of malfunction in any part of that system might be failure to operate under extreme cold (or extreme heat). To mitigate cold, some systems might incorporate trace heating.

In the UK, December 2010 was the coldest December for 100 years. For winter overall, the years 1947 and 1963 were especially harsh. The lowest temperature recorded in Britain has been −27.2 °C in Braemar, Scotland in 1982 (Met Office, 1982). But contrast that with the winter of 2017–2018; in January 2018 in a Siberian village, reputedly the coldest inhabited place on Earth, the external thermometer broke as temperatures sank to −67.7 °C. Over the same period, the US East Coast was also extremely cold. In such conditions, populations just cannot survive without infrastructure protection.

10.4.8 Extreme snow

Many countries need to cope with severe snowstorms or snow accumulation as a matter of course. In structural design, the immediate interest is the unit weight of snow applied to surfaces. But the melting phase is also a consideration. Snow melting slowly with drain blockage might lead to progressive ponding, a sagging roof and a worse loading state. On sloping roofs (or surfaces) local roof slides can take place. In countries where snow build-up is common, roofs are often fitted with guards to contain the slipping snow and prevent masses crashing to ground. In mountainous areas of very high snow fall, it is sometimes necessary to provide protective roofs over roads to limit the risk of blockage (and impact on traffic) of local avalanches. Avalanches can destroy whole villages. One of the deadliest appears to have occurred in Peru in 1970, triggered by the Ancash earthquake. This generated an avalanche mass which moved 11 miles at very fast speed and buried about 20,000 people. Perversely, lack of snow can also be

Box 10.6 Snow in the UK (2010)

There were very severe snow falls in November/December 2010. Many motorists were stranded and in Scotland, the Forth Road Bridge was closed for over ten hours. Many airports had to close, including Heathrow, which was shut for days just before Christmas, creating chaos. An extreme low of −21 °C was recorded in the Highlands.

The Secretary of State estimated that travel disruption costs were £280 million per day; and the overall costs of the cold spell have been put at £1.6 billion.

a problem: the severe North American drought of 2012 was partially due to a lack of winter snowfall, combined with several years of below average rainfall.

The pattern of snow fall differs across countries and in the UK, there is codified guidance on intensity levels along with advice on potential drift build-up. Nevertheless, there is necessarily a good deal of uncertainty and patterns of roof failure under excess snow build-up do occur (Chapter 8 gives examples). A factor of interest is not just the total amount of snow, but the rate at which it can fall, since if the rate is excessive, the emergency services have little time to react, e.g. for road clearing (Box 10.6). Coping with snow on the roads generates its own problems; a significant cause of decay in bridge parapets has been linked to using de-icing salts, which accelerate rebar corrosion.

10.4.9 Extreme ice

The risk of ice build-up on exposed structures is a concern. Excessive ice on power lines and towers has created enough extra weight to bring them down. Less obviously, ice build-up can affect wind response and has been responsible for some major disasters on cable structures. It is a factor to consider on any cable-stayed structure as either rain, snow or ice accumulation can affect cable dynamic response under wind conditions and thus the risk of oscillation via vortex shedding (see Box 8.29 and the Emley Moor TV mast). As with many natural phenomena, it pays to think beyond the immediate consequences and look at wider implications. Ice storms in Canada in 1998 brought down transmission lines and were responsible for days of power loss over the whole eastern area, and this at a time when populations were dependent on power for heating. Even in the British Isles, storms frequently damage power supplies.[3] In parts of the world where low temperatures are common, a design risk to waterside structures (say bridge piers) is that of ice floe collision.

10.4.10 Freezing rain

Freezing rain is a separate phenomenon, and rare in the UK. When it does occur (as in January 1996) it causes objects to be coated in ice. Precipitation initially falls as snow

3 In the winter of 2021–2022 Storm Arwen caused power losses over parts of northern Britain for well over a week.

at high level, then melts into rain/drizzle but is supercooled en route and refreezes at ground level, glazing objects with ice. Power lines and telephone lines are most affected, and the increased weight can bring cables down. The ice coating can be so thick it has been known to bring down whole transmission towers. Freezing rain is common in parts of Canada and in 1998, a particularly severe ice storm created deposits up to 120 mm thick, damaging power lines all over the country (power outages lasted for weeks in some places). Costs were estimated at $5–7 billion.

10.4.11 Hailstorms

Hailstorms can be so destructive that farmers insure against them. Studies have been made in the UK which show that there is a pattern of behaviour, with most storms occurring in the summer months and crossing the middle of the country. There are several records of hailstones large enough to damage cars, roofs and greenhouses. At Ottery St Mary in Devon, in 2008, a very unusual hailstorm persisted for two hours, causing extensive damage.

10.4.12 Lightning

The hazards of lightning strikes on buildings have been known for centuries but were not solved until Benjamin Franklin 'invented' the lightning conductor. All tall structures are required to have such protection; nevertheless, even today, lightning consequences have not been entirely eliminated. The dreadful fire at York Minster in 1984 was ignited by a lightning strike (Chapter 9, Section 9.4.2) and many regional fires in dry areas which threaten property and cities are initiated by lightning. As with some other natural hazards, a significant effect on structures/infrastructure might arise consequent on plant malfunction initiated by a strike. Strikes can lead to power outages and 'loss of power' is a generic hazard to consider in any system. Strikes can also knock out electronic systems. In special structures, like the London Eye, there is a need to consider passenger protection directly, first, because the Eye is a tall steel structure where lightning strike is an obvious hazard. The second reason is to ensure that passengers can be evacuated and that in turn demands that drive capability is maintained with high reliability. Since that reliability is governed by a computerised control system, all the on-board and ground-based electronics need proper surge protection. Another Eye design hazard was to ensure that the main bearings (enclosed within the steel structure hub) would not be affected by lightning strike, partly for commercial reasons and partly to ensure that the wheel would not be prevented from turning. This is an example of a 'system as a whole' and a need to consider a generic hazard (lightning); how that might affect any part of the system and the consequences of damage to any part of the system on the safety strategy (summarised as the ability to evacuate under all credible circumstances). Local codes exist which define the risk

> **Box 10.7 Wenzhou high-speed train crash, China (2011)**
>
> This was a catastrophic failure of the new Chinese high-speed train. Two trains collided on a viaduct and four cars fell over the parapet, killing nearly 200 passengers. The ability of two trains to be on the track at the same point was thought to be linked to a failure of the control system, which in turn was thought to be due to a malfunction after a lightning strike. As with many other disasters, the blame and political repercussions were immense (see also Chapter 2).

of lightning attack in any part of the world. Singapore, for example, has a very large number of annual lightning storms.

In 1977, a lightning strike knocked out power transmission lines near New York. As a result, power was off for 25 hours, trapping people in subways and elevators. See also Box 10.7.

10.4.13 Fire

Fire is a domestic hazard in any building or plant (Chapter 9). But as a natural hazard, it is a phenomenon that might evolve into regional fires sufficiently fierce to threaten whole communities. TV pictures of such fires in Australia, California and southern Europe are common. In 2009 Greece was very badly affected and smoke plumes across the whole Attic peninsula could be seen on satellite images. Fires began about 25 miles from Athens but spread to its suburbs, engulfing numerous towns on the way. Thousands of residents were forced to flee. The fires lasted for days. Likewise, in the very hot and dry 2022 summer many European countries were badly affected by fierce regional fires. The United States loses thousands of acres of forest and many homes annually in wildfires; those in California at the end of 2017 and 2018 were the most destructive on record. In the 2018 Californian fires, some towns, were utterly destroyed, and overall fire costs are set at billions of dollars.

As a separate but related hazard, fire pollution seriously degrades air quality. In the very hot Russian summer of 2010, many peat fires developed around Moscow, producing awful pollution across the city: it wasn't possible to see across Red Square (these fires followed severe drought and temperatures above 40 °C). Pollution was five times greater than normal and the polluted atmosphere badly affected people with heart conditions and asthma. About 5,000 deaths above normal levels were registered in July across the Moscow region. The UK does not suffer from such severe fires, but in 2018, fires on the moors on the eastern outskirts of Manchester were hard to control, and pollution spread across parts of the city.

10.4.14 Pollution

Pollution problems such as the ones in Moscow are exacerbated by high air temperature / high pressure conditions. In plan, air continually circulates around the centre of the high pressure area, and so the supply of fresh and clean air is diminished towards the centre. In elevation, air temperatures usually reduce with height but cold air sinking in high pressure systems leads to inversion, with air temperatures being higher at upper levels. This condition then creates a trapping of pollution and smoke below the inversion layer; industrial pollution cannot escape, and air becomes stagnant. Industrial pollution is a major issue in developing countries and was a point of great concern before the Beijing Olympics (2008) (at the time, Beijing was ranked the 13th most polluted city in the world). Because of pollution concerns, the authorities closed down building sites and restricted car usage.

Pollution is an issue all over the world. Fog and its derivative 'smog' were at one time a defining London image (so rightly a man-made disaster). One of the worst events occurred in 1952, when the city was blanketed with smog for four days, causing about 4,000 deaths. Visibility was badly reduced and traffic of all kinds brought to standstill. In parallel with concern about smog was concern about acid rain; such rain occurs naturally, in the form of very dilute carbonic acid and is the cause of caverns and sinkholes in limestone (which can be a natural hazard in themselves). But the acidity increases when industrial sulphur and nitrogen fumes are absorbed. The effects on structures suffice to cause exacerbated steel corrosion and, in particular, severe weakening of stone. Many of the UK's historic limestone cathedrals have been badly eaten away and weakened by acid rain. Absorption of atmospheric CO_2 is a cause of carbonation in the cover of reinforced concrete and thus one of the causes of reinforced concrete decay. Government action followed the Great Smog of 1952: it was realised that something had to be done, and that 'something' was the gradual introduction of the Clean Air Acts and the switch away from burning coal in power stations. The introduction of flue gas desulphurisation technology in power stations to limit SO_2 emissions was another step towards pollution reduction. The world has still not solved its power supply problems, but a return to coal burning is not a safe option.

10.4.15 Underground hazards

In discussing natural hazards, we tend to think primarily of effects above ground. But dangers also lurk below ground, with the dramatic surface breaking of sinkholes being one. News reports often focus on large holes suddenly opening which might be related to old mine shafts. True sinkholes are caused by eroded limestone caverns breaking out up to the surface. In 2013, one opened in Florida, swallowing a complete house; one sleeping occupant was never recovered. Holes might also be caused by erosion of ground due to leaking drains and there have been spectacular examples of this around the world

Box 10.8 Abbeystead, Lancashire, UK (1984)

The valve house for this system was underground, set into a hillside. On the day of the incident, 44 people were present to see a demonstration of water pumping. Just after pumping started there was a flash and an explosion, causing severe damage. Sixteen people were killed and all others present were injured. The explosion was caused by the ignition of accumulated methane, which might have been dissolved in water leaking into the system. During design, the hazard of gas accumulation had not been considered. (See also Chapter 5.)

(Figure 10.4). SCOSS has issued an Alert on sudden loss of ground support; on roads these holes may cause severe accidents; elsewhere, they might undermine foundations of buildings (SCOSS Alert, 2017). Underground gas is a major hazard and the history of coal mining is littered with examples of gas explosions. It is not just mining that presents a risk; a notorious incidence of methane explosion occurred within the Abbeystead valve house in 1984 (see Box 10.8).

A major concern since Fukushima has been that of radioactive water escaping into the ground and reaching the aquifer. Radiation in the form of natural radon seepage upwards is a general health hazard, increasing the risks of cancer. Radon is a colourless, odourless gas, formed by the radioactive decay of the small amounts of naturally occurring uranium. The risk varies across the UK according to the local

Figure 10.4 Loss of ground (Dreamstime 186058221)

geology, but it does figure in the Building Regulations, which require a risk assessment and implementation of preventative measures in building design.

10.4.16 Landslide/mudslide

Landslides and mudslides are the motion of large masses of earth down slopes. They might be precipitated by heavy rain or earthquakes and they have the ability to cause terrible problems. Hong Kong, being a city with many steep-sided slopes and tropical downpours, has been badly affected in the past. In 1972, a series of landslides occurred, causing multiple deaths; since then, a long-term programme of slope stabilisation been underway. Mudslides seem particularly horrific. Motion can be fast and akin to a flood. Slides can occur anywhere in the world. Bog slides occurred in Ireland in 2008, extending over 4 km and engulfing two bridges and a section of road. Landslips disrupted UK rail traffic significantly in 2012 after heavy rain. Some events are truly awesome, for example, the Vargas tragedy in Venezuela in 1999. This killed tens of thousands and led to widespread infrastructure collapse; whole towns disappeared. Another bad event occurred in Indonesia in 2006, when a mud volcano was caused to erupt by a blowout of natural gas following well drilling. At its peak, 180,000 m^3 of mud per day was spilling out. The flow, though now reduced, is expected to continue for decades.

In the UK one of the most traumatic events in living memory was the slide of a waste tip at Aberfan in South Wales (1966) which engulfed an entire school, killing 116 children and 28 adults (albeit this is better classified as a man-made disaster (Box 10.9).

10.4.17 Other hazards

There are other, rare, natural hazards. In early 2013, a large meteorite struck parts of central Russia, destroying several structures. Meteorite events are so rare they would generally be discounted on probabilistic grounds. However, damage from severe hailstorms is not that uncommon. Damage to structures from dust storms is uncommon; but an issue to consider is of dust potentially damaging mechanical plant and thence consequential failure of engineering systems (examples are known of structural overload from accumulated dust in coal plant). Perhaps surprisingly, dust storms do

Box 10.9 Aberfan, South Wales, UK (1966)

Aberfan was a coal mining village and for many years waste had been dumped in tips beside the town. The waste had been stacked up over porous rock containing many springs. Just before the tragedy, there had been several days of heavy rain. The tip was unstable and on the fateful day, about 150,000 m^3 of slurrified material flowed down at speed straight into the village. En route it destroyed a terrace of houses but worse was the burying of Pantglas Junior School; the children inside were killed by the impact or suffocated (McLean and Johnes, 2000).

occur in the British Isles, and Scotland experienced some in 2013 so severe that driving visibility was obliterated. Volcanic eruptions occur (such as the mud volcano cited earlier) and we are all aware of the potential devastation exemplified by the burying of Pompeii and Herculaneum by a Vesuvius eruption in AD 79. More recently (1995) the Caribbean island of Montserrat had to be half abandoned when a volcano erupted; the island's capital was buried under 12 metres of mud. Although volcanoes might not seem relevant to the UK, it was only in 2010 that the whole of European air space closed down for five days due to the ash cloud generated upwind by eruptions in Iceland.

10.5 Disaster prediction

Engineers need to be interested in the natural hazards that can exist and in their causes. The ultimate objective is to predict the likelihood and intensity of an event and so design infrastructure accordingly to prevent failure or to mitigate the consequences of disaster. It is worth pointing out immediately that the word failure here needs to be carefully interpreted since many hazards are so intense that structures cannot be designed to survive intact with normal safety margins (Section 10.7). Instead, it is necessary to define a strategy of survival rather than expect structures to remain undamaged (Chapter 5). Another reason for understanding cause is to do with preparedness. Some events like earthquakes are essentially random, whilst others are semi-predictable. Communities need preparedness for all types of natural disaster, which includes issuing warnings where practicable, and they always need preparedness for post-disaster recovery.

It is well known that in the US hurricanes will develop in the summer months over the Gulf of Mexico and then track north. Typically, there will be five to a dozen such storms developing from June to November, since the energy required for generation can only be gained when sea temperatures rise. The development of individual storms is monitored via satellite and other means, and predictions are made of path and timing. Thus, populations at risk are warned and can either be evacuated or take shelter. Both those precautions require engineering intervention, first to ensure the feasibility of moving large numbers of people at short notice and second, by acquiring the skills of shelter design. Such action requires considerable organisation on behalf of the authorities. To assist preparedness, predictions of likely damage from hypothetical events may be required.

Other broad patterns of weather disturbance are understood, such as the El Niño effect. This is a band of unusually warm ocean water that occasionally develops off the South American coast, influencing Pacific and world weather patterns. There is intense research worldwide to try and understand the phenomenon of global warming, which seems to be associated with climate change and a general shift in weather patterns and a general increase in more erratic weather. These are of interest to engineers whose structures have to cope. There is, for example, a gradual increase in sea levels due to polar ice melt, exacerbated because warmer water expands above colder water. The Committee on Climate Change (2017) states that sea levels globally and around the UK have risen by 150–200 mm since 1900. Future predictions of sea-level rise are even greater and enough to be factored into designs for dockside structures and coastal protection.

In countries such as the UK, both normal weather and extreme events have been studied over long periods, so good records exist as a basis for future prediction. Furthermore, in terms of atmospheric weather, techniques for prediction via computer modelling have been massively enhanced over the last 50 years. Even so, it is still not possible to forecast more than a week or so ahead or even be confident that a broad pattern of winter or summer behaviour can be determined (partly due to the perverseness of the jet stream since the UK is on the fringe of some more stable weather patterns). Nevertheless, by utilising records of many years' activity, such as rain, wind and temperature, it has been possible to prepare probabilities of likely data sufficient for design purposes. Thus, civil engineers are able to work from codes which suggest wind speeds which are unlikely to be exceeded over a given period, customarily 50 years (a probability of 10^{-2}). We have, as a profession, to accept the concept of likelihood and probability. That begs the question of possible exceedance and in safety-related structures, where a very low probability of failure is desired, it is normal to work with data such as the likelihood of wind/snow/rain, etc. with a return period of 1 in 10^{-4} (colloquially referred to as the '1 in 10,000 year' event). Techniques exist to do this and there are accepted values for extreme wind, extreme snow and so on over the UK (see also Chapter 4, Sections 4.5.12 and 4.5.13).

More broadly, understanding weather patterns in the UK allows for consideration of combinations of events that might put populations at risk. Many of our severe storms are the tail end of hurricanes from across the Atlantic. The severe storm of 1987, which caused significant damage in southern England, was one such example. Another severe event is that which caused flooding in East Anglia and the Netherlands, in 1953 when more than 300 people died after sea walls were breached and floods progressed inland over many miles. About 40,000 were made homeless. The cause was a combination of events: a full moon to cause a high tide, a deep depression to raise the sea level even higher and a northerly storm to push waters yet higher still (a storm surge as described in Section 10.4.2). That combination of events might not be predictable at any one precise time in the future, save to note that it will happen at some time. The risk can be anticipated if forecasters see that combination of phenomena developing. In the long term, that understanding drives the need for the Thames Barrier and the Dutch sea defences. In the short term, it can be fed into emergency response preparedness. The 1953 combination did recur in 2014; due to better preparation, consequences were minimal.

No one can predict if a meteorite or asteroid will strike. But the probability of any individual building being hit is so low as to be ignored. Such rare risks are best dealt with by insurance. A much more difficult problem, yet one of direct significance to engineers, is dealing with earthquakes. Clearly very many parts of the world are at risk to a greater or lesser extent. It is possible to identify areas at risk and suggest design values, but there is considerable uncertainty. The New Zealand Christchurch earthquake in February 2011 was set off by an unknown fault and the earthquake that caused so much damage in Kobe in 1995 was a very rare event. In 2009, an earthquake at L'Aquila in Italy caused a great deal of damage and about 300 people died. Italy has an advisory centre about likely intensity of earthquakes and it appears that on this occasion, the group's judgement was that significant increased activity prior to the main event was likely to be dissipating strains in the Earth's crust, rather than be a forewarning of worse to come. This judgement proved to be wrong and there was a significant earthquake. In the aftermath, the group were criticised by the authorities and a judge sentenced some eminent scientists to six years for manslaughter, as he put it: 'for them giving a "generic and ineffective" risk assessment'. Such a reaction is hardly likely to encourage others to be objective in pursuit of prediction. All that can be said is that there is a likelihood of earthquake activity at some stage of some unknown (but probable) intensity. The likelihood might be greater if the locations are close to known fractures in the Earth's crust such as the Californian San Andreas fault, or if the region is known to be close to the edge of one of the tectonic plates, such as regions close to the 'Pacific Ring of Fire'. Earthquake forecasting remains elusive.

The level of uncertainty is much more than the question of whether there will be an earthquake or not. Design engineers are interested in the detailed characteristics of earthquake motion such as peak acceleration, signal frequency content and duration of ground shaking, all of which have their own uncertainties. In countries where earthquake activity is more commonplace, design codes contain criteria for earthquake activity and these will normally be region-specific. In Eurocode 8 (BS EN 110108), it will be seen that the design standards for earthquake activity in, say, Italy or Greece exceed those in northern Europe. The described ground motion is some sort of envelope of several motions regarded as typical.

A key question for all extreme events then is that, given the high level of uncertainty both in value and frequency of disaster occurrence, how should structures be designed for safety? What should the design strategy be to cope with uncertainty? This is discussed in Section 10.6. But to complement physical design, part of the strategy overall is preparedness both for semi-predictable events and for those that are less predictable. Most countries thus have facilities for preparedness and for disaster relief. Engineers can feed into these plans with information on what might happen; on post-event response and on retrofit of damaged structures. Our technology has improved to permit identification of weaknesses/vulnerability in overall infrastructure and lifeline systems. What we know for certain is that natural events causing disaster will recur.

10.6 Emergency preparedness and response

10.6.1 Preparedness

If we look at the last century, it becomes apparent that until comparatively recently, natural disasters and their effects were just more or less accepted. Countries could do little about them. They could do little because the political structures did not exist, because the costs looked so huge and because the technology was hardly understood. But since the Second World War, changes have taken place. Richer countries have come to accept their responsibility in terms of overseas aid. The growth of television and the power of those images to convey horror have had a major effect: the unfolding events of famine or Fukushima could be seen in real time. Society now recognises the inevitability of annual disasters, often in poorer counties and that leads to two demands: first, preparedness and second, response. Simply put, we must prepare for the inevitable.

Most countries now have some sort of agency whose mission is to anticipate the type and likelihood of disasters in their applicable region and to prepare response plans. For example, all inhabitants in Japan are trained on what to do if an earthquake strikes and there are national and regional plans of preparedness for rescue and so on. It can be anticipated that after an earthquake, buildings will need to be shored up and heavy lifting equipment will be required in search and rescue missions. Search and rescue teams require special training and equipment.

In the USA, emergency preparedness is under the control of FEMA (Federal Emergency Management Agency) (FEMA (undated). In the UK, there is national control under the government's Civil Contingencies Committee, known as COBRA due to its usual meeting place (Cabinet Office Briefing Room A), and much information on dealing with all manner of emergencies can be obtained via the Cabinet Office website (Cabinet Office, undated). In local areas, isolated incidents are dealt with by the Fire and Rescue Service. The Meteorological Office (Met Office) has an alert system of severe weather warnings, whilst regionally, the Environment Agency operates a system of flood alerts over the TV, radio and internet. In the Scottish Highlands, there is an avalanche warning system.

Most public utilities have their own domestic plans to cope with potential disaster. Highway authorities plan with stockpiles of salt and equipment to keep roads clear in winter, though they don't always get it right. In December 2010, very heavy snow storms closed Heathrow Airport, causing transport chaos (Box 10.6). Since that time, more investment has been made in snow-clearing equipment and preparedness.

10.6.2 Response

In early times, relief was left to church organisations. But in 1863 the Red Cross was founded as a direct response to suffering in war. The proposal was for the creation of national relief societies, made up of volunteers trained in peacetime to provide neutral and impartial help to relieve suffering in times of war. Early non-governmental organisations (NGOs) such as Oxfam (1942) were founded to provide relief to civilians in Europe. There are now many NGOs with specific interests trying to protect populations in the aftermath of natural disasters. The UN itself has a major support role in dealing with relief and humanitarian crises. RedR (Register of Engineers for Disaster Relief) is of special interest to engineers. It was founded in 1980 to provide a system to deploy engineers post disaster (RedR UK, undated).

10.6.3 Resilience

Resilience is the term used to encompass the whole of emergency preparedness and response but including having a strategy of recovering quickly from any disaster. The UK government has a national infrastructure resilience programme: 'to enable public and private sector organisations to build the resilience of their infrastructure, supply and distribution systems to disruption from all risks (hazards and threats) as set out in the National Risk Assessment'.

The Cabinet Office has published Sector Resilience Plans and the 2012 version included sections on assessing existing resilience and building resilience. Some illustrative extracts are:

> **Energy**: … Building Resilience … Priorities include:
>
> **Upstream Oil and Gas**. Assessment of the risk to oil and gas beach terminals from fluvial and coastal flooding;
>
> **Electricity Generation**. Assessment of the risk to power stations from fluvial and coastal flooding;
>
> **Electricity Networks**. Assessment of the risk posed by severe space weather; and completion of the electricity networks vegetation management programme.
>
> …
>
> **Nuclear**: … Assessment of Existing Resilience …
>
> **Hazards**. Where necessary, sites must erect flood defences to resist a 1 in 10,000 year flood; and under Building Resilience: (4): On 11 March 2011, Japan suffered its worst recorded earthquake. The resulting tsunami severely damaged Fukushima Dai-ichi nuclear power site triggering a national and international nuclear emergency. (5) Findings from Her Majesty's Chief Inspector of Nuclear Installations report, examining lessons from the Fukushima accident to enhance the safety of the UK nuclear industry, will support the National Framework.

> ...
>
> **Transport**: ... Assessment of Existing Resilience
>
> 1. The major risks the transport network faces are severe weather, flooding, power outages, reduced fuel supplies and volcanic ash. To maintain essential services, transport operators have: [several measures are described]
>
> ...
>
> **Water**: ... Building Resilience ...
>
> (7). In the previous price period, Ofwat made £400m available to companies to improve the resilience of assets and systems to flooding and other hazards. Companies have been instructed to once again consider the resilience of assets to major risks in their business plans for the period 2010–2015. Immediate priorities will be to respond to and recover from the current drought.

Clearly, all these plans have as their aim keeping the country running against numerous potential threats. Equally clearly, civil engineers have a major part to play in providing the essential resilience (Proceedings of the Institution of Civil Engineers, 2012).

10.7 Design criteria for disasters

A wide range of disasters occurs. Although we can make coarse predictions of which region is likely to be affected by what type of disaster, predictability is still relatively crude, so engineers have to cope with uncertainty. We must do this both regionally and in respect of individual buildings. Regionally, we need to look at what disasters might strike and then also try and assess the effects on infrastructure and the likely kinds of damage and likely cost of mitigation. It is an unpalatable fact that communities just cannot afford to make every structure or infrastructure system 'disaster proof' against the worst credible events that nature can throw at them. Nonetheless, some systems are more important than others. Engineers need to identify lifeline systems, identify applicable threats and then see what can be done about them, all rather as set out in the UK Sector Resilience Plans.

For extreme events, design is about damage control and mitigation, accepting the inevitability of 'failure', but seeking to ensure that the consequences are relatively tolerable.[4] Design encompasses concepts of selective and partial failure coupled with an ability to recover. Hence the design process is both conceptual and numerical. We

4 What is tolerable changes with time. Hereto water companies have been allowed to discharge raw sewage into water courses in times of exceptional rain. By summer 2022 this was causing much public agitation

cannot stop fires breaking out, but measures can be taken to minimise the risk of fire spread or we can detect fire quickly and provide measures for suppression. We cannot stop natural events that might cause flooding, but we can perhaps make individual structures temporarily watertight or we can arrange drainage systems such that the land that does flood offers least risk. We cannot stop tornadoes and cyclones, but we can provide strong shelters and we can evacuate communities to safer areas.

Numerical approaches encompass concepts of controlling modes of failure to tolerable ones, such as, for example, making sure extreme scour does not destroy a bridge's foundation but accepting partial damage to bridge superstructure. Numerical design can also adopt lower safety margins to correspond with loads having low probability of occurrence, or it can trade on a structure's ability to absorb energy, accepting that after an event the structure may have distorted but will not have collapsed. These strategies cannot be presumed effective unless structures are detailed accordingly (Chapter 4).

10.8 Examples

There are plenty of examples to illustrate the form of civil engineering works that have been set up to try and avert disaster and the challenges presented. In historical times populations have often tried to protect themselves against floods, sometimes by careful siting of settlements or by providing high refuges for times of inundation. Today is no different: it is foolish to site key infrastructure facilities in places or at levels that might be at risk of flood or adjacent to active faults. Sometimes the protective effort has been by the clearing and dredging of rivers to remove impediments to flow, often supplemented by attempts to confine flow by the building of banks and levees. Defences of this type have existed in the Netherlands for centuries, and huge flood defences have been built in other parts of the world too. In the US, the Mississippi levees stretch for 3,500 miles along the lower river reaches. Large parts of New Orleans, located on the delta, lie nominally below sea level, as do two-thirds of the Netherlands: both are regionally protected. Venice floods quite regularly under certain climatic conditions, although this situation is beginning to be felt intolerable.

Just before the Second World War, the Netherlands constructed the first of what were to be its modern defences against flooding. The first major task was blocking off the Zuiderzee to form the Ijsselmeer, now the largest freshwater lake in Europe. The constructed dyke was 32 km long. However, even at this time, it was realised that sea defences were weak in the south-west where the rivers Rhine, Meuse and Schelde have their estuaries. Four million people live in this low-lying area. Around 1950, massive new protection works were started (and given added urgency by the 1953 events). Certain estuaries were closed off, with a plan to ring the whole area and shorten the number of dykes required. The target flood defence height is set against a 1 in 10,000 year event for population protection but lower returns for agricultural land, and that target also differs with respect to freshwater flooding and the more damaging saltwater flooding. Nothing is ever static and current assessments of climate change and sea-level rises are imposing demands for a one-metre increase in protective height. The delta defensive works consist of dykes, dams and storm surge barriers. One of the most massive is a floating gate across the Meuse (the Maeslantkering) which is designed to temporarily restrain a storm surge rise. The Netherlands also has a national warning system in place.

Fortunately the UK does not have quite the same potential problems as the Netherlands, but it does have a major risk of London flooding; if inundation occurred, the commercial losses would be immense. To counter this, the Thames Flood Barrier was constructed, being operational from 1982. This iconic moveable structure is the world's second largest (the larger one being in the Netherlands) and its function is to protect London when Thames estuary waters rise dangerously under the combination of climatic conditions described in Section 10.5. In very high tides, the barrier can be raised to hold the waters back, but it must be lowered again at low tide to let river water discharge.

The barrier's original design target was a life of 70 years and protection against a 1 in 1,000 year flood, and allowance was made for rising sea levels over time. But the predicted level of high water is increasing, driven partly by climate change, and partly by slow land mass changes that are still taking place as a consequence of the ice removal after the last Ice Age (southern England is slowly sinking). As a measure of change, the trend of number of gate closures per year is seeming to increase. In terms of reliability, it has to be ensured that the barrier will function on demand, so there is a monthly test. Nowadays the controller is assisted by computer modelling and automatic feed-in of weather and marine conditions, such that there is a grace period of about 36 hours' warning of a potential closure need. In thick fog in 1997, a dredger crashed into the barrier, dumping its 3,000 tonnes load of gravel and preventing the gates from closing. This is a good example of how any system needs to be considered, in total, for all the hazards that might cause malfunction, not least as bad weather was a contributory factor to the dredger incident.

Another example of a huge civil engineering project partly initiated to control flood is the Three Gorges Dam across the Yangtze in China. Annual flooding of the Yangtze has been the cause of immense misery over the centuries. The dam is expected to reduce major flooding frequency from once every 10 years to once every 100 years. Moreover, it is hoped to mitigate the effects of periodic super-floods.

Another project that has been under discussion for many years is a barrier to prevent Venice flooding. Venice has always flooded and, as in London, bad flooding requires a coincidence of high tide, a depression and strong winds to push waters northwards up the lagoon. Again, as with London and the Netherlands, some very long-term physical changes are making the flooding more frequent, in this case, erosion and gradual subsidence (plus global warming). In 1966, there was a catastrophic flood, the highest ever recorded, reaching nearly 2 metres above datum: the entire city was flooded. In 2018, floods reached 1.5 metres deep, the worst for 22 years.

Not all structures are required to be massive to protect against disaster. Many lives have been saved in Bangladesh by the provision of cyclone shelters, which are often public buildings specially strengthened to protect inhabitants in the short term against fearsome storms. The structures need to be strong enough to withstand wind and storm-water flow and high enough to lift people above surge levels. Windows may be shuttered to keep out the worst effects. In tornado-prone areas of the US, many houses incorporate small underground bunkers to provide immediate shelter while the storm passes by. In 2005, during Hurricane Katrina, thousands of evacuees sheltered in the Louisiana Superdome.

10.9 Chapter summary

Engineers face few safety tasks more demanding than protecting populations against natural disasters. That task is a core activity. Indeed, it is one embraced in the Royal Charter of the Institution of Civil Engineers: 'being the art of directing the great sources of power in Nature for the use and convenience of man'. There are always going to be severe natural events but whether these turn into disasters is partly controlled by the efforts society puts in to mitigate the consequences. Thousands of people die every year when buildings collapse in storm or earthquakes, or when floods devastate. Yet the risk to an individual is much less in the developed world than it is in the developing world, simply because of better quality engineering, better resourcing and better organisation. That reality is not easily overcome. It is a challenge simply because the developing world lacks resources. Thus, although building gigantic and sophisticated flood defences or carrying out complex dynamic analyses or probability studies are fascinating and worthwhile activities, it is equally important to devote much effort to creating low-tech structures using cheap indigenous resources that have the best chance of protecting most of those at risk. Much remains to be done, especially if we acknowledge collective responsibility.

References

BS EN 110108, Eurocode 8: Design of structures for earthquake resistance.

Cabinet Office (undated), www.gov.uk/government/policies/emergency-planning.

Committee on Climate Change (2017), 'UK Climate Change Risk Assessment 2017. Synthesis report: priorities for the next five years', www.theccc.org.uk/wp-content/uploads/2016/07/UK-CCRA-2017-Synthesis-Report-Committee-on-Climate-Change.pdf.

Da Silva, J. (2008), 'Coping with natural and man-made disasters', *The Structural Engineer*, **86**(14).

FEMA (undated), Federal Emergency Management Agency, www.fema.gov.

House of Commons Transport Committee (2010), 'The impact of flooding on bridges and other transport infrastructure in Cumbria', TSO, https://publications.parliament.uk/pa/cm200910/cmselect/cmtran/473/473.pdf.

McLean, I. and M. Johnes (2000), *Aberfan – Government and Disaster*, Welsh Academic Press.

Met Office (undated), www.metoffice.gov.uk/research/climate/understanding-climate/uk-and-global-extreme-events-heavy-rainfall-and-floods.

Met Office (1982), Historical Weather Factsheet, www.metoffice.gov.uk, https://tinyurl.com/2mvjz534.

NOAA (undated), US National Oceanic and Atmospheric Administration, www.noaa.gov.

Proceedings of the Institution of Civil Engineers (2012), Special Edition, 'Infrastructure Resilience', **165**(6), November 2012.

RedR UK (undated), www.redr.org.uk.

SCOSS Alert (2017), 'Sudden loss of ground support', www.cross-safety.org/sites/default/files/2017-07/sudden-loss-ground-support.pdf.

The Epic of Gilgamesh (1964), trans. A. George, Penguin Classic Books.

World Economic Forum (2018), World Economic Forum Annual Meeting 2018, www.weforum.org/events/world-economic-forum-annual-meeting-2018.

11 Man-made disasters

11.1 Introduction

Chapter 10 described natural events precipitating what we commonly regard as disasters. Engineers cannot avoid such events, but need to mitigate their impact. Equally, there have been many engineering failures which themselves have consequences so serious that the incidents could be regarded as man-made disasters. The definition of a disaster is loose, but generally it is either the scale or consequences of the failure that are determining.

The HSE have recognised that catastrophic events occur within the UK construction industry and accordingly they commissioned a report on this topic which was published in 2011 (HSE, 2011). Their rationale stemmed from a belief that: 'the industry may not be sufficiently aware of the potential for it to be associated with more major events (those involving multiple deaths and/or significant damage to property and infrastructure)'. These were referred to as events of: 'low probability' but 'high consequence'. Some lessons are reported in Chapter 13. The HSE's report scheduled more than 50 incidents that had taken place within the decade 2000 to 2010, hence showing that events qualifying as disasters, or near disasters, are not rare.

For an incident to qualify as a disaster, the consequences must be severe. Hence, an important lesson is to consider the abstract nature of safety margins weighed against the consequences and this is one lesson from the HSE report (HSE, 2011). In one sense, a design is 'safe', or not, in absolute terms. But in reality, we can observe that 'safety' is not an absolute commodity. Figure 11.1 shows two identical retaining walls. One holds back 1 metre of water and the other holds back 1,000 metres. In both cases the water pressure and forces on the walls are exactly the same. But if (a) fails there may only be a splash, whereas if (b) fails there will be a deluge.

Another key word is uncertainty; there is always uncertainty and engineers need to ensure there is 'more safety' or their designs are 'safe enough' if there is more uncertainty in their design or if the consequences of any failure are disproportionate. This is a bal-

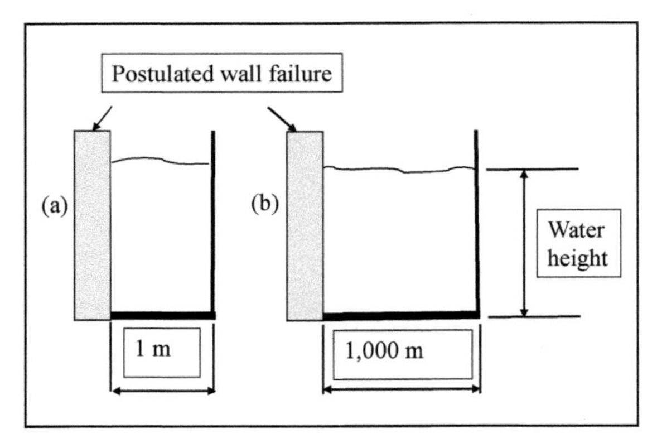

Figure 11.1 Consequences of water-retention failure.

ancing act. 'More safety' usually comes at a cost and it is for society to decide if that cost is worthwhile. In the example of Figure 11.1, many engineers would instinctively feel that more care should be taken with wall design (b) and would feel much more comfortable if its safety margins were higher than for wall (a). (Section 11.3.3 will illustrate some dramatic consequences of dam failure.) In some codes, and against some loading, this decision-making is done explicitly via inclusion of a numerical 'importance factor' built into code-defined equations (Chapter 3).

Part of the design process is therefore to consider all the hazards involved plus the consequences and weigh up the design approach, taking account of all factors. This is not easy, but many examples illustrate what can happen when things go wrong or where design has just been done by rote. Failure consequences may be severe inconvenience, direct risk to health or life, financial or environmental. Critical failure of any part of the infrastructure is a prime candidate for consideration as a potential disaster. There is the immediate cost of the failure itself but thereafter, potentially much larger costs linked to delay and disruption. Engineers should not just carry out their work mechanically; they must have regard for the wider implications of potential failure. Not doing so can be seen as one factor in many historical incidents.

Man-made disasters often involve fire and examples are discussed in Chapter 9. Many disasters have occurred during the construction phase and Chapter 8 is devoted to them. Examples of man-made failures in service are considered below.

11.2 Man-made disasters

There is no shortage of historical precedent with disaster classics such as those Brunel endured or the Tay Bridge failure of 1879 (Tay Bridge disaster, 1879; Martin and MacLeod, 1995). Even a cursory search on the internet will reveal incredible failures of engineering systems continuing up to the present day. Society has had to accept huge environmental consequences linked to the search for oil. Another surprising fact is that it would be imagined that the engineering expertise assigned to some of these mega projects must have been of the highest quality. With such teams on board, the question must then be: 'How could matters have gone so wrong?'

In studying the disasters described in this chapter, it will be observed there are a myriad of reasons for their occurrence. But a very common one is the team not having proper regard for safety management and all too often concentrating on capital costs and continuing production income at the expense of 'safety'. When things do go wrong, the costs are usually orders of magnitude greater than the perceived costs of any temporary production halt. And those additional costs are not just financial; there are costs to reputation, public outrage to endure and prosecution for negligence. There are costs to the environment which we all have no choice but to bear.

In a number of the failures discussed, there might at first seem to be no direct connection to civil/structural engineering. However, there are counters to this perception. First, all professional engineers need to have a wide perspective on 'safety', not least as many disasters stem from lack of comprehension across technical boundaries. Second, civil/structural engineers contribute to the project. Unless they have the right perspective on hazards and consequences, they are less likely to formulate a design strategy appropriate to control and minimisation of risk. Looking at examples, the source of initiating events is extremely varied, ranging from vandalism and terrorism to simple error. Unless we learn from what has gone on before, we are not likely to be informed enough to identify future hazards correctly. Overall, part of safety education is to understand the background to the many regulations now in force since many of these have been drafted or strengthened in response to bitter experience.

11.3 Disasters involving structures

11.3.1 Buildings

Rana Plaza, Bangladesh, 2013

In spring 2013, a nine-storey clothing factory, the Rana Plaza building in Dhaka, Bangladesh, failed more or less instantly; see Figure 11.2 (Rana Plaza building collapse, Dhaka (2013). The first pictures showed an utter collapse though the full implications did not emerge for some days. Five thousand occupants were dragged down and eventually the death toll from this single event was confirmed at over 1,100. Reports suggested that cracks had previously occurred in the building but had been ignored and the owners had provided assurances as to safety (the structure was reinforced concrete). It appears these assurances were false, and failure seems to have been linked to gross overloading by installing additional (and illegal) storeys, perhaps exacerbated by machinery vibrations. By any standards, the structure was not 'robust', and its mode of failure was exactly of the type aimed to be avoided by codes of practice implementing the lessons learned in the UK after the failure of Ronan Point (Chapter 8). Newspaper reports shortly afterwards stated that surveys confirmed 65% of comparable buildings in the locality were at similar risk.

Figure 11.2 Rana Plaza collapse, Dhaka (Shutterstock 797806171)

In this disaster, search and rescue was hampered by lack of training. The Register of Engineers for Disaster Relief (RedR UK, undated) issued an appeal for support as they provide assistance in such situations.This is not the first time that disastrous failure has occurred from overloading exacerbated by vibration, as shown below.

Sampoong department store, Korea, 1995

The Rana Plaza case might seem to be a one-off failure attributable to criminal negligence. But it bears similarities to the failure of the Sampoong department store (Korea) (1995). This was a devastating structural collapse, killing 502 people and injuring 937, making it the largest peacetime building failure in Korea. Some three months before failure, cracks appeared but no action was taken. On the day of the failure, cracking increased significantly, and the top floor was evacuated but incredibly, the building remained open for trading. Several hours before final failure, load bangs were heard, and the building vibrated noticeably: this was linked to air-conditioning units which were then switched off in response. The top floor then began to sink, but even an hour before final disaster, the store was still trading. Eventually the roof gave way bringing with it the heavy air-conditioning equipment. Columns then sequentially failed and the whole structure pancaked into the basement. Almost instantly, 1,500 people became trapped and 502 were killed.

The store's structural frame was made of concrete, incorporating columns and flat slabs, and a common mode of failure (indeed one that usually governs slab thickness) is punching shear at the slab/column interface. By design, it is required either to have a large column (giving a bigger shear perimeter) or a thicker slab, giving more shear area. The investigation revealed that the slab shear design was quite inadequate, a mistake compounded by having only half the required number of rebars fixed on site. The bars were placed 100 mm below the top surface instead of the intended 50 mm (which clearly must have been a detailing or fixing error). Local reductions in column diameter had been made (i.e. weakening) by cutting into them to allow installation of escalators and a protective fire shield. Presumably these items also added considerable load. Certainly, a fifth-floor restaurant had been added, along with its equipment, so, collectively, the structure was overloaded, weakened and had a faulty design. As to the failure, as so often happens, warning signs were ignored. The final straw seems to have been repositioning of three air-conditioning units, following complaints of noise. Initial installation caused cracking and every time the units operated their vibrations caused crack propagation. Another theme that has been repeated many times in this book is a failure brought about by unauthorised change (Chapter 6).

This was structural failure, but, as with many failures, the consequences were far-reaching. There was public outrage characterised by demonstrations, criticism of safety standards and a level of public corruption was exposed. The store's owners and managers were charged with negligence, received jail sentences and forfeited their assets.

Summary

The generalised safety issues featured in these two example incidents are: modes of failure; brittle modes of failure; speed of failure. Chapters 3 and 4 promoted the notion of preference for ductility, for robustness and for avoidance of brittle modes. Any engineer experienced in flat slab design would understand the critical reliance on column head shear capacity. Under overload, it is far preferable for a slab to fail in bending before it fails at its supports in shear. Another general theme scheduled in Chapter 6 was to avoid unauthorised change: a feature here in the overload, in-service weakening and relocation of plant. In construction, another safety attribute is to be confident that what is built matches design intent; here it did not. Structures should show signs of failure before it is too late. Both these structures did, but warnings were ignored. That is unacceptable in terms of safety management and a commonly recurring theme in many of the disasters described later. Equally, worldwide failure reports regularly point to the need for structures to be under some form of regulation. Weak regulation enforcement is a feature in many of the disasters discussed later.

11.3.2 Factories

Factories are places where industrial processes are carried out and many people gather to work. Historically the processes themselves present many hazards so scandals over workers' health during the Industrial Revolution were a major driver in the progressive development of UK health and safety law (see Chapter 12). Factories are also major emitters of pollution, again potentially affecting the health of the wider population and damaging the environment. Gaining control over emissions has been a significant factor in the development of legislation. Fire has been an ever-present risk and the need for 'fireproof' mill buildings underpinned early developments in structural engineering such as the use of iron pillars and concrete floors (see Chapter 9).

Although incidents still occur in the UK, there is a culture of control and of addressing safety actively by looking at the potential hazards within a factory (or office) environment and then acting to minimise the risks. Indeed, this is a legal demand. In other parts of the world, standards are less rigorous, and deaths still occur. The Dhaka incident described earlier is an extreme example of the hazards workers might face. In the garment factories of Bangladesh there have been many fires and deaths due to smoke, blocked exit routes and so on. Safety in mines remains weak around the world, especially where there is unlicensed activity.

Bhopal gas tragedy, India, 1984

The event in Bhopal remains one of the starkest examples of what can happen. The plant was used to produce pesticides, which involved use of a toxic gas (methyl isocyanate) and other chemicals. One night there was a leak and gas permeated over a large area of densely populated neighbourhoods. The immediate official death toll was over 2,250 but other estimates reach multiples of that figure. It resulted in severe illness for large numbers, including blindness. A significant issue after the event was the difficulty of holding the plant owners to account.

The Seveso disaster, Italy, 1976

It should not be presumed that severe accidents only occur in developing countries. In 1976, there was a very bad accident at Seveso, a small town north of Milan. The chemical plant was used to produce herbicides and plant operation required some reactions at high temperature. On the occasion of the incident, operations were being shut down for the weekend but at the particular stage of the process that had been reached, this action caused certain temperatures to rise too much, leading to chemical decomposition that progressively developed into a runaway reaction. To cope with the pressures generated, a relief valve opened, permitting about six tonnes of material, which included dioxin (a deadly poisonous chemical), to be discharged and dispersed over a large area. Dioxin was normally present at tiny levels (parts per million), but on this occasion, because of the temperature, it was more concentrated.

Shortly afterwards many animals in the vicinity died and emergency slaughtering was ordered to prevent contaminated meat getting into the food chain (80,000 animals were killed). Fifteen children were hospitalised, and many adults treated for skin disorders. The local population was advised not to eat locally grown produce. The government had to spend huge amounts on land decontamination. As a response to the incident, the EU drafted new industrial safety regulations, known as the Seveso II Directive or, in the UK, as the COMAH Regulations (COMAH = Control of Major Accident Hazards).

Armley asbestos factory, Leeds, UK, 1960s and long term

An asbestos plant in Leeds (UK) caused contamination of its immediate surrounding area extending over about 1,000 houses. Over a period, at least 300 employees appear to have died from asbestos-related illnesses or cancers: all apparently due to emissions from the factory. Asbestos was a material once widely used in the building industry and still encountered as lagging in demolition and refurbishment works. It is now regarded as highly dangerous (asbestos may still be encountered when demolishing older buildings).[1]

Foot and mouth outbreak, UK, 2007

In 2007, the UK suffered an outbreak of foot and mouth disease which led to the slaughter of about 2,000 animals. A second outbreak occurred later in the same year which was traced back to pathogens escaping from the drain (a civil item) of a laboratory studying the disease. The drain was apparently leaking, disrupted by tree roots. The safety principle here is that a risk assessment could have predicted that a secure disposal route was required. Assessment would then have considered what factors might interrupt that route (i.e. invalidated design assumptions) and further considered how it could be ensured that the route remained intact. If there was no monitoring, it would not be known and that would be unsatisfactory if the consequences of a leak were unacceptable, as they were. The same logic will apply to the failure via explosion of the Stockline plastics factory in Glasgow (see later at Section 11.4.8). The safety issue at Stockline was that gas escaping from the buried pipe would lead to unacceptable consequences. There were predictable modes of failure in the supply pipe (corrosion) but no way of inspecting the pipe or monitoring its condition. Hence the system's safety was unacceptable.

Sayono-Shushenskaya power station, Russia, 2009

The Sayono–Shushenskaya power station was a giant hydroelectric plant, in fact the largest in the country. In 2009, one of its turbines spectacularly, failed ejecting its 920 tonne rotor. This missile destroyed the turbine hall, which then flooded. Nine of the plant's ten turbines were destroyed and the entire 6,400 MW plant was put out of action, leading to a blackout. A total of 75 people were killed. The cause seems to have been fatigue damage of the mountings, driven by turbine vibration. After the accident, some bolts were found to be missing and others were fatigue damaged. The plant staff had known that the turbine was vibrating for some time and vibrating much more than other turbines. The failure pictures are dramatic (Figure 11.3).

1 In 2022 it was reported that the Department of Education had confirmed that the vast majority of Northern Ireland schools still contained asbestos.

Figure 11.3 Sayono–Shushenskaya power plant failure (Alarmy 2E76KYP)

Summary

As a generality, all plants have hazards associated with their operations and those may demand safety containment via the civil engineered structure or there may be restraint interfaces with the civil engineered structure. Plants generally produce solid, liquid or gaseous products as waste and the collection and disposal of such waste is often via the civil engineered structure. A safety issue is therefore to consider the possibilities and consequences of unwanted emissions and their effects on health and the environment. When such emissions are intolerable, a robust containment policy is required. Part of that policy is at least to ensure that a leak is detectable before the emitted quantity becomes intolerable. There is commonality in this with nuclear containment demands and to the disaster at Buncefield, where fuel escape went unchecked until it was too late.

11.3.3 Dams

It is not difficult to imagine the devastating consequences of a dam failure. During the Second World War, dams in Germany were targeted by the RAF in famous raids in the Ruhr valley (1943). The dams destroyed on the Möhne released a huge volume of water which travelled quickly as a wave about 10 metres high, and reaching some 50 miles downstream. The flood waters destroyed two power plants and caused significant industrial damage. Agricultural land was washed away and about 1,600 people drowned.

The Brumadinho Dam failure in Brazil in 2019 was similarly catastrophic (Brumadinho Dam disaster, 2019). The failure can be seen, as it happened, on the Web.

Other dam failures have occurred 'by accident'. Failures have been generated by extreme in-flow leading to water levels too high or by impounded water overtopping with subsequence erosion. Failures have also occurred due to erosion caused by leaks (e.g. in piping), particularly in earthen dams; and through geological instability, or poor foundations.

Dolgarrog Dam, North Wales, UK, 1925

The worst dam accident in the UK occurred in 1925 in the Conwy valley. High above Dolgarrog village, two linked reservoirs fed an aluminium works on the valley floor. After a period of heavy rain, the upper dam failed releasing a flood down into the lower dam, which then also failed. The released waters poured down the mountainside into Dolgarrog, killing 16 inhabitants. Following this disaster, laws were passed governing the safety of reservoirs. Current requirements are covered under the Flood and Water Management Act of 2010. For many years, the effect has been to ensure that reservoir safety is assessed by specialised engineers (panel engineers) with particular attention being given to reservoirs of the largest capacity (British Dam Society, undated). There are about 3,000 dams within the UK.

Malpasset Dam, Fréjus, France, 1959

The Malpasset Dam above Fréjus collapsed in 1959, killing 423 people and causing huge financial losses. Failure was in a catastrophic mode, through disintegration of the concrete wall, releasing a torrent of water some 40 metres high and travelling at speeds great enough to destroy two villages in its path before eventually reaching the sea. The failure cause was put down to an undetected rock fault, but it might have been precipitated by very heavy rainfall that filled the dam to the brim and, for various reasons, the discharge valves (used to reduce water height) were not released in time.

Issues here are the mode of failure (which was so rapid as to form a deluge) and the role of operators in not responding to early warnings and not taking control action in time. There are reports of inadequate geological investigation too.

Teton Dam, Idaho, US, 1976

A significant failure occurred on the Teton earth dam in Idaho in 1976 with failure occurring as the dam was being filled for the first time. Failure was catastrophic, killing 11 people and 13,000 cattle. The dam's build costs were about $100 million and compensation claims about $300 million, with damage estimates of maybe $2 billion. During construction, large fissures were discovered in the base material but were not grouted up as being outside the shear key. Where grouting had been carried out, much more material was required than originally allowed for, which the inquiry suggested should have raised doubts about the likely success of any grouting operation. The first indications of a service problem were from leaks containing sediment, suggesting material erosion. The rate of discharge increased over a period of hours, and could not be plugged. Gradually, areas of ponding got out of control as the pace of failure accelerated. Local residents were evacuated. Thereafter the dam crest sagged and its main wall collapsed, releasing a torrent and emptying the reservoir completely. The cause was put down to permeable material in the core and fissures in the abutment, allowing water to seep through and develop piping. Whatever the exact cause, in all it was a combination of material factors and design error, with catastrophic consequences.

Carsington Dam, Derbyshire, UK, 1984

Another UK failure occurred during the construction of Carsington Dam in Derbyshire and this has similarities to the Teton failure. Slope stability/slip failures occurred during filling via shearing of the core through material that was too weak and through weak material in the foundations. The earth dam was intended to be about 37 metres high. Initially, tension cracks were seen in the crest and then, over a period of hours, a 400 metre length slid forwards by up to 15 metres. Although there were criticisms of the design approach, there also seems to have been some link to the manner of compaction (i.e. related to the compaction equipment and rate of filling). Hence, overall, this might be said to be a failure of both design and construction. Eventually the whole dam had to be replaced. Given the scale of the event it is perhaps fortunate that the failure occurred when it did rather than the dam failing later under operational water pressure. One of the lessons reported is that instrumentation ought to be added which could detect displacement as a precursor to failure. This would have fulfilled one of the safety objectives defined in Chapters 2 and 5 that, wherever possible, safety requires advance warning of something going wrong.

Another recommendation was that for such major projects, a design report needs to be prepared and reviewed by a panel of experts or other advisors and consultants: this is, in effect, the third-party review advocated elsewhere along with recommendations that such a review is not limited to dams (Chapter 13). The Carsington failure also illustrates that there is risk in construction itself and that there is significant interaction between design and construction. At both Carsington and Teton, the process of verifying the design assumptions ought to have been carried out. Although at Teton there was at least some warning of collapse, the value is questionable if the pace of failure is such that timely effective action cannot be taken. A key factor in all such structures is that whatever confidence there is in design, having an evacuation plan in case of failure is prudent (Chapter 13).

Kingston Fossil Plant: coal fly ash slurry spill, Tennessee, USA, 2008

In this failure, 4.2 million cubic metres of coal ash slurry was released, the largest release in US history. The spillage covered surrounding land up to 2 metres deep and clean-up costs were estimated at up to $975 million.

The direct cause of spillage was collapse of an earth retaining wall. Reports dispute the reasons, though one claims fundamental instability in the lower layers of sludge which was never identified and went undiscovered for decades. It was suggested that internal warnings were ignored. Overall charges of a corporate lack of safety culture were levied. Whatever the truth, no one seems to dispute that it was a terrible environmental disaster not just in terms of its scale, but also because the ash was laden with contaminants such as selenium, mercury and arsenic.

Ajka alumina plant accident, Hungary, 2010

The consequences of failure are no better exemplified than by the failure of this tailings dam. The dam retained thick alumina sludge. The event was the collapse of one reservoir corner which released about 1 million cubic metres of red sludge, discharged as a wave up to 2 metres high. The consequences were 10 deaths, 150 people injured, and sludge spread over about 40 square kilometres, with some reaching the Danube and polluting it severely. Unlike water, sludge does not drain away and so it lay as a blanket over much of the surroundings. Hence, apart from the need to observe that the consequences of dam breach are a flood, any risk assessment of consequences needs to assess such factors as rate of discharge (as at Teton) and whether the longer-term effects will be 'harmless' (fading away) or a residue of longer-term land pollution needing clean-up. In all cases, where the consequences are critical, a required design decision concerns the amount of safe volume to be impounded, i.e. reservoirs such as this could be subdivided (at additional cost) to limit the amount of accidental discharge, should one ever occur.

Oroville Dam spillway, California, USA, 2017

The Oroville Dam, completed in 1968, is the tallest in the US and it provides water to the arid southern part of California. Early in 2017, a large crater appeared in the main spillway, which was closed for repairs, but had to be quickly reopened to avoid overtopping of the dam. Before such repairs could be completed, the main spillway then began to erode further, and the emergency spillway was also damaged. Emergency plugging repairs had to be executed but about 200,000 people had to be evacuated in case the dam failed.

Summary

Several safety themes emerge from these notes of dam failures; these are largely universal in their application to all civil projects (see Chapters 4 and 5).

- Identify modes of failure and in the case of dams, how they will release contents.
- Look at the consequence of failure, not just in terms of flood volume, but in terms of rate and longer-term consequences. Consider ways of controlling the volume of discharge should

the worst happen (controlling sequential spread is also a factor to be watched as at Dolgarrog, or at Buncefield (Section 11.4.7) where the initial problem in a single tank progressed to other tanks).

- Always validate design assumptions. That process includes site investigation work and the need to modify designs when site conditions deviate from those anticipated.
- Note the role of construction methodology on quality. So, validate design assumptions during construction. Assess design construction proposals independently.
- Have a monitoring inspection system in place to give advance warning of trouble (the value of this in relation to the Whaley Bridge dam (2019) can be seen in Chapter 10).
- Always have management and inspection plans in place and an evacuation plan ready, linked to anticipated warning time.
- Always have an independent review of the total design.

11.3.4 Offshore disasters

Some of our most impressive, challenging and expensive engineering structures are those required for offshore oil or gas extraction. Rigs are massive structures subject to difficult operating conditions and subject to a range of plant hazards linked to potential explosion and fire. They operate in hostile waters where storm and wave action apply (and possibly earthquakes too). Oscillation of sea states means that much of the rig structure will be subject to fatigue. There are fundamental safety conflicts in that to operate, rigs must be manned and support overnight accommodation. This means that workers must co-exist with hazardous operations for extended periods. In case of disaster, rigs might be hard to evacuate, and the consequences of spillage are highly polluting.

To try and minimise such risks, rig risers incorporate automatic cut-off systems. However, despite the engineering care that might be expected in rig design, there have been any number of high-profile disasters. In several cases, commercial pressures to maintain production or operator error / human fallibility have been major causal factors in the disasters.

Sea Gem, UK, 1965

Sea Gem was the first British offshore oil rig in the burgeoning 1960s North Sea boom but unfortunately also the first British offshore rig disaster. The rig's structural form included a suspended platform and this collapsed completed precipitated by hanger failure (see Box 7.8 for technical cause) collectively there was both a progressive and disproportionate collapse. 13 workers died .

Alexander Kielland, Norway, 1980

The Alexander Kielland was a Norwegian drilling rig that capsized whilst it was being used for offshore worker accommodation. The rescue boat took an hour to reach the scene and rescued no one; the capsize had killed 123 men. The accident's direct cause was fracture in one of the bottom stabilising braces (see Box 7.10 for technical cause).

This failure offers many lessons. First, it was clearly a disaster losing an entire rig and killing so many workers. A second lesson is the scale of catastrophic failure that can occur consequent on even a small defect. It is a prime example, highlighting the need for engineers to build robustness into their structures. The features of initial defect, poor weld profile and plate tearing all suggest use of an inadequate material, an inadequate QA regime and an inadequate QC regime in failing to pick up the defect at fabrication. The failure may also be an example of the need for all changes to be authorised (i.e. the addition of the hydrophone fixtures) so as not to unwittingly undermine any of the initial design parameters (Chapter 6). It might be speculated that the Alexander Kielland failure, as with the Sea Gem failure, was at least partially brought about by engineers who were not conversant with the severe limitations that fatigue loading brings about. But the Alexander Kielland incident also illustrates the need to have in place a proper command structure to deal with crises when they arise.

Piper Alpha, UK, 1988

The Piper Alpha event was one of the most defining in engineering history (Figure 11.4). It has major interest to those who study 'safety' and ought to be a fundamental 'case' studied by all engineers. Piper Alpha was a large North Sea oil platform that started production in 1976. It was connected via oil and gas pipelines to two other installations and thence to shore. On 6 July 1988, there was a massive gas leak, which ignited in an explosion leading to large fires. The heat ruptured the gas pipeline riser from another installation which then also exploded into a massive fireball, engulfing the whole platform. Photographs of this event are iconic. All this took just 22 minutes. The scale of disaster was enormous: 167 people died; 62 people survived. The total insured costs were about £1.7 billion but a further £2 billion are said to have been spent on subsequent North Sea safety improvements.

The initial leak came from pipework connected to a pump. A safety valve had been removed from the line for overhaul and maintenance and the pump removed. There were two lines, one out of use and the alternative. Something went wrong with the alternative and pumping was erroneously switched back to the out-of-use pipe whose parts had been removed (there appears to have been a mix up in understanding exacerbated by a change in shift; see Chapter 6). As soon as the pipeline was pressurised, the pump blanking plate blew off, releasing gas, and the chain reaction of disaster was initiated.

An adequate safety response was lacking and commercial pressure for production was intense. Thus, even after the initial event on Piper Alpha, adjoining rigs remained in production and were, in effect, back-pumping fuel into the now stricken Piper Alpha rig, making conditions even worse.

Lord Cullen was commissioned to chair a public inquiry into the cause, leading to publication of his key report (Cullen, 1990). He made 106 recommendations, all of which were accepted, many being a direct result of industry evidence. Responsibilities

Figure 11.4 Piper Alpha blowout (Alarmy P6KMX8)

for implementation were spread between the regulator and the industry. One key recommendation was the requirement to prepare and submit safety cases. Such documents should be comprehensive and embrace full details of the arrangements for managing health and safety and controlling major accident hazards. Operating companies ought to have safety management systems in place; to have identified risks and to have reduced them to as low as reasonably practicable (ALARP) (Chapter 13). There must be management controls, provision of safe refuges and provisions for safe evacuation and rescue. Preparing such safety cases is a major task and, to be of use, they must be kept up to date. It will be observed that the safety case contents should cover many of the attributes (or lack of them) that have featured in so many other failures.

Sleipner, Norway, 1991

The Sleipner platform was a large offshore rig with its top deck supported below water by a large multi-celled concrete structure. The scale was massive, with the top deck weight alone being 57,000 tonnes, and providing accommodation for 200 staff and support for huge weights of drilling equipment. Below water there were 24 concrete cells (vertical tubes) each 12 metres diameter, and the whole structure was 210 metres high.

The structure failed catastrophically and sank during construction. Collapse was total and losses were up to $700 million. The cause was a design error that led to a gross underestimation of shear stress across the shell walls under hydrostatic pressure; the walls were just not thick enough (Chapter 6). The failure occurred during installation; the walls cracked, allowing water to pour in at such a rate that the rig became unstable. This failure was a 'classic' in the sense of being attributable to poor finite element modelling. It is also a classic example of 'common cause failure' (an error in one place repeating many times) and of lack of robustness: one error leading to catastrophic global failure. It is also a good case highlighting the significance of loading during the construction phase.

Deepwater Horizon, Gulf of Mexico, 2010

Piper Alpha was an undoubted disaster. Deepwater Horizon was by some measures even worse (excepting loss of life: 165 died on Piper Alpha and 11 on Deepwater Horizon.). The rig was a semi-submersible drilling rig operating in the Gulf of Mexico when there was a blowout. Just before the main accident, a geyser of sea water erupted from the riser, pushed by mud and methane, which then ignited into a huge explosion and fireball. The blowout preventer had failed. The fire could not be extinguished in the short term and the rig sank, leaving oil gushing out at seabed level, so creating the largest oil spill in US history and one having crippling effects on wildlife, fisheries and tourism. Eventually, after huge efforts, the well was capped off.

The investigating US Coast Guard laid blame on numerous causes; on engineering and on operations. Their report suggested that the rig had serious safety management system failures and a poor safety culture. It was claimed there was too much pressure on production at the expense of safety. Technically, the Coast Guard concluded that electrical equipment causing the igniting spark was in a bad condition and seriously corroded. Other deficiencies were alleged, such as improperly assembled gas detectors and emergency equipment; audible alarms switched off because of nuisance false warnings; complacency with fire drills; and poor preparation for dealing with a well blowout. All were cited as contributory factors, though all of this was contested by the owners. The full facts will most likely not emerge for some time yet but there are strong suggestions that the root cause was a poorly cemented riser seal (it ended up being off-centre). This had two effects; first, there was a potential upwards leak path for gas and second, the off-centre location prevented the blowout closer acting properly: an example of a single error knocking out two safety barriers. So far, BP has endured 11 counts of manslaughter, a felony count of lying to Congress, fines of $4.5 billion and projected settlement costs of $65 billion (2018 estimate). Partially in response, BP has filed lawsuits against former partners such as the company that cemented the well (improperly) and the manufacturers of the failed blowout preventer.

Petrobras rig, Brazil, 2011

The Petrobras rig was at the time the world's largest semi-submersible oil platform costing nearly $500 million to build, yet it sank progressively over five days after it was damaged by explosions. Support was derived from two large floating pontoons which in turn supported four columns and the upper deck. Failure began after an emergency drain tank located in one of the columns had been shut down and supposedly isolated. However, leakage of fluids and gases through the closed off valve overpressurised the tank, which then ruptured. A seawater service pipe adjacent to the tank also ruptured. Alerted by alarms, firefighting teams activated the seawater service pipe, but instead of supplying water for firefighting, this pipe was now flooding the support column. Gases from the burst drain tank then exploded, killing 11 firefighters. Separately, the flooding of the column also short-circuited a seawater pump located at the column's base within the pontoon so the valve failed 'open', which then allowed uncontrolled flooding of both the column and pontoon. In turn, that flooding tilted the rig submerging one corner progressively until the whole rig sank.

This failure highlights the importance of looking at engineering systems in total and examining the consequences and often complex interactions that can occur from individual failures, potentially culminating in disaster. In any mechanical system, it is always wise to assume that malfunction will occur at some stage and then to look carefully at the tolerability of likely consequences. Chapter 13 will describe the HAZOP and failure modes and effects analysis (FMEA) techniques which are ways of understanding such interactions. As discussed in Chapter 4, engineering systems need to be robust. It should not be the case that any single failure can initiate a chain of events escalating to disaster (see also NASA Challenger failure in Chapter 6).

Summary

Many of the features in these disasters echo generalised safety issues summarised earlier. There is a need to understand all modes of failure and it seems some early rig designs just failed to appreciate the hazard of fatigue damage or appreciate the needs for robustness. Some failure modes relate to loading conditions that only exist during construction or adjustment, so may be erroneously overlooked. Sleipner highlighted a current theme of concern, which is total reliance on computer modelling: 'safety' demands validation of the approach (Chapter 13). Piper Alpha brought home the lessons of having to look at safety overall, which includes recognition of all hazards; the way these might be controlled; the link between safety and maintenance; and the need for clarity of communication. It also highlighted the need to document the safety approach for external scrutiny. The role of human error stands out (Chapter 6). Deepwater Horizon reinforces the lesson of needing to know that what is built matches design expectation and the potential for common mode failures where one error might negate multiple safety barriers. Petrobras is a good example of the need to look at systems as a whole, perhaps by FMEA (Chapter 13) and to follow through chains of events sequentially across different technical boundaries consequent on some mechanical malfunction. Another lesson is that in any analysis, a starting point is to presume the mechanical device will fail, or alternatively, seek to understand why it is going to have its presumed reliability. It also reinforces the lesson to look at any system and identify where there is a possibility of one failure leading to catastrophe; these are sometimes called 'single point failures' (Chapter 5). Where these exist, if the consequences are intolerable there may be a need to make an 'incredibility of failure' argument (Bullough et al 2001).

11.3.5 Disasters involving structures: summary

The overall failure of any structure can be regarded as a disaster in its own right. But many historical incidents have had much wider implications when the consequences of structural failure have led on to the release of toxic substances, to floods or to widespread pollution. A key lesson therefore is that notional safety does not suffice; rather the amount of 'safety' provided should be tailored to uncertainty and to the potential consequences of failure.

Another key lesson is to build in the attribute of time. Knowing that something is going wrong and having time to react, to mitigate or to evacuate should be part of the overall strategy. Hence building in provisions for inspection and monitoring and automatic shutdown are all measures that should be considered.

One possible mitigation is to site hazardous industrial processes well away from more densely populated areas. This is wise when feasible, but in a crowded world not always possible. Where distancing is impossible, that imposes even more imperatives for proactive design and management of safety.

11.4. General disaster causes

11.4.1 Crowds

Structural engineers are used to adding live load to structures to account for the weight of people acting on the structure. In almost all cases the assumption is that the load is static though the notional load allowance in codes (say up to 5 kN/m^2) does allow some margin for impact effects. There have, however, been several cases when such notional loading has not sufficed, leading to disaster and great loss of life. There are two broad situations: those of crowds moving in rhythmic motion close to the supporting structure's natural frequency, which then produces amplified response; second, loads that arise from crowds in panic and stampedes and these introduce lateral loads on walls and barriers.

The knowledge of how to deal with crowd rhythmic response has benefited from our increasing ability to assess natural frequencies more easily and from studies of crowd behaviour (Chapter 8 gives background and examples).

The worst case for dynamic response is when stands are used for pop concerts, since then the crowd is 'trained' by the music to move in synchronisation and at a set beat (~ 2–3 Hz or so) which might easily coincide with the support's own frequency. An example of what can go wrong with crowds is the motion experienced by the Millennium Bridge in London on its opening day in 2000 (Chapter 8, Box 8.26) (Dallard et al., 2001).

Actual disasters have occurred several times from crowd crushing loads. An early case occurred at a Bolton football match in 1946. A worse event occurred at Bethnal Green Tube Station in 1943. During an air raid, people rushed into the station for shelter. In the rush (and in poorly lit conditions) a woman and child fell over, creating a blockage which escalated as more people tried to enter. Altogether 173 people were killed. More recent tragedies have occurred at the Heysel Stadium in Belgium (1985) and at Hillsborough (UK, 1989). In other parts of the world, almost unbelievable crowd numbers congregate at religious festivals, funerals and political rallies, so mass crowd behaviour is an ever-present risk. For example, about 30 million visited Allahabad for a Hindu festival in 2013. One stampede at the railway station killed 42 people. Crowd stampedes have occurred regularly at the annual Mecca Hajj. In 1990 more than 1,400 pilgrims were trampled to death in a tunnel. This was followed year on year by more deaths with about 2,000 being killed in a crush and stampede in 2015.

Burnden Park, Bolton, UK, 1946

A crowd of over 85,000 fans descended on the stadium to see an FA Cup match. The stadium was packed and overcrowded and as the crowd surged forward en masse, two metal barriers broke, leading to crushing, with 33 dead and 400 injured. In a similar 1902 incident at the Ibrox Stadium in Glasgow, a wooden terrace collapsed during a major match and 26 fans died.

Ibrox Stadium, Glasgow, UK, 1971

In another disaster at Ibrox, 66 Rangers fans were killed and over 200 injured in a crush during the last minutes of a match. As at Bethnal Green Tube Station, it appears that someone fell on the staircases while exiting and in doing so initiated a chain reaction of falling and pile-up. A similar incident occurred at the Luzhniki Stadium in Moscow in 1982, as fans exited early. Again, someone fell, there was a pile-up and the official account said that 66 people had been crushed to death (although eyewitness accounts suggest this figure was woefully short of the reality).

Heysel Stadium, Brussels, Belgium, 1985

There was riot between rival fans, some of whom were pushed towards a wall. As fans tried to escape by climbing over it, the wall collapsed, killing 39 people.

Hillsborough, Sheffield, UK, 1989

This is regarded as a defining tragedy that led to widespread changes in stadium design and crowd safety control. After the event, there has been much controversy and legal action between the victims' families and South Yorkshire Police over responsibility. Consequent on poor crowd management, too many fans entered the stadium, clashing with those already inside and 96 Liverpool fans died in a crush. Lord Taylor produced a report (Home Office, 1990) which introduced major changes; all-seater grounds, better barriers, removal of perimeter fencing and better crowd control. It was also a spur to understand better the dynamics of grandstands (especially those for dual use in pop concerts) and that in turn has led to the Guide to Safety at Sports Grounds (Home Office / Scottish Office (1986)).

Love Parade, Duisburg, Germany, 2010

In assessing crowd response, pinch points such as exits and staircases are especially important. In this festival, 21 people were crushed to death in a 240 metre long tunnel leading into the entry point. The tunnel became overcrowded, but perhaps the disaster was set in motion by people falling over and being unable to recover.

11.4.2 Impact generally

When anything moves there is danger of impact with nearby structures. Deliberate impact is a factor to be considered in any building which requires security control. Motorway crashes onto bridge piers, car crashes into buildings, train crashes onto structures and aircraft impacts have all occurred. The Chinese high-speed train crash of 2011 described in Box 10.8 is one such incident. Whatever the cause of that crash, one of its consequences was the train hitting the parapet. Observations such as that reveal that a general design objective is to recognise crash as a generic hazard and to try and contain the consequences: first, to minimise the risk of passengers within the vehicle incurring even more harm by plunging over an edge, and second, to make sure that the effects on the structure are not disproportionate (an allied safety issue is also to ensure passenger containment: Chapter 6).

The energy of impact is 0.5 mv^2, so normally a heavy vehicle travelling fast has the most energy. However, that is not the full picture, since damage caused is also a function of how much energy absorption capability there is within the 'missile'. An airplane, for example, has many parts that are quite 'soft' (the engines are the main missiles) so the effects of the impact itself during the attack on the New York Twin Towers or the Pentagon (2001) were containable and it was the subsequent fire which went on to create the total catastrophe. The New York Empire State building has been struck by a light aircraft in the past and survived relatively unscathed. Impact by ships tends to be severe because, although moving slowly, their mass may be huge. The risk is ever-present, as demonstrated by the Titanic hitting an iceberg. In 2013, the Queen's Jubilee barge Gloriana lost power on the Thames and collided with a bridge.

11.4.3 Vehicle impact

Vehicular impact is serious but tends not to cause disasters on the scale of other events described in this chapter. Nonetheless the hazard should not be taken lightly. Statistics show that certain bridges are habitually hit. One in London at Tulse Hill is reported to have been hit 92 times since 2009, accruing total repair costs of £23 million. But the UK record appears to be held by a bridge in Cambridgeshire which has been hit 113 times in eight years.

Apart from obvious dangers of damage to the vehicles or bridge, a more serious concern is the risk of completely dislodging a bridge deck and this has certainly happened. At Leybourne in Kent, in the UK, a lorry impact caused an M20 footbridge to collapse in December 2017. Fortunately, no lives were lost, though costs were put at £1.5 million. Vehicles coming off roads and impacting trains can cause much more serious damage, as at the Great Heck rail crash in 2001 (Chapter 2).

11.4.4 Ship impact

Ship impact incidents are characterised by enormous damage. Ships are slow moving, but their mass is so heavy that the energy they impart during impact is massive.

Tasman Bridge, Hobart, Tasmania, 1975

The Tasman Bridge disaster occurred when a large carrier collided with parts of the high-level bridge, precipitating complete dislodgement and collapse of the central deck. Several drivers were killed as their cars plunged off the free edge, unaware that the bridge had collapsed. The incident cut Hobart off from its suburbs, causing massive disruption and economic loss. A lesson is that a relatively easy way exists of adding an electronic detection system to warn of a bridge deck break and if that system had been in place at Hobart, at least it might have been possible to alert drivers and avert some of the deaths. Again, the safety principle is that if there is going to be a failure, the earlier the warning, the better.

Erskine Bridge in Glasgow, UK, 1996

Bridge bashing is the result of vehicles trying to pass beneath bridges that are too low for them to clear. It occurs on roads and in bridges over water. Some eye-watering incidents of the former, of motorway bridges being struck, can be seen on YouTube. An example of the latter occurred when an oil platform was being towed under the girders of the Erskine Bridge in Glasgow with its mast too high. The result was a catastrophic rip in the bottom flanges of the bridge deck, with repair costs put at £4.25 million. Heavy bridge traffic was prohibited for four months afterwards.

Genoa Harbour Control Tower, Italy, 2013

The harbour control room was a 50 metre tall concrete tube positioned on the corner of a jetty. During manoeuvres, a cargo ship rammed the jetty, causing a complete tower collapse. (Taking account of this failure and the Tasman and Erskine Bridge catastrophes, it was probably a wise decision to abandon any idea of bridging the English Channel and opting for the alternative of a tunnel.)

11.4.5 Aircraft impact

Aircraft impact is a generic risk since almost invariably flight paths cross cities. In the UK, Heathrow's main flight path lies all the way across London and planes cross every few minutes. There is no real protection beyond reliance on the high safety standards of airplanes and the control system on approach; hence the probability that any individual property will be damaged is statistically very low. Nevertheless, accidents do happen. In Manchester in 1957, a BEA Viscount landed short and crashed into a row of houses at one end of the runway. Remarkably only two residents were killed. Plane reliability has improved immensely since 1957, but remains imperfect. In Amsterdam in 1992, an Israeli plane crashed just after taking off from Schiphol Airport, directly impacting buildings and killing 43 on the ground, the damage being worse because the plane exploded and started a fire. Likewise, in the Twin Towers (2001) attack, the hazard was aircraft impact, but worsened by fire, which is an almost inevitable coupling. In Heathrow in 2008, a Boeing passenger plane landed just short of the runway (probably because of icing up in its fuel supply). There is always the risk of terrorist attack as in the Twin Towers or as at Lockerbie, Scotland, in 1988 when Pan Am Flight 103 was destroyed by a bomb and came down in the town, killing 11 residents. There is always the chance of pilot error; in 2013, a very experienced helicopter pilot crashed into a central London tower crane in fog. The potential for aircraft impact is a considered threat for high-security facilities such as nuclear power plant, partly protected by design and partly by enforced no-fly zones. For other tall structures, the standard demand is to ensure robustness to try and limit the scale of collapse should any impact occur. All tall structures should, of course, have aircraft warning lights.

11.4.6 Train crashes

A whole book could be written about train crashes; their cause and effect are worthy of a dedicated study on their own. From a safety study perspective, the crashes are of interest since they all feature the interaction between potential failures of mechanical and control systems overlain by the chance of operator error and compounded by maintenance error. Crash consequences frequently involve impact with civil engineering structures. The Chinese high-speed train crash (2011) has been described earlier (Box 10.8). In the UK, even in recent years, there has been a spate of bad crashes: Hither Green (1968); Southall (1997); Ladbroke Grove (1999); Hatfield (2000); Great Heck (2001); Potters Bar (2002); Grayrigg (2007).

Eschede, Germany, 1998

This accident occurred on Germany's high-speed train line. The train shed a wheel following fatigue cracking. The train was then derailed at points after which it collided with a road bridge; 101 people died and 88 were injured.

Viareggio, Italy, 2009

In this incident, a freight train derailed within the railway station. On derailing, the train hit the platform and overturned. Some of the wagons were transporting LPG (liquefied petroleum gas), and subsequently exploded. A large area of central Viareggio was damaged and 32 people died.

Lac-Megantic, Canada, 2013

On 6 July 2013, a runaway train filled with crude oil derailed in the small eastern Quebec town of Lac-Megantic. The tragedy began when a fire broke out in the main locomotive after the train had been parked. Firefighters extinguished the flames and turned off the engine; an action which inadvertently cut the air brakes. An hour later, the train rolled into the town along a stretch of track which was the second steepest incline in Canada. There the train derailed and exploded in a huge blaze. Forty-seven people were killed and much of the town centre was destroyed. When looked at as a system, this incident exemplifies the chain of events scenario discussed in Chapter 5. It also a safety risk inherent on discipline interfaces where apparently it was not appreciated that turning off the engine would create another hazard. For any moving system, maintaining control in all circumstances is a tenet that underpins system safety (Chapter 5). A HAZOPs-type study (Chapter 13) ought to have predicted this type of accident.

11.4.7 Explosions

There are many forms of explosion. In former times, explosions of gunpowder or munitions occurred, and many underground mine explosions have taken place where methane accumulation has been the hazard. During the early development of steam power, many boilers exploded (Chapter 2) and later on, when power was linked to supplies of gas and oil, there have been significant incidents in storage depots and refineries. Some industrial processes involving dust (wood, sugar, flour, etc.) all have explosive potential. A more recent threat is that from terrorist bombs. The consequences of explosion can be direct loss of life or might be the destruction of infrastructure or buildings.

Notable disasters still happen. In July 2011, a munitions dump blew up next to a Cyprus naval base. This was the island's worst-ever military accident and perhaps the fourth largest accidental explosion ever. The cause was self-detonation of 98 containers of explosives that had been surface stored in the sun for 2.5 years. The blast was large enough to damage hundreds of nearby buildings including a power station responsible for supplying over half of Cyprus's electricity. Much of the island then went without power. Costs were estimated to be 10% of the entire Cypriot economy. Dramatic city explosions have occurred from other causes, such as in China (Tianjin, 2015) and Jakarta, Indonesia (2017), but none has been so bad as the Port of Beirut explosion (2020) of stored ammonium nitrate, which left about 300,000 people homeless and is estimated to have caused $15 billion of damage.

Miners have always known the principle of detection and their ancient use of canaries for gas detection is well known. Matters improved in 1815 when Davy invented his safety lamp, a device that gave light but could not ignite methane. Despite this improvement, mining gas explosions were never eliminated. In 1910 at Pretoria Pit (England) 344 miners were killed. In 1913 at Senghenydd (Wales) 439 miners died. In the Blantyre mining disaster of 1877 (Scotland) 207 miners were killed. These explosions remain some of the worst incidents in UK mining history. Current UK risks are much reduced simply because coal mining is no longer a major industry. Nevertheless, gas explosion in any underground workings remains a generic risk exemplified by the gas incident at Abbeystead in 1984 (Chapters 5 and 10). The lesson is to be conscious of the hazard/risk in any area below ground (including any underground workings and sewers).

A common explosion cause in the UK is from poor installation of domestic gas supplies; failures in low-rise domestic property recur on a regular basis. About 40 occurred in the year 2019 to 2020 albeit that seemed to be an exceptional number, utter destruction and death can occur. The Ronan Point disaster (Chapter 8) was initiated by one such leak and dramatically showed the potential consequences (hence the incident led to the banning of gas supplies in similar tall structures). Figure 11.5 shows an example of blast consequences on a multi-storey building. Any fire in locations storing gas cylinders

Figure 11.5 Structural consequences of a gas explosion (Dreamstime 134326307)

is likely to be dangerous (Chapter 9). Another, albeit rare, domestic cause has been the explosion of heated water cylinders that are not vented to allow hot water expansion. Carbon monoxide leaking in houses is a generic risk though not an explosive one.

Fireworks factories

It might be imagined that fireworks factories are an obvious source of risk and they are; several have blown up. At Enschede in the Netherlands in 2000 a fireworks factory blew up with such intensity that 23 people were killed, 400 homes were destroyed and 1,500 buildings damaged.

Buncefield, UK, 2005

Buncefield was the site of a large storage depot holding about 60 million gallons of fuel. In 2005 it exploded, and the subsequent fuel fire has been described as Britain's largest since the Second World War: the blast was so large that ground motions could be measured. The incident started in one tank with the initiating event being tank overfilling, followed by ignition of the escaping fuel. The scale of the event was defined by the spread of ignition to about 20 other tanks.

The incident happened on a Sunday, which was fortunate, since many nearby buildings were empty when they were destroyed by the shock wave, so incredibly, the scale of injuries was quite slight. Inquiries afterwards suggested that in similar facilities, safety needed to be based on ensuring fuel could not escape or that, if it did, the scale of failure needed to be limited. Sometime after the event, claims were made that fire retardants used had contaminated local water courses. At the trial, the judge criticised sloppy practices and inadequate site risk assessment. The owner (Total) admitted exposing the public and its staff to risk and admitted allowing water below the depot to become contaminated.

Two aspects of safety stand out. Petrol being pumped in overflowed as the device measuring fill levels seems to have failed. This looks like an example of a control system failure (see Chapter 8). In looking at the safety of an entire system, an FMEA would have to consider the modes and consequences of any system failure. In this case, a conclusion might have been that the safety integrity level (SIL) required of the control system should be very high. Design rules are available for configuring control systems to meet such levels of reliability. The second aspect is one of containment and spread. Given that one tank might fail on a storage farm, a design objective should certainly be that disaster does not spread to other tanks in a chain reaction. Clearly in this case a chain event did take place. There are similarities to another event in 2009 at Cataño (US). This was an oil refinery fire where there was an initial explosion spreading to other storage tanks, culminating in a huge explosion. At the beginning of January 2018, a fire started in one car stored in a Liverpool multi-storey car park (Chapter 9). The fire rapidly spread to other cars and eventually about 1,400 cars were destroyed, along with the car park itself (SCOSS Alert, 2018).

11.4.8 Chemical plant explosions

There have been numerous chemical explosions. Examples are:

Flixborough, UK, 1974

This disaster occurred in a chemical plant. Pictures of the aftermath are dramatic and there was formal inquiry. Sometime before the explosion, a crack had been discovered in part of the plant and to permit continuing operation it had been decided to install a bypass pipe. On the day of the incident, this pipe ruptured, releasing a large amount of vapour which then exploded destroying the entire plant and causing widespread damage in the vicinity; the explosion also killed 28 workers and injured 36 more. Around 1,800 buildings were damaged within a mile radius. The event was the country's biggest explosion until the Buncefield storage depot incident in 2005.

The initiating event allowing gas to escape was the pipe failure and the inquiry concluded that its design and construction had been seriously defective. The people who designed and installed the pipe were unqualified and had no experience of high-pressure piping work; no drawings or calculations had been produced and the pipe was so badly supported that it was able to twist under pressure (other theories as to initiating cause have been proposed). Effectively an unauthorised and uncontrolled change had been made. Following the event, the UK COMAH regulations were amended: these cover the safety of hazardous industrial processes.

Pepcon, Nevada, USA, 1988

This was an industrial plant disaster. The plant produced fuel components for missiles and the space shuttle. For some reason, a fire started and spread to drums storing chemical components. A fireball was created leading on to a series of massive explosions. Two workers were killed, more than 300 were injured, and damage costs were put at more than $100 million.

Stockline factory, Glasgow, UK, 2004

Stockline factory was a plastics factory. In 2004, there was a terrible explosion, the worst in Scotland since Piper Alpha. The factory was destroyed, nine workers were killed and 40 were injured. The cause was later established as a gas leak from a corroded pipe laid under the factory. £400,000 fines were levied for breaches of health and safety law. (See comments in Chapter 5.)

Texas City refinery explosion, USA, 2005

In March 2005, the Texas City refinery exploded causing 15 deaths, injuring 180 people and forcing thousands of nearby residents to take shelter. The cause was an escaping vapour cloud which exploded. After the event, the operator (BP) faced severe criticism for safety mismanagement, much of it centred on 'cost-saving cuts in maintenance budgets'. BP pleaded guilty to violation of the Clean Air Act, and was fined $50 million. Even worse, in 2010, BP agreed to pay a further $50 million fine for continued violation of emission standards. The state prosecutor cited 72 pollutant emissions that had occurred after the 2005 explosion.

11.4.9 Pipeline rupture

Another generic explosion form can be linked to pipelines. Many serious accidents have occurred. Failure may be via fatigue, via corrosion, vandalism or even stupidity. In Alaska in 2001, a drunken man shot a hole in the Trans-Alaska Pipeline and released 285,000 gallons of crude oil; clean-up costs were in the region of $13 million. In Nigeria, people seeking to illegally tap off fuel have initiated disasters. Ruptured lines are common after the severe distortion imposed by earthquake ground movement and are one reason why fires occur in conjunction with seismic events (pictures from Kobe and Fukushima show such fires). All such incidents are illustrative of the general need to consider the location of hazardous facilities in relationship to inhabited areas. Spillages have also been responsible for huge environmental disasters. In 2001, an Iowa spillage during maintenance work killed 1.3 million fish in the Des Moines River. Any search of the internet will reveal an extensive list of incidents such as in the 'List of pipeline accidents in the United States' (undated) in the References.

Ufa train disaster, Russia, 1989

The explosion occurred when a leaking natural gas pipeline created a highly flammable cloud which was ignited by sparks from two passenger trains passing nearby. The result was train destruction and 575 deaths (many of them children) with 800 injured. There was another significant explosion on the Trans-Siberian pipeline in 1982.

Prudhoe Bay oil spill, Alaska, 2006

This leak was the largest in Alaska with potential for huge environmental consequences to wildlife. Fortunately, damage was contained since the leak occurred in winter when animals were absent. A feature of the event was that the leak persisted for five days without detection during which time more than 200,000 US gallons of oil escaped. The owners, BP, were prosecuted due to their neglect of corroding pipes. After the leak (which was from one hole) inspectors found several more zones of pipe wall thinning.

Puebla oil pipeline explosion, Mexico, 2010

This explosion occurred within the city of San Martín Texmelucan de Labastida, and it occurred when drug cartel thieves tried to siphon off oil. There was a gas explosion and oil fire resulting in a huge blaze extending over a 5 km radius. The explosion killed 29 people and injured many more. The blast destroyed homes and required the evacuation of thousands of residents. Environmental impacts included pollution of reservoirs and the local river (see also Section 11.4.10 on vandalism).

11.4.10 Vandalism

The case of a drunk shooting and puncturing an oil pipeline in Alaska is one example of vandalism creating disaster. But there are many others (Box 11.1). On railway lines, thieves have stolen cables for their copper content. In Derbyshire in 2012, there were a few incidents of vandals leaving concrete blocks on the track and in one instance a train hit them. In the same year, someone threw a brick over a motorway parapet which hit a van (as an aside, all bridge parapets should be secured down to minimise the risk of them being easily pushed over). In Manchester in 2007, youths were found to have tried hacksawing through support cables for a footway bridge over the M60, leading the Highways Agency to say it was the worst case of vandalism they had ever seen. A similar case occurred in Wales in 2013, when thieves took stainless steel wire forming part of a pedestrian bridge parapet. An even worse case was reported from British Columbia (Canada) in 2020 when for the second time someone deliberately cut through a cable supporting a cable car system, sending many cabins crashing to the ground.

Box 11.1 Falkirk Wheel, Scotland (2002)

The Falkirk Wheel is an award-winning rotating boat lift elevating boats 24 metres between the Forth and Clyde Canal and the Union Canal. Naturally there is a waterway at the top. Shortly before opening, in 2002, vandals managed to overcome the upper sluice gate and in so doing flooded the entire central control tower with all its machinery and electrical equipment, delaying the opening for a month.

Arson is a form of vandalism and a generic threat (arson in the royal dockyards remained a capital offence up to 1971!). The scale of threat can be gauged by observing that half of all UK fires are started deliberately. The bulk of school fires are arson, whilst hospitals, churches and public buildings are regular targets. Damage to property and the economy runs at about £2 billion p.a. One of the most tragic arson fires was that of York Minster in 1829 (see Chapter 9, Section 9.4.2). A considerable amount of arson damage occurred in the 2013 Hackney riots in London.

In early 2011, vandals set fire to a scrapyard that happened to be under the M1 motorway north of London. The fire weakened the overhead concrete bridge beams significantly; the motorway had to be closed (just before a bank holiday) and traffic was severely disrupted, at huge cost.

Protection against vandalism is partly a matter of looking at systems to assess their vulnerability and adding precautions as appropriate, potentially by added robustness or enhanced security. As buildings and structures become more sophisticated, that demand is likely to continue. In 2013, a woman stole a train from a depot outside Stockholm. She drove it for some distance before it derailed and crashed into an apartment block. Maybe this seems an incident rare enough to be ignored; but unlawful access to site yards and joyriding construction plant is not unknown. Security is required. Security is now critically important for any engineering project that includes computerised control, since hacking is an ever-present threat (see Section 11.7 on cyber-attacks).

11.4.11 Terrorism

Terrorist attacks on structures are a phenomenon the modern world is still coming to terms with. In the last half of the 20th century, there were many bomb attacks in Ireland and on the UK mainland perpetrated by Irish terrorists. One of the most spectacular was the bomb detonated in Manchester in 1996, which destroyed a good part of the city centre and required its redevelopment. Continental Europe has also suffered bomb attacks from various groups (for example, the 2004 incident in Atocha Station, Madrid). In the US in 1995, a bomb attack in Oklahoma City led to a spectacular collapse of the whole front face of the Alfred P. Murrah Federal Building after a column was blasted out. Unfortunately, the beam overhead had no bottom rebar continuity over its support,

Box 11.2 Twin Towers, World Trade Center (11 September 2001)

This attack by suicide bombers is probably the worst terrorist attack of all time, certainly in terms of its effect on a structure. Awareness is all the more highlighted because of worldwide TV pictures showing the calamity live as the tall towers disintegrated, falling in on themselves into huge rubble piles. Many lessons have been learned on the resilience of structures, their modes of failure and the need for an integrated design approach. Amazingly the plane impact did not precipitate immediate failure; rather it was the prolonged exposure to burning fuel that brought about the scale of damage. A tragedy is that the towers stood for a while, yet it proved impossible for many occupants to escape. The need to learn lessons is urgent as many similar tall structures are planned throughout the world (Institution of Structural Engineers, 2002).

so had no chance of survival once its prop had been removed. Equally, the attack on the US Embassy in Dar es Salaam in 1988 resulted in total building failure. However, no attack has had greater impact both in actuality or in political terms than the suicide plane attacks on the World Trade Center in New York in 2001. The iconic Twin Towers, 502 metres high, were both completely destroyed, with a death toll of around 3,000 (Box 11.2).

All these events have led to much debate across the engineering community on methods of defence and how to protect and design structures with the objective of at least minimising life loss. General principles revolve around making structures inherently robust, keeping potential threat vehicles well away from facades, making windows shatterproof and protecting occupants from flying glass. A forensic investigation of incidents has supported rules for detailing to achieve robustness (Institution of Structural Engineers, 2010, 2013).

There are no indications that threats are likely to be eliminated. The Lockerbie bombing of 1988 remains a deep scar. Glasgow Airport was targeted with a car bomb in 2007. London has endured many bombs, notably those on the underground in 2007 and Boston (US) suffered bomb attacks during its annual marathon run in 2013. There were bomb attacks in Manchester (UK) in 2017 (Box 11.3).

Box 11.3 Manchester Arena bomb attack, UK (2017)

On 22 May 2017 a suicide bomber left a device in the Manchester Arena concert venue. The bomb exploded as the pop concert ended, killing 22 people. Fortunately, the building survived, or the death toll would have been higher. This incident illustrates the importance of major public buildings having a good measure of robustness to cater for the unknown.

11.4.12 General disaster causes: summary

There are many causes of man-made disasters, some linked to standard structural engineering and error, some due to relatively new phenomena such as suicide bombers. But generally, civil engineering structures only supply the containment envelope or support for other activities, be they industrial processes or transport of some kind. Each type of usage brings specific hazards linked to plant malfunction or operator error. If the project as whole is to be safe, it remains vital to identify all the hazards and their probability of occurrence; to be skilled enough to identify modes of failure and to provide mitigation. Skill in identifying hazards and modes of failure comes partly from dialogue across the whole project team but especially from observing what has gone wrong before on similar projects.

All the subheadings listed in this chapter are associated with standard hazards which should be considered in the civil/structural engineering design. In popular crowded venues, the loading demands on support structures or barriers can be very high and if barriers fail, the consequence can easily be multiple deaths. Many types of structure are potentially subject to impact from missiles, varying from lightweight cars up to the enormous mass of ships. Failures that have occurred point to what might happen and highlight the need to reduce impact risk as much as possible or to configure designs such that impact can be absorbed short of catastrophic consequence.

Designers need to stay abreast of new hazards that arise over time. No doubt there has always been vandalism, but some of the examples described in this chapter illustrate what terrible harm can result from malicious acts. Hence in any risk assessment, the heading of 'vandalism' as a hazard ought to be included; likewise, the very difficult topic of 'terrorism'. Some government buildings are obvious targets, for which explicit preventative measures are warranted. For many other structures, the perceived threat level might be low, but a measure of protection created by robustness is always prudent.

Consideration needs to be given to potential coincidence. At Viareggio, it seems to have been incredibly bad luck that there was (a) a derailment, (b) it occurred within a station, and (c) the trains were carrying LPG (unless that was a regular cargo). It might also have been low probability that the Eschede train shed a wheel just before a bridge (unless there were many bridges). In contrast, the incident at Great Heck (described in Chapter 2) could be regarded as having quite a high probability. The incident at Lac-Megantic might be thought of as a low probability event, but it is not the only case where control has been lost of moving vehicles, or moving structures. It serves to highlight the essential demand to assess systems as whole, looking for potential chains of events.

11.5 Nuclear disaster

11.5.1 Introduction

Some of the worst man-made disasters, or at least some of the disasters people fear the most, are those linked to the nuclear industry. The fear is of explosion, fallout, radiation and contamination and their effects on health and, not inconsiderably, just a fear of the unknown. For whatever reason, public trust in nuclear power is low and partly justified since the industry and governments have not always been as open as they should have been. Nuclear disasters have the potential for being very long lived, for causing environmental catastrophes and for being no respecters of national boundaries. It is therefore no surprise that some of the greatest challenges faced by engineers are those of crafting a safety case against a potential failure and of persuading populations about nuclear safety. This is an essential skill since many believe that the world's energy problems, and the world's defence against global warming, cannot be resolved without some reliance on nuclear power within the energy supply mix.

11.5.2 Containment

Civil engineering boundaries play a significant part in the containment of radiation and contamination; it is therefore important to be aware of potential failure modes. These include mitigation of plant failures directly, plus containment of consequent discharges. The boundaries might be required to contain pressure or active waste products. The direct effects of radiation on the stability and durability of engineering materials are a significant consideration. Given the extreme reliability demanded from plants, allied with a demand for very low probabilities of failure, the expected design forces and durability demands on the civil structures are often extreme.

11.5.3 Waste control

Following the demonstration of energy release via the dropping of atomic bombs on Japan in 1945, many countries made efforts to harness the enormous energy potential of nuclear fission for peaceful purposes. The first functioning plant producing power was at Calder Hall in Cumbria, UK, which went on stream in 1956. Thereafter, other plants provided power in the US and Russia, with various types of reactor being tried out. In the UK, much early effort was hasty, so control of waste in terms of inventories and storage was not good. Nor were the containment design standards acceptable to modern thinking and little consideration was given to capability to withstand accidents and extreme events. Since that time, clean-up, making safe, and decommissioning have absorbed huge resources over many years. Safe waste control and disposal remain major issues for the community.

11.5.4 Failures

There have been a few major accidents all characterised by losing control of plant; by inadequate regulation; and by inadequate safety thinking. The consequences have been land contamination plus radiation releases with their direct threats to health.

The Windscale fire, Cumbria, UK, 1957

This event is reckoned to be the worst nuclear accident in the UK. Windscale in Cumbria was a plant producing plutonium for the UK's atomic bomb project. In 1957, plant activity was taking place during which some of the uranium fuel ignited, creating a heat so intense that it threatened the integrity of the concrete biological shield. Part of the cause was due to scientists raising the reactor temperature beyond its original design specifications and so unwittingly altering the internal core heat distribution to such an extent that it created hot spots. These spots remained undetected because thermocouples were not positioned in the hottest regions. By the time problems were noticed, the fuel fire was out of control and there was no effective way of putting it out. Efforts to cool the core by increasing air flow only supplied more oxygen, so intensifying the fire and pushing radiation up the chimney. It was only by heroic staff efforts and some risky undertakings that the fire was eventually damped down; the two chimneys were then sealed, but decommissioning was not started till 2012.

During the incident, much surrounding land was contaminated and milk from grazing cows had to be destroyed. At the time, the significance of the event was not publicised, and no evacuations were carried out.

Three Mile Island, Pennsylvania, USA, 1979

The Three Mile Island plant incorporated a pressurised water reactor (PWR). An accident there in 1979 was the worst in the history of the US nuclear industry and involved partial meltdown of the reactor core, an event that designs seek to avoid at all costs. Mechanical failures allowed reactor coolant to escape but, apparently due to human error, the operators did not recognise the true nature of the consequences of the incident because some radioactive elements along with a large amount of contaminated water were released into the local river. During the incident, local evacuation took place. Clean-up costs were reported to be about $1 billion.

Chernobyl, Ukraine, 1986

This incident is thought to be one of the worst disasters of all time. During the accident, which involved an explosion and fire, a significant portion of the plant was destroyed, and large areas of surrounding land rendered uninhabitable. About 350,000 people had to be evacuated and resettled and a local city was abandoned. Fallout from the incident drifted all over northern Europe, though the effects in terms of directly caused deaths (of plant workers and locals) and longer-term health have been much disputed.

The incident was initiated by a catastrophic power increase which itself was a consequence of plant operation being unstable in certain conditions (the sequence of events is complex). The power surge caused the rupture of a reactor pressure vessel and generated a series of steam explosions with the damage collectively exposing the core to air. The core then ignited, sending a huge amount of radioactivity into the atmosphere where the active particles drifted with the prevailing wind over a vast area. The effects were noticeable even in the UK some 1,300 miles to the west, where restrictions on grazing sheep entering the food chain were imposed. The incident was bad enough, but its significance was exacerbated by the secretive nature of the Soviet government, which attempted a cover-up. Paradoxically, the accident took place during an experiment which was being carried out to test core cooling as an essential safety feature. There is evidence that the experiment was being improperly controlled because certain safety features had been overridden: the plant was operating outside its intended design envelope. Furthermore, the operators were unaware of a fundamental flaw that had been revealed by a previous incident in Leningrad in 1975. As so often there was a precursor to the disaster.

Given the dangerous nature of the plant, it was rapidly enveloped to contain the remaining debris. However, that envelope itself decayed and a subsequent major civil engineering project eventually encased the whole in a 'sarcophagus'. Given that safe access did not exist, the scheme built the cover (a very large, arch-shaped hangar type construction) at one the end of the damaged plant and then slid it over. The engineering effort to achieve this was enormous and can be appreciated in the Chernobyl arch (2016) video.

Fukushima nuclear disaster, Japan, 2011

The spread of disasters from Windscale in 1957 to Fukushima is 2011 suggests that the problem of safety in plants is not resolved and can only ever be contained by vigilance and by studying past failures. Clearly no engineers working today have contemporary knowledge of an event that happened more than half a century ago. One key difference between the two incidents is that whereas the public were largely ill-informed about Windscale as it happened, Fukushima played out live on TV in front of the whole world. Moreover, in contemporary society there is a greater scepticism, a more intrusive press and a wider desire to know the full facts.

At Fukushima there was loss of cooling to the reactor, which overheated, causing core meltdown and thereafter a release of radiation. Local populations were evacuated (about 140,000 people) with a significant life loss during the evacuation process (about 50 deaths – but none directly related to radiation). A few workers received very high doses, but it is too early to say what the effects will be on their long-term health. As with previous disasters, the cause was failure of mechanical systems (in the cooling circuit) and then overheating; but of interest in safety terms was the failure of the backup systems through causes and in scenarios that were foreseeable, and indeed known about, but had not been rectified. Thus, there were violations of the safety requirements and inadequate enforcement by the regulator. The direct cause of the incident was the tsunami that struck the coast and overtopped the defensive walls (see Chapter 10). In civil engineering terms, there was a lack of containment. Moreover, concrete tanks that stored fuel under water for shielding and cooling cracked and leaked, allowing the fuel to become uncovered and overheat. Serious though Fukushima was, it could have been far, far worse (see also Chapters 5 and 6).

11.5.5 Nuclear disaster: summary

Four serious nuclear incidents have been cited. In each case part of the civil engineering structure was affected, always in its role as a containment boundary. Furthermore, in the abstract study of safety there is much to learn. First, the spread over time shows that successive generations of engineers really do have to understand and learn from what has occurred before. Second, it might be observed that the four incidents all occurred in different countries: Britain, USA, Ukraine and Japan. It is not therefore the case that the incidents can be blamed on laxity or ignorance isolated to any one country. We must all learn on an international scale not least because, in nuclear incidents, the consequences are global.

There is commonality between the failures. In each case the plant could be looked at as a complex system with significant interaction consequent on local failure. In each case, 'knowing what was happening', or not, was an important feature in responding to the incident. In each case, instability or the failure getting out of control was a characteristic. In each case, there was a failure of plant operatives to fully comprehend what they were doing; so, training, competence and leadership are vital. In each case, there also seems to have been some failure on the regulation side. Each incident also suggests that whatever confidence there is in safety, there needs to be contingency planning for responding to potential failures and there need to be contingency plans for protecting local populations at short notice (all very similar to the response needed for other disasters). All these ingredients are in place in the modern nuclear industry and latest reactor plans are designed to be 'fail-safe' and to include containment for the core even if it does melt. But as Fukushima shows, there needs to be rigour in studying safety plans, for the prime incident initiating the event also knocked out the backup systems (a type of common cause failure; see Chapter 5). The lesson about knowing what is going on via instrumentation and so on is one that has wider application in the safety of all systems.

11.6 Environmental disaster

11.6.1 Introduction

In one way or another industry has been responsible for much industrial pollution, whether in solid, liquid or gas form. In earlier times, engineers paid scant attention to the consequences their endeavours might have on the environment and during the Industrial Revolution, the results were truly dreadful. Engel's description of Manchester in 1844 (Engels, 1892) included the following description:

> The view from this bridge, mercifully concealed from mortals of small stature by a parapet as high as a man, is characteristic for the whole district. At the bottom flows, or rather stagnates, the Irk, a narrow, coal-black, foul-smelling stream, full of debris and refuse, which it deposits on the shallower right bank.
>
> In dry weather, a long string of the most disgusting, blackish-green, slime pools are left standing on this bank, from the depths of which bubbles of miasmatic gas constantly arise and give forth a stench unendurable even on the bridge forty or fifty feet above the surface of the stream. But besides this, the stream itself is checked every few paces by high weirs, behind which slime and refuse accumulate and rot in thick masses. Above the bridge are tanneries, bone mills, and gasworks, from which all drains and refuse find their way into the Irk, which receives further the contents of all the neighbouring sewers and privies.

There is much more in this vein; mercifully Manchester is not like that now! Nevertheless, many years later we're still faced with clean-up and remediation of contaminated rivers, sea and land. In the 1840s there were cholera epidemics in London caused by foul drinking water, though that was not understood at the time. The city had no proper sanitation and all manner of industrial and human waste littered the land or was dumped in the Thames, with the river water frequently being recirculated for human consumption. The air was notoriously foul and the streets filthy, all of this having a measurable impact on the lives of the poor. Charles Dickens's contemporary novels give graphic descriptions of poverty and pollution. Enlightened reformers, such as the secretary of the Poor Law Commission, Edwin Chadwick (publisher of a report, 'The Sanitary Condition of the Labouring Population of Great Britain'), grappled with the problem and a large Victorian sanitary movement evolved, trying to improve matters. In 1843, Prime Minister Peel created the Royal Commission on the Health of Towns and the Public Health Act came into law in 1848; some drainage improvements followed. But it took the hot summer of 1858 for conditions to reach a state so acute that they gave rise to an overwhelming stench ('The Great Stink'), a smell so bad, that even work in the House of Commons was interrupted. Thereafter Parliament passed a bill for the building of a new London sewer scheme. Engineers then played a huge and positive role in improving health conditions for the general population, exemplified by the great London civil engineering works of drainage and sewage disposal by Joseph Bazalgette

(1819–1892), which began the cleansing of the Thames and the elimination of cholera. Since that time, 'public health engineering' has been a key specialisation within the civil engineering profession.

Water pollution is both an environmental disaster and a health hazard, since many other diseases, as well as cholera, are transmitted via foul water. Yet it is really not so very long ago that the quality of water in our water courses was still way below acceptable standards and it was horribly contaminated with pathogens. Engineering work to improve quality is a long-term task. Section 11.3.2 referred to the disasters of factory failures dumping pesticide chemicals onto land and people (Seveso and Bhopal) with awful health consequences. Even if such disasters cause no immediate health problems, they can destroy our environment: the sludge release by at the Kingston Fossil Plant (Section 11.3.3) has been described as a huge environmental disaster. There cannot have been a more dramatic example of the potential for engineering failure to pollute than the dumping of oil in the Gulf of Mexico after the Deepwater Horizon rig failure in 2010.

The generic safety issue now is always to consider the consequences of waste products and their safe treatment and disposal. The general assumption should always be that any discharge is a potential health/environmental hazard and engineers have both legal and moral responsibilities to limit risks. Nevertheless, the topic of dumping untreated sewage into the UK's watercourses remains one that arouses much contemporary public and government concern.

News reports in 2018 revealed that industrial air pollution in India is having severe effects on the population. It is so bad, and includes so much sulphur dioxide and other contaminants, that even the colour of the Taj Mahal is changing from white to yellow.

11.6.2 Emissions and contamination

Just about any industry creates some degree of pollution and in redeveloping any site, legacy contaminated ground is to be expected (this poses health risks to workers). Waste from old industries, mines, gasworks, dye factories, bleaching works, chemical works and so on have left us with legacies we have to cope with. Wastes posing direct health risks are heavy metals, poisons such as cyanide and arsenic and buried asbestos (see the case of Armley asbestos factory, Section 11.3.2). A notably bad case is that of Wittenoom, Australia. At one time this was Australia's only source of blue asbestos; in fact mining asbestos was the town's only reason for existence. Working conditions were extremely poor, with workers breathing in asbestos dust. The health legacy has been dreadful. Organic products that decay to produce methane are also hazardous, as are those that contaminate aquifers or water courses. New threats emerge, such as in 2018 with the worldwide realisation that waste plastics are polluting the oceans with unknown consequences. Some case histories are bad enough to figure on the 'disaster scale', and illustrate what can happen.

Poisoning

Chemical discharge (liquid, solid or gaseous) may directly poison populations. Some notable cases are given below.

Lead poisoning (1900 to 2000)

Lead is widely used in construction. Before the introduction of copper and plastic pipes it was commonly used as a water piping material and used in paints. Early in the 20th century, lead was used in petrol and was not phased out until the end of the century; in the UK leaded petrol was still available until 2000. However, lead is highly toxic and the discharge of compounds in car exhaust gases generated widespread health impacts: lead is absorbed in the bloodstream and is linked to brain damage. Indeed, phasing out lead has been credited with helping cut crime, as one of its adverse effects seems to have been promotion of aggressive behaviour. Domestically, the greatest exposure in households is from swallowing or breathing in lead paint chips and dust (many older houses will still have lead-based paint on woodwork; the paint can become a hazard during renovation). Lead in drinking water is known to have adverse effects, particularly in infants, by delaying their physical and mental development and it can increase blood pressure in adults. There have been a severe problems in Nigeria where children have been poisoned by lead in the gold-mining areas by breathing in dust or ingesting it from contaminated soil. Estimates of about 400 child deaths and perhaps 2,000 permanent disabilities have been published.

Japanese environmental disasters (1900 to 1970)

In Japan there have been some particularly graphic illustrations of the effects of toxic waste on human beings. Mass cadmium poisoning has occurred (up to the 1940s), which softened the bones and caused severe joint pain and disability. The cadmium was released into rivers by mining companies, with that water eventually being used to irrigate paddy fields, so the source of ingestion was rice. Much more recently, up until the end of the 1960s, there have been bad cases of mercury poisoning, causing all kind of terrible reactions. In the 1950s and 1960s epidemics, the source was mercury, released in waste water from chemical factories, which entered the food chain via fish and shellfish and then afflicted several thousand victims. There have also been epidemics (and resultant deaths) of pulmonary disease, bronchitis, emphysema and asthma consequent on breathing excess sulphur oxide pollution from petrochemical processing facilities (up to the 1970s). The only solution was to introduce flue gas desulphurisation to clean the discharge. Although successfully solved in Japan, air pollution and allied health problems remain in much of the industrialised world. Chapter 10 discussed some of the effects of air pollution in the UK that are linked to gaseous emissions from diesel engines, which will translate into engineering demands for cleaner transport systems.

Lead poisoning is not some distant problem. In 2014, the town of Flint in Michigan (USA) switched its water supply and the new supply was badly contaminated with lead.

A national scandal ensued (President Obama intervened) and the required fix was reported as costing $28 million.

The Sandoz chemical spill, Switzerland, 1986

This major environmental disaster was caused by a fire and its extinguishing at a chemical store in Switzerland. Tonnes of pollutants entered the River Rhine, turning it red. The consequences were the destruction of fish and plant life downstream, and disaster for many populations along the river who rely on it for drinking water. It should be recalled that the measures for extinguishing fires might have consequences as harmful as the fires themselves. After Buncefield (Section 11.4.7), there were claims that the fire retardants used had polluted groundwater supplies.

Love Canal disaster, USA, 1970s

This disaster shows just what can happen when waste dumps pollute groundwater. The event was so bad it has become a symbol of improper storage and lax regulation. The 'canal' was never a true canal; rather it was part of an abandoned project of which one remnant was a large ditch. Up to the 1950s, the ditch was used as a dump for all manner of chemical wastes; about 21,000 tonnes in volume. The waste was initially buried deep and covered over. But after some time, contaminated groundwater seeped into nearby basements; the ground cover was eroded, exposing some of the waste and plants began to die. Local residents complained of birth defects and other health problems. The Carter government then declared a state of emergency, designating the issue a human-caused environmental disaster. The clean-up was declared finished in 2004, with costs up to $400 million. As a follow-up, thousands of other hazardous waste dumps, were assessed with many hundreds being declared dangerous.

Cornish mines, UK, 1990s

Mining has been carried out in Cornwall ever since pre-Roman times. At one time Cornwall was the world's major source of tin. Nowadays the mines are abandoned and flooded but these flood waters are potential sources of contamination (for example, from the Wheal Jane group). Historically the water had been acidic and high in metal content: iron, zinc and cadmium. In the past arsenic, copper, silver and zinc were also worked. In the 1990s, acidic waters from Wheal Jane overflowed and drained into the Carnon Valley and thence to Falmouth Bay, killing some wildlife. In response, settling ponds were constructed, with treatment costing many millions of pounds.

Oil spillages (1960s to 2013)

Oil spillages at sea are tremendously damaging to wildlife, fisheries and tourism. Spillages are an ever-present risk from wrecks of giant tankers or from an offshore oil well's blowout. Some notable cases are:

The Torrey Canyon was wrecked off Cornwall in 1967 and in the process dumped around 35 million gallons of crude oil in the sea. The cause has been variously reported as navigational error, or human folly. It has also been reported that a contributory factor was a design error in the steering system. At the time, the Torrey Canyon was one of the largest wrecks ever. But in 1978 the Amoco Cadiz ran aground off Brittany, dumping 69 million gallons of oil so that incident then became the largest oil spill to that date.

The Exxon Valdez spill in Alaska in 1989 was up to 120,000 m^3 (about 11 million gallons) and considered to be one of the most devastating-ever human-caused environmental disasters. The remote location made clean-up exceptionally difficult, so although in terms of volume, the Exxon Valdez spillage ranks well below the worst, its environmental effects were some of the most severe on wildlife.

The Prestige oil spill off northern Spain (2002) dumped around 20 million gallons of oil and polluted a huge length of coastline, creating a disaster for both fishing and tourism. The cause was rupture of one tank during a storm. It was defined as the largest environmental disaster in the history of Spain and Portugal and it also affected France. Heavy coastal pollution meant that offshore fishing was suspended for six months.

All these previous events have been dwarfed by emissions from the Deepwater Horizon blowout in the Gulf of Mexico in 2010 (Deepwater Horizon, 2010). Leakage over the duration of the crisis has been estimated at about 206 million gallons, an order of magnitude larger than previous spills. The environmental impact and compensation claims are gigantic. Estimates of oil dumped into the Persian Gulf during the Iraq War are even higher than the amount released by Deepwater Horizon, and there are many other incidents of extreme oil spillages.

Air pollution

Air pollution is a disaster almost entirely man-made. It is a condition that used to badly afflict the UK before the introduction of the Clean Air Acts. Chemical smogs and hazes still affect many cities around the world, causing severe breathing problems. Air pollution in Beijing was a serious concern just before the 2008 Olympics. Occasionally there are really bad incidents caused by industrial spillage. One such case occurred in Alberta (Canada) in 1973, when a plume of poisonous hydrogen sulphide escaped from oil drillings and spread over a wide area, requiring mass evacuation. Chemical concentrations were reported to be 40 times higher than normal and many people suffered burning eyes and throats. Occasionally the pollution comes from 'natural events'. In summer 2013, the atmosphere in Singapore was so bad that people had to wear protective masks as the city was enveloped in smoke haze from forest fires in nearby Indonesia. This was reminiscent of Moscow in summer 2010 when surrounding forest fires polluted so badly that tourists could hardly see across Red Square. Newspaper reports early in 2013 reported residents of Beijing struggling through 'airpocalypse-smog' (The Guardian, 2013). On one day pollution levels were 30 times higher than those deemed safe by the World Health Organization (WHO). For most of January, the air quality was worse than that of an airport smoking lounge. And all linked to industrial emissions. The air pollution in Delhi is notorious. In the UK, ever increasing attention is being given to pollution from water courses, traffic and industry. All this will affect civil designs in the water, transport and energy sectors.

Environmental destruction

Aral Sea, Kazakhstan/Uzbekistan

Some regard the destruction of the Aral Sea as one of the greatest environmental disasters of all time. It used to be one the world's largest lakes, but after diversion of feeding rivers for irrigation, it is now only around 10% of its original size. Local ecosystems have been destroyed, there is heavy pollution and public health problems. Where there was once sea there are now plains covered with salt and chemicals and the toxic dust is dispersed by wind. In some ways this echoes the 1930s US Dust Bowl (Dust Bowl, 1930s): both disasters brought about by man's lack of understanding.

11.6.3 Global warming and climate change / climate emergency

Global warming is high on the world's agenda for action. Despite controversy over causation, most scientists seem to agree that warming is a reality with strong evidence of cause being linked to atmospheric pollutants released by mankind. If so, perhaps it will be the greatest disaster yet to come, for changes in weather patterns can affect our very ability to survive. Recognising this, the UK Institutions of Civil and Structural Engineers have both adopted the term 'climate emergency' and are expending great efforts to embed decarbonisation into construction activities. This is having profound effect on infrastructure design and construction.

In January 2018 the World Economic Forum at Davos issued its annual report spelling out global risks (World Economic Forum, 2018). The report makes sobering reading in its descriptions of the year's climate challenges and its forecast of trends. In 2017, the US (as a single country example) had suffered a record year from wild fires, hurricanes and other related disasters with total costs estimated to be $306 billion ($90 billion higher than the previous worst year in 2005). So these effects, although apparently 'natural' might in effect be 'man-made' (Chapter 10). One telling quote under the section heading 'Our planet on the brink' is:

> However, the truly systemic challenge here rests in the depth of the interconnectedness that exists both among these environmental risks and between them and risks in other categories—such as water crises and involuntary migration. And as the impact of Hurricane Maria on Puerto Rico has starkly illustrated, environmental risks can also lead to serious disruption of critical infrastructure.

The warning about disruption of critical infrastructure is a challenge engineers need to heed. Man-made or natural, these inescapable changes portend more erratic and extreme weather events which will be highly significant for their effects on the built environment Hence the construction industry is facing demands both to contain the degree of warming and to contend with the consequences.

11.6.4 Environmental disaster: summary

Self-evidently the risk of environmental disaster is real. Mankind seems reluctant to stop searching for oil, cannot stop mining or running industries, and cannot eliminate energy use. What engineers have to do is realise that virtually any project they undertake will involve solid, liquid and gaseous emissions. Any of these three can be devastating in terms of pollution, environmental impact or their effects on health. One case described earlier in the book is that of the Aberfan (UK) disaster in 1966: it was a solid waste tip from coal mining that slipped and entombed the village school; in the case of the Ajka alumina plant accident (Section 11.3.3) it was retained wasted sludge that flooded out. Overall, any engineering activity needs to take seriously its environmental impact study and consider how the risks of unwanted emissions and dealing with waste can be controlled. The challenges that climate change poses to the professions are formidable.

11.7 Chapter summary

There is no shortage of man-made disasters to use as examples of what can go wrong. Events are not that rare and remain an ever-present threat with potentially huge safety and commercial consequences.

Although some of the examples cited in this chapter might seem to be one-off incidents, they do fall into patterns. First, they can be grouped under specific headings such as 'dams' or causes such as 'crowds'. Second, it is not hard to find repeat examples within each group. Generally, whatever the form of the failure type, it has almost certainly occurred before, so repeats should come as no surprise.

Possibly some disasters are unavoidable, and cases have been cited of unfortunate coincidence (such as at Viareggio). Conversely, in many other cases, elementary mistakes or poor communications or poor management initiated or contributed to the event. New threats arise such as terrorism and, for example, in 2005, a cyber-attack on Brazil's power grid blacked out an area of north Rio de Janiero. Cyber-attacks on our infrastructure are seen as developing threats. 2017 saw a cyber-attack having serious implications for many UK hospitals. (World Economic Forum (2018) notes that cyber-attacks are becoming normal and it cites the 2015 attacks on Ukraine's power grid which shut down 30 substations.)

Mankinds needs are not going to change. We will continue to need buildings, transport, power and industry. Hence, we must do what we can to minimise the risk of disaster. The underlying themes generating disasters are nearly always the same (see Chapter 2):

- Most engineering enterprises are systems involving different engineering disciplines. Ensure there is one coherent safety philosophy for the whole and that the sub-demands on each separate engineering system are understood.
- Don't put profit before safety.
- Identify the hazards.
- Identify probabilities.
- Learn the techniques of prediction as HAZOPs and FMEA for investigating interactive systems (see Chapter 13).
- Identify modes of failure.
- Ensure the consequences of failure modes are understood and tailor the level of safety provided accordingly.
- Make sure as many mitigations as possible are in place.
- Create a safety case (watch out for chains of events with interlinked failure modes or causes).
- Have imaginative, independent, third-party reviews.
- Make sure what is built is what was intended to be built.
- Proactively manage durability and through-life safety.
- Make sure any changes are properly authorised.
- Consider ways of monitoring plant and system functionality to give advance warning of things going wrong. The objective is always to keep control and try to make systems 'fail-safe'.
- Make sure there are procedures for reacting to warnings.
- Consider the worst, and plan for it (including evacuation).
- Consider all the emissions from any system and ensure safe containment and disposal of wastes: solids, liquids and gases.
- Be aware of relevant legislation. It was probably drafted as a response to previous disaster.
- Above all, learn from what has gone wrong before.

References

British Dam Society (undated), https://britishdams.org.

Brumadinho dam disaster (2019), https://en.wikipedia.org/wiki/Brumadinho_dam_disaster.

Bullough R., Burdekin F.M., Chapman O.J., Green V.R., Lidbury D.P.G., Swingler J.N., Wilson R. (2001),'The demonstration of incredibility of failure in structural integrity safety cases'. International Journal of Pressure Vessels and Piping 78 (2001) 539-552)

Chernobyl arch (2016), www.youtube.com/watch?v=n7aMcKinrWY.

Cullen, W. D. (1990), *The Public Inquiry into the Piper Alpha Disaster*, Vols 1 and 2, HMSO.

Dallard, P., T. Fitzpatrick, A. Flint, A. Low, and R. Ridsill Smith (2001), 'The Millennium Bridge, London: problems and solutions', *The Structural Engineer*, **79**(8).

Deepwater Horizon (2010), https://en.wikipedia.org/wiki/Deepwater_Horizon.

Dust Bowl (1930s), https://en.wikipedia.org/wiki/Dust_Bowl.

Engels, F. (1892), *The Condition of the Working-Class in England in 1844*, Swan Sonnenschein & Co.

Home Office (1990), 'The Hillsborough Stadium Disaster, 15 April 1989', Final Report by Lord Justice Taylor, HMSO.

Home Office / Scottish Office (1986), *Guide to Safety at Sports Grounds*, HMSO.

HSE (2011), 'Preventing catastrophic events in construction', Research Report RR834, prepared by CIRIA and Loughborough University for the Health and Safety Executive.

Institution of Structural Engineers (2002), *Safety in Tall Buildings*, IStructE Ltd.

Institution of Structural Engineers (2010), *Practical Guide to Structural Robustness and Disproportionate Collapse in Buildings*, IStructE Ltd.

Institution of Structural Engineers (2013), *Manual for the Systematic Risk Assessment of High-Risk Structures against Disproportionate Collapse*, IStructE Ltd.

List of pipeline accidents in the United States (undated), https://en.wikipedia.org/wiki/List_of_pipeline_accidents_in_the_United_States_in_the_21st_century.

Martin, T. and I. A. MacLeod (1995), 'The Tay Bridge disaster – a re appraisal based on modern analysis methods', *Proceedings of the Institution of Civil Engineers*, **8**(2).

RedR UK (undated), www.redr.org.uk.

Rana Plaza building collapse, Dhaka (2013), https://en.wikipedia.org/wiki/2013_Savar_building_collapse.

SCOSS Alert (2018), 'Fire in multi-storey car parks', www.cross-safety.org/sites/default/files/2018-02/fire-multi-storey-car-parks.pdf.

Tay Bridge disaster (1879), https://en.wikipedia.org/wiki/Tay_Bridge_disaster.

The Guardian (2013), 'Chinese struggle through "airpocalypse" smog', 16 February, www.theguardian.com/world/2013/feb/16/chinese-struggle-through-airpocalypse-smog.

World Economic Forum (2018), *The Global Risks Report 2018*, World Economic Forum, www.weforum.org/reports/the-global-risks-report-2018.

12 Occupational health and safety

12.1 Introduction

Other chapters of this book vividly illustrate why engineers must take the safety of our built and rural environment seriously. These include a moral imperative, compliance with a code of conduct as typically set out by the major professional institutions, business implications, contractual requirements, duty of care, and last but not least, because the law of the land usually requires it. Whereas some of these may vary in necessity or detail, the latter is always a constant feature where harm to people may result from a work-related activity and is derived from statute.

This chapter considers the influence of occupational safety legislation on the approach and attitude to safety and also describes the obligations relating to the occupational health of those at work or who may be affected as a consequence. In this chapter, 'safety' is used as an encompassing term to include 'health', unless the circumstances clearly indicate otherwise. It records the history of occupational safety legislation, charting the emergence of concern for the worker. The development of modern legislation is summarised to illustrate how health and safety requirements in construction derived from this desire and how its framework lends itself to the management of uncertainty within the design and construction process more generally. Thus, the chapter concludes with a description of how this proportionate risk management can not only satisfy the law, but also provide a sound basis for good business and effective working.

12.2 Background

For much of the long history of human development, building techniques (in their widest sense) have progressed through trial and error until what is deemed to be a satisfactory solution is achieved. From the earliest of civilisations – Greek, Roman, Byzantine, Aztec and others – lessons were learnt the hard way and passed down through the generations and over the centuries, often reaching a level of sophistication at which we still marvel.

This was the period of the art of structural engineering as practised by master craftsmen. The science was to come much later and started to emerge during the Enlightenment (17th century), beginning the still-continuing process of scientific understanding of how it is that structures behave, resist applied loads, and fulfil their function. It was during this early period of the Enlightenment that, for example, Hooke (1635–1703) and Newton (1642–1727) identified and then set out many of the fundamental understandings and rules on which we rely. (Chapter 1 has further information.)

However, even during the late 18th, and 19th centuries, that unique period of industrial revolution and influence throughout the emerging British Empire, there remained a large amount of trial and error in the design and construction process as new boundaries were pushed. Isambard Kingdom Brunel (1806–1859), one of the UK's most respected engineers of the Victorian era, suffered numerous failures (Chapter 1). His eminent Victorian colleagues (Smeaton, Stevenson, et al.) had similar setbacks. Setbacks have continued during the 20th, and now the 21st century, when, instead of a lessening of failures, examples continue to abound.

Throughout most of this long period of human endeavour (certainly up to and including some of the 20th century) the greatest emphasis has been on the asset. Safeguarding those constructing or maintaining the facility was not uppermost in the minds of sponsors, constructors or indeed such authority that existed. After all, constructing things was dangerous, but perhaps not so different from many other activities at that time, or life in general; things often went wrong; hence people frequently were ill, injured or killed. This mirrored the attitude towards workers generally; we are all aware of the appalling working conditions of the majority, which reached its apogee in the workplaces of the newly industrialised cities, and indeed exist in many worldwide locations today.

Safeguarding the asset during operational use was of course a concern (and knowledge was often limited in this regard): but again, the concern was generally centred around the business consequences of failure, rather than those persons who might be individually harmed. In mitigation, the natural robustness of much engineering design at this time did mean that in-life maintenance and repair were often minimised.

12.3 Emerging concern for people

It was in the powerhouses of the Industrial Revolution, i.e. the mills and the mines, that the first measures emerged, specifically designed to afford some degree of protection to workers. However, these initial attempts were very limited; they only applied to women and children (mostly the latter) and initially only with regard to the hours of work, although later with some attention to ventilation of workplaces. Nothing was proposed in respect of the physical dangers of the workplace itself. Death and maiming were just risks of the job. Notwithstanding the measures to improve ventilation, workplace ill health was even further removed from consideration: partly because many causal aspects of ill health were unknown, e.g. dusts and industrial fumes, and partly because of the overriding prevailing attitude, which did not include worker well-being as part of the business model. There were some employers, such as Gregg of Quarry Bank Mill, in Styal, Cheshire, who were recognised at the time as benevolent; in today's terms, of course, the mill workers' conditions would still horrify us, but it is unfair perhaps to measure those who were trying to make a difference then against current attitudes.

Legislation to protect the worker is generally reckoned to have begun with the Health and Morals of Apprentices Act 1802. This was followed by a series of Factories Acts, leading up to the consolidating Factories Act 1961, which was the modern precursor to contemporary legislation (and still partly in force but of no practical relevance to contemporary engineering disciplines).

Since 1961 there has been an accelerating drive to improve health and safety. Partly this is because it is just, and partly in recognition of the enormous cost attendant on injury and ill health. In the UK (2018/2019) these accounted for some 28 million working days lost with annual costs of £15 billion; while 1.4 million working people suffered from a work-related illness (Health and Safety Executive, undated); note that this covers the entire workforce, not just the construction sector. In 2016, the Health and Safety Executive (HSE) published their 'Helping Great Britain work well' initiative (HSE, 2016). The target was to build on existing practice but maintain the impetus to get all sections of the workforce involved. To quote from the pamphlet:

> There is broad agreement among those with whom we have engaged that preventing harm to workers and the public is integral to businesses being successful and achieving sustained growth. Our collective challenge is to ensure that this is known, understood and becomes embedded firmly in everyone's thinking – and in all of our actions.

The HSE's pamphlet lays out their guiding principles as:

- Those who create risks have a responsibility to manage those risks – placing the ownership of risk in the right place.
- Action should be proportionate to the risks that need to be managed – which means we need well-thought-out measures to be applied that are tailored to each business, to the nature of the work undertaken and the people who work there.

And the HSE add: 'Too much complexity and bureaucracy has built up around health and safety.' The target should be 'keeping things as simple and straightforward as possible', with the promise that 'If we can all come together to help achieve these things, maintain the gains made in safety, and seize the opportunity to give health the same priority, it will help improve productivity, keep business costs down, help keep workers safe and well, and protect members of the public.'

Six themes are presented, one of which is: 'Tackling ill health: Highlighting and tackling the costs of work-related ill health'. Another two are: 'Managing risk well: Simplifying risk management and helping business to grow' and 'Keeping pace with change: Anticipating and tackling new health and safety challenges'.

For engineers to respond to these challenges, a starting point is to have a good perspective over the most common hazards and most common causes of work-related ill health in the construction sector. Construction remains a high-risk industry yet whilst there are significant number of fatalities, the industry also accounts for a high number of injuries and wider health issues. An overall picture can be gleaned from the annual HSE statistics. In 2018/2019 these were:

Table 12.1 Number of fatalities: 30 (this remains broadly similar year on year)

Causes	Percentage
Falls from height	49%
Trapped by something	14%
Struck by moving vehicle	11%
Struck by moving object, including flying/falling object	10%
Contact with electricity/electrical discharge	5%
Other	11%

Table 12.2 Non-fatal injuries: 54,000

Causes	Percentage
Slips, trips, or falls on same level	25%
Injured while handling, lifting or carrying	20%
Falls from height	18%
Struck by moving object, including flying/falling object	12%
Other	25%

Table 12.3 Work-related ill health: 79,000 workers suffering (new or long standing)

Causes	Percentage
Musculoskeletal disorders	62%
Stress, depression or anxiety	21%
Other	17 %

Note: These figures are self-reported and broadly similar year on year. There are reasons to suspect true figures are much higher.

The statistics reveal that many more working days are lost due to work-related illness than to injuries. They also reveal that construction workers have a high risk of developing diseases such as cancer (from asbestos, silica and fumes); breathing and lung problems from dusts, fumes and gases; skin problems such as dermatitis (from contact with hazardous substances); physical health risks from lifting or vibration; or deafness from prolonged noise exposure. Apart from that, the same basic hazards causing the most severe effects recur year on year.

In very recent times the stigma around mental health has been partially lifted and the raw statistics reveal much tragedy. Of the 79,000 workers suffering from ill health, 17,000 were suffering some form of mental stress and that is just on reported figures. The Office for National Statistics (Office for National Statistics, undated) has reports for the five years up to 2015. The figures are complicated (as are the causes), but the suicide rate for construction workers is about one per day. The risk of suicide among low-skilled male labourers in construction is about three times higher than the male national average. Whatever the number, it is multiple times the death rate linked to 'falls from height'.

12.4 Construction regulations

The first comprehensive regulations of the modern era, specifically targeted at the construction industry, were the 1961 Construction Regulations, made under the 1961 Factories Act. These covered general provisions and lifting and were of a prescriptive nature. However, the benefits of having construction-specific legislation were outweighed by the increasingly unwieldy and ineffectual format.

12.5 1972: A new era begins

A review of legislation in place around this period indicated the growing muddle and confusion. In 1961 there were, as examples:

- Sandwith Anhydrite Mine (Lighting) Special Regulations 1962 S.I. 1962/192
- Non-ferrous Metals (Melting and Founding) Regulations 1962 S.I. 1962/1667
- Construction (Working Places) Regulations 1966 S.I. 1966/94

and many more, each trying to legislate on a specific activity, by setting out *prescriptive* obligations.

Thus this prescriptive nature of the law resulted in legislators trying to cover all industries or situations. This inevitably led to a plethora of regulations which became unwieldy, inflexible, impractical to keep up to date with new industrial practices and hence, ultimately, unworkable.

It was against this backdrop that the government commissioned an enquiry, to be led by Lord Robens, with a view to improving the situation. At this time the construction industry alone was killing some 250 workers a year.

12.6 The Robens Report 1972

The Robens Report, as it became known, was and remains a seminal stage in the history of workplace well-being. It emanated from the committee set up 'to review the provision made for the Safety and Health of persons in the course of their employment, other than transport workers, and to consider whether any changes are needed in the scope or nature of the major relevant enactments, or the nature and extent of voluntary action' (Committee on Health and Safety at Work, 1972). The report was well received and the recommendations led directly to the Health and Safety at Work etc. Act 1974 (HASWA) (see web address in References).

12.7 Health and Safety at Work etc. Act 1974

The introduction of the Health and Safety at Work etc. Act in 1975, as a direct consequence of the Robens Report, ushered in a new approach to occupational health and safety. The drafting was far-sighted and even today, some four decades on, it remains one of the classic Acts of the post-war years, ranking, in the author's view, alongside the 1944 Education Act, and the various anti-discrimination Acts enacted between 1965 and 2006. It was designed for the UK, although Northern Ireland has its own equivalent, but its style and approach have also been adopted elsewhere.

To quote from the HSE's booklet, 'Thirty years on and looking forward' (HSE, 2004):

> HSE's first director general described the HSW Act as 'a bold and far-reaching piece of legislation'. It marked both a watershed in health and safety regulation and a recognition that the existing system had failed to keep up with the pace of change and was trailing behind industrial and technological developments.

The key to the thinking behind the Act was to place the responsibility for deriving a safe place of work with those who created or controlled danger (the risk), leaving the solution in their hands, subject to some overarching requirements and industry norms. In this manner it was no longer necessary for the legislators to determine workplace solutions for every industry and each circumstance. This not only provided flexibility but also, crucially, inferred that standards would rise in line with new techniques, processes and indeed societal expectation. So it has proved.

In simple terms the Act requires[1] employers (and the self-employed) to have regard to the health, safety and welfare of:

- employees (section 2);
- those who are not employed, but who may be affected by the undertaking (section 3).

This non-prescriptive approach is now referred to as 'goal setting' legislation and can be summarised by three facets:

- The responsibility for managing 'danger' lies with those who create or control the risk.
- The solution must take account of the particular circumstances.
- The solution must take account of technological progress.

Although this approach provided flexibility as noted, it also introduced a degree of uncertainty, as the 'acceptable solution' was often not defined by any authoritative source. The Act was framed whereby, once a case to answer has been proved, the duty holder (a defendant once in court) was responsible for demonstrating compliance. This is known as the reverse burden of proof (described in section 40 of the Act) and remains controversial.

The Act sets out one level of obligation, and two levels of qualification, in terms of the effort that should be made to eliminate or reduce risk. These are described below in Section 12.15.

The Act introduced the Health and Safety Executive[2] as the government agency responsible for overseeing the delivery of a safe workplace, supported by local authorities. The HSE's current mission statement is 'to prevent death, injury and ill-health in Great Britain's workplaces – by becoming part of the solution'.

As Hansard states:

> The 1974 Act's record speaks for itself. Between 1974 and 2007, the number of fatal injuries to employees fell by 73 per cent. The number of reported non-fatal injuries fell by 70 per cent. If we look behind changing employment patterns during that period, we see the same picture. Between 1974 and 2007, the rate of fatal injuries per 100,000

1 The Act requires a number of other things to happen but in terms of this chapter, HASWA sections 2 and 3 are the key requirements.

2 The Health and Safety Commission, also introduced at the time, is now subsumed into the HSE to form one body.

> employees fell by a huge 76 per cent. Britain had the lowest rate of fatal injuries in the European Union in 2003, the most recent year for which figures are available. The EU average was 2.5 fatalities per 100,000 workers; the figure in the UK was 1.1. (Hansard, 2008)

However, no one would suggest we can be complacent; commercial pressures, self-employment, home-working, foreign workers, and new technologies all present continuing or new challenges. In particular, occupational ill health is proving a difficult nut to crack.

12.8 Regulations and guidance

HASWA allows for Regulations to be made under the Act, providing more specific requirements either on an activity, e.g. manual handling, on a process, e.g. construction design and management, or on materials, e.g. lead. The European Union (EU) has had a significant influence on the introduction of many of these Regulations, stemming from EU Directives. Directives are implemented by each member state in their own style and particular detail. Thus, there can be many different interpretations around the 27 member states.[3] For example:

Table 12.4

Temporary and Mobile Work Sites Directive 92/57/EEC	
UK	Republic of Ireland
Construction (Design and Management) Regulations 2007	Safety, Health and Welfare at Work (Construction) Regulations 2006

12.8.2 The key EU influence, as far as the construction industry is concerned, has centred around those Directives outlined in Table 12.5.

Table 12.5 Legislation (mostly) originating from the EU

Framework Directive 90/269/EEC 89/391/EEC
Display Screen Equipment Regulations
Manual Handling Operation Regulations
Management of Health and Safety at Work Regulations
Personal Protective Equipment Regulations
Workplace (Health, Safety and Welfare) Regulations
(commonly referred to, collectively, as the 'six pack')

3 EU Regulations on the other hand, are implemented verbatim without any Member involvement e.g. the Construction Products Regulation.

Temporary and Mobile Work Sites Directive Construction (Design and Management) Regulations (CDM 2007) (Note 1)
European Council Directive 2001/45/EC Work at Height Regulations (2005)
Workplace Regulations (These replace much of the Factories Act) Regulatory Reform (Fire Safety) Order (originating in the UK)

Note 1: when first introduced in 1994, these Regulations were supplemented by the Construction (Health, Safety and Welfare) Regulations 1996, which, taken together, implemented the EU Directive. In 2007, the latter Construction Regulations were revoked, and the contents were subsumed, substantially unchanged, into the revised CDM Regulations (CDM 2007; these are now at a 2015 edition).

The Regulations overall, and those of direct relevance to the construction industry, may be placed into logical groupings, shown in Table 12.6, which allows for readier access for those who may not have day-to-day knowledge of them.

Table 12.6 Key construction industry regulations

	Category	Regulation
A	Management	Management of Health and Safety at Work Reporting of Injuries, Diseases and Dangerous Occurrences (RIDDOR) Construction (Design and Management) Personal Protective Equipment (PPE) First aid Confined spaces
B	Materials & Substances	COSHH Asbestos, Lead
C	Health	Manual Handling Noise Vibration
D	Safety	Work at Height Electricity Fire (Note 1)
E	Plant & Equipment	Provision and Use of Work Equipment (PUWER) Lifting Operations and Lifting Equipment Regulations (LOLER) Supply of Machinery

Note 1: Post Grenfell (Chapter 9) it can be anticipated a number of changes will be introduced.

Of the Regulations in the table, it is the Management of Health and Safety at Work, CDM, Workplace Regulations, and Regulatory Reform (Fire Safety) Orders that relate to built infrastructure safety and hence are of interest in this chapter and are discussed again later.

There are of course many other sets of Regulations applicable generally, but which may not be driven by the EU or relate to structural safety.

12.9 The obligation of duty holders

As noted above, HASWA creates one level of absolute obligation, and two levels of qualification to an absolute duty to safeguard people. In other words, the amount of effort that is required to make a situation 'safe', and compliant, varies. The three levels are:

Table 12.7 Obligations and duties

Obligation	Comment
An absolute duty:	
Strict liability (SL)	You shall do ... Generally, there is no defence for not doing so.
An absolute duty, qualified by:	
To do something 'so far as is reasonably practicable' (SFARP)	This is the most common obligation and the basis for the qualification in sections 2 and 3 of HASWA. It is generally represented by 'good contemporary industry practice' although the legal definition is not helpful in a practical sense and confusion abounds (Health and Safety Executive, undated, provides guidance, geared to the larger 'process' or similar industries, but provides no answers for mainstream practitioners; Carpenter (2021) provides background and is written to provide clarity for designers); see also Section 12.16.2).
To do something 'so far as is practicable' (SFAP)	This occurs in higher risk situations (as specified in certain Regulations). To satisfy this obligation, if elimination or mitigation is technically or managerially feasible, action must be taken. Cost is not a consideration. However, it too can be satisfied by 'good contemporary industry practice'.

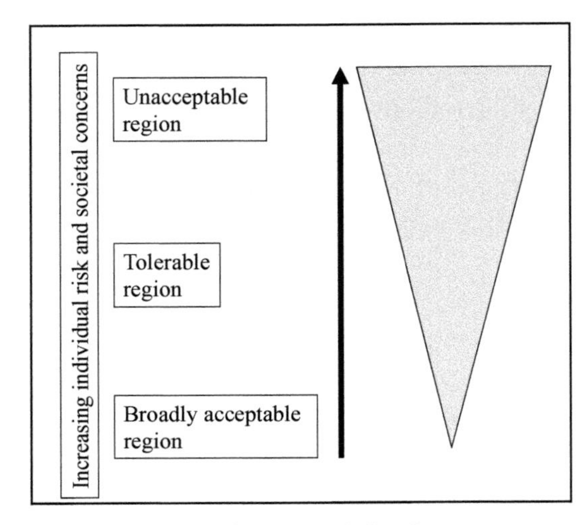

Figure 12.1 HSE framework for the tolerability of risk

This can be thought of visually as a triangle (Figure 12.1), with three layers; the higher up the triangle, the more stringent the test, but (as the triangle narrows), the fewer instances of risk occur (HSE, 2001).

The implementation of these definitions is described in Section 12.16.

12.10 Jurisdictions

Legislation for workplace safety applies to Great Britain and Northern Ireland[4] separately, although the detail is essentially the same. However, this is complicated further when considering other relevant legislation alongside HASWA and its Regulations, for example, the Building Regulations, which also have a 'safety' remit. The latter are structured separately for each UK jurisdiction (England, Wales, Scotland and Northern Ireland) although, as to be expected, there is much commonality.

12.11 Case law

The law of Parliament is given practical interpretation by the courts. It is an ongoing mechanism, sometimes seeing reversals of interpretation, and often extending over many years in respect of a single issue. The higher the court, the more influential the decision, with the Supreme Court (previously the House of Lords) making 'decided' law.[5] However, even it can change its mind.

Table 12.8 Case law interpretations

Issue	Case
The meaning of 'So far as is reasonably practicable'	*Edwards v National Coal Board*
Involvement of engaging party	*R v Associated Octel*
Foreseeability	*R v HTM*
Proportionate action	*Baker v Quantum*
Real risk	R v Science Museum

4 Where, instead of HASWA, Northern Ireland has the Health and Safety at Work Order 1978.

5 For criminal law, Scottish appeal cases are heard by the High Court of Justiciary.

Common law also develops in the same way and it has provided some of the most critical interpretations over the years, also influencing the interpretation of statutes. Case law has been an important element in safety legislation. Examples of key issues are shown in Table 12.8.

12.12 Manslaughter

Construction remains a dangerous industry despite the significant reduction in fatalities and accidents over the years. Accordingly, a commensurate amount of care must be taken to safeguard those who may be adversely affected by work activities. Where it is considered by the prosecuting authority that there has been a particularly serious breach of this duty, a charge of manslaughter may follow. Manslaughter investigation is the responsibility of the police.

In this situation we are referring to either a charge of gross negligence manslaughter (levied against an individual), or a charge of corporate manslaughter (levied against an organisation). The former is a common law offence, the latter, since 2007, a statutory offence. The necessary tests to prove guilt are different.

For corporate organisations, the requirement under the Corporate Manslaughter and Corporate Homicide Act 2007 (CMCHA) is to ensure that the manner in which the organisation is managed and organised by its senior managers does not fall 'far below that to be expected in the circumstances'. In effect this means that it is essential that there are effective systems in place, that they are monitored and reviewed, and that actions are implemented.

As may be expected, the penalties are severe. Individuals are likely to receive a custodial sentence; organisations may receive a range of penalties: significant fines, 'name and shame' measures, disqualification of directors.

The message is clear; seek to ensure that one's personal professional work and organisational operations are well organised and managed. Seek competent advice if there is uncertainty. Above all, manage the risks associated with, or generated by, your business.

12.13 Human factors

Chapter 6 is devoted to the topic of 'human error' since this facet normally plays a part in every failure. In this chapter, 'human error' might refer to inadequate risk assessment or inadequate management of the whole project. Harm to individual workers is frequently the consequence of carelessness or disregard for elementary precautions. It is important, however, to bear in mind that the law does not require us to be perfect. People are expected to behave reasonably and to show a degree of competence commensurate with the task being undertaken.

12.14 Major events

Unfortunately the UK has experienced a number of truly catastrophic events over the decades, illustrating a failure to manage workplace safety and hence falling under the auspices of HASWA. Examples such as Piper Alpha (Chapter 11), King's Cross Underground fire (Chapter 9), and the *Herald of Free Enterprise* (Chapter 6) have all led to considerable loss of life, but, as is often the case, also led to actions by industry and the HSE to improve the safety regime.

12.15 Occupational safety requirements relating to construction industry asset failure

The development of occupational safety law as described above means that those working in the construction industry must consider the safety of the asset, not only from a business perspective (and the prospect, for example, of being sued by the client), but also from the statutory obligation to have regard to the well-being of those who may be adversely affected if it fails. Contractors have been in the enforcing agency's sights for many years and there is substantial experience of cases coming to court. However, for designers this is still very much 'work in progress' in terms of the necessary compliance levels as, by comparison, very little has reached the courts.

The relevant safety legislation, for designers and contractors working on facilities that could harm people if they collapsed (taken from Table 12.5), include:

Health and Safety at Work etc. Act 1974

The overarching Act for safety legislation. It provides for obligations of the employer towards employees (section 2) and to others, not in the employment of the employer, but nonetheless affected by the work activity, i.e. the undertaking (section 3). It also places specific obligations on individuals (section 7) and directors (section 37). For defined construction work, enforcement falls to the Health and Safety Executive.

These obligations should form the basis of good working practices. However, the groups of persons to whom an employer has an obligation varies depending upon the role in a construction context, as shown in Table 12.9.

Management of Health and Safety at Work Regulations 1999

These give detail to the Act by requiring employers to, inter alia:

1. undertake a 'risk assessment' on all hazardous operations;
2. co-operate with other employers;

Table 12.9 Corporate responsibilities

Designer Has some responsibility for:	**Contractor** Has some responsibility for:
Own employees	Own employees
Public	Public
Contractors	(Other) Contractors
User	
Those who maintain	
Those who implement foreseeable repair	
Those who demolish/decommission	

Note that, with the exception of employees, who by definition have a contractual link with the employer, in all other cases the contractual relationship is irrelevant.

3. have a source of competent advice;
4. have appropriate arrangements in place;
5. provide training, information and supervision.

These requirements relate to the specific actions that the undertaking (as a contractor or designer) must take in order that their obligations may be discharged. Forward-looking organisations will see these obligations as an integral part of a good business model, bringing efficiency and improved productivity, as well as targeting Parliament's narrower intent of ensuring a safe place of work.

Construction (Design and Management) Regulations 2015

These regulations set out what are no more than sensible and recognised management principles for all construction projects. They require:

- the engagement of competent persons (individual or corporate);
- adequate information flow;
- appropriate co-operation and co-ordination;
- suitable risk management.

Workplace (Health, Safety and Welfare) Regulations 1992

Clauses 4 and 5 describe duties. Clause 4 covers duties of people other than employers if they have control of a workplace to any extent. Clause 5 sets out how duties are limited to the matters which are under the party's control.

12.16 Safety risk management

12.16.1 General

As we have seen, the corporate management of safety risk is a legal requirement covered by the Acts (HASWA and CMCHA) and Regulations. There are also other drivers which require competent management of risk, for example, business success, image, employee morale, and the common law duty of care.

Typically there are three levels of risk management that should be covered (Table 12.10).

Table 12.10 Board risk management levels

Level	**Requirement**	**Comment**
Strategic	The board sets the style and culture for the organisation; it sets and reviews corporately identified strategic risk. The board leads by example.	Culture is key, and has to be generated from the top. Advice is available from a number of sources (see Chapters 6 (Section 6.8) and 13).
Operational	Sets procedures for running and reviewing the business sectors, within the strategic framework set out above, and the means of communication.	
Project	Implements operational procedures at a project level.	

All listed companies are required to consider strategic risk at board level, with that consideration being made known to shareholders.

The key to ensuring safety against failure, either during construction, or thereafter, is to implement a proportionate risk management system, that:

- identifies the safety issues;
- eliminates or mitigates them;
- communicates the necessary information to those who need to know;
- implements the requisite control measures;
- monitors the measures;
- amends them as required.

The responsibility for this rests with designers and contractors, as appropriate and illustrated in Table 12.9, and also the owner/tenant/operator over the operational life of the building or structure.

If the failures mentioned throughout this book are examined, it will become apparent to the reader that risk was not managed as required. The obligation to manage risk is governed by safety legislation, as outlined above, but there is also a general duty of care, and a recognition that asset failure is not good business. Failure in all three of these aspects can have a devastating effect not only on those directly affected (and killed, injured or otherwise harmed) but also on organisations, in terms of financial cost, morale and image and on the managers and other individuals involved, under whose watch the event occurred. (The severe consequences to BP after the Deepwater Horizon disaster have been described in Chapter 11. A further contemporary example is the consequences to Boeing after their problems with the 737 MAX. Although not finalised, it is clear that the ramifications following the Grenfell Tower fire in 2017 will be very significant, and will affect much of the construction industry.)

12.16.2 Suitable and sufficient risk management

The question arises as to 'What does a suitable and sufficient risk management system look like?' As for all matters covered by HASWA and goal-setting legislation, it will be for the court to decide. CMCHA has not yet bedded-in with sufficient case examples to provide guidance. However, there are pointers available which provide good advice as shown in Table 12.11.

Table 12.11 Corporate attributes

Attributes	Sources	Comments
Competency		
	CDM 2015 Guidance L153	PAS91:2013+A1:2017 (Construction Prequalification Questionnaire) provides a framework for assessment, a set of standardised questions, categorised into modules.
	Accreditation schemes	Belonging to a scheme which itself belongs to safety schemes in procurement (SSIP).
Management	Advice from HSE and Institute of Directors	'Leading health and safety at work'.
	Advice from others	The Institution of Civil Engineers has published guidance on the management of strategic risk, and the Institution of Occupational Safety and Health has done likewise.
		There is a wealth of literature on this topic.

In few cases will it be possible, or desirable to manage risk using quantitative methods (see Chapter 13). Qualitative methods will be the norm. In this regard, whether as contractor or designer, the use of ERIC (Carpenter, 2021) will allow a pragmatic approach. ERIC is, of course, an acronym:

Table 12.12 ERIC

E	**E**liminate	Eliminate Hazards (SFARP). If not feasible:
R	**R**educe	Reduce the risk from remaining hazards (SFARP). For significant risks which remain:
I	**I**nform	Provide relevant information to those who may need to know.
C	**C**ontrol	Those who implement the task introduce controls based on the information and other considerations.

The advantage of this simple procedure is that it is easy to remember, broadly mirrors that set out in CDM 2015, Regulation 9, and reminds the user of the order of approach, for example, just supplying information to others of 'danger' does not satisfy the law as no attempt has been made to eliminate or reduce it. One must start at the top, with E.[6]

It also has one other crucial advantage; ERIC may be used on any hazard, whether it relates to safety or not. In this way, the barrier that is often seen between 'safety' risk and other 'project' risk, is removed. All can be handled in this manner.

Most organisations will have management systems in place. However, few have specific procedures for identifying the potential catastrophic event. A study by the HSE in 2011 ('Preventing catastrophic events in construction', HSE, 2011) showed that the industry was generally ill-prepared.

Catastrophic events would be those having the following potential consequences:

- potential for multiple deaths and serious injuries in a single incident and/or
- serious disruption of infrastructure (e.g. road, rail) and/or services (e.g. power, telecoms).

In addition, such events may well have the following features:

- ability to adversely affect organisations commercially, either directly or through loss of reputation.
- creation of public demand for action, possibly leading to demand for a public enquiry and/or changes to relevant legislation.

6 Unfortunately, CDM2015 omits reference to 'Hazard', contrary to other established advice and practice. It is retained here.

12.16.3 Links with other legislation

Although safety legislation (stemming from HASWA) stands on its own, there are occasions when it runs alongside other legislative requirements which also are safety-based. The prime example is the Building Regulations (Building Regulations, 2010) and specifically Approved Document A (Approved Document A, 2010) which accompanies them.[7] The purpose of the Building Act 1984, from which the regulations originate, is to protect persons 'in and around buildings'.[8]

The Building Regulations only apply at certain times in a building's life, are limited in their requirements to accidental scenarios affecting the structure, and generally set out a 'de minimis' standard that experience shows will satisfy for most buildings, most of the time. Safety legislation takes a different approach as shown in Table 12.13 below (and discussed in Institution of Structural Engineers, 2010 and 2013).

Table 12.13 Relevance of Building Regulations over structure's life cycle

Stage/Aspect	**Safety Legislation**	**Building Regulations**
Construction phase	Obligation to have regard to those who may be affected by any collapse	No requirement
On completion		Must comply with Building Regulations
During life of building (generally)		No requirement, but Building Control may take action if a structure is dangerous
During structural repair or maintenance		Must comply with Building Regulations
During demolition		No requirement, but Building Control may take action if a structure is dangerous
Actions	Foreseeable accidental and malicious actions	Foreseeable accidental actions

Although designers are not duty holders with regard to the Regulatory Reform (Fire Safety) Order 2005, this also plays a part. Designers can influence the ability of the responsible person under the Order (usually the landlord or tenant) to comply with its requirements, through careful consideration of the consequences of their design, and the information provided. In fact it is a requirement of Part B of the Building Regulations (clause 38) that 'The person carrying out the work shall give fire safety information to the responsible person', where 'fire safety information' means information relating

7 Or the equivalent in Wales, Scotland and Northern Ireland.

8 Following the Grenfell Tower fire, these will be joined by the Building Safety Act (yet to be enacted), which, as noted above, will have a major influence.

to the design and construction of the building or extension, and the services, fittings and equipment provided in or in connection with the building or extension that will assist the responsible person to operate and maintain the building or extension with reasonable safety.

12.17 Chapter summary

The health and safety of all those who work in the construction industry is a matter to be controlled so that risks of harm are minimised. The law prescribes duties that are applicable to different sectors within the industry so everyone should be aware of what those duties are. Originally, matters of health and safety were controlled by prescriptive rules. But this changed significantly with the introduction of the Health and Safety at Work Act. Nowadays, what is prescribed is the outcome. So, duty holders are supposed to manage affairs under their control so that the outcome is achieved. The aims are quite simple as defined by HSE

- Those who create risks have a responsibility to manage those risks.
- Action should be proportionate to the risks that need to be managed.

The industry is not static. New hazards emerge and society's expectations of safety change with time. For example, much contemporary concern is being given to mental health; a topic that used to be taboo. Proper management of risk and properly thought out risk assessments can cope with these changes. Our overall endeavour is to keep people safe and to do that proactively.

References

Approved Document A (2010), Structure: Approved Document A: Building regulation in England covering the structural elements of a building.

Building Regulations (2010), Approved Documents, published 2010, as updated, www.gov.uk/government/collections/approved-documents.

Carpenter, J. (2021), *Designing a Safer Built Environment: A Complete Guide to the Management of Design Risk*, ICE Publishing.

Committee on Health and Safety at Work (1972), *Safety and Health at Work: Report of the Committee, 1970–72* (Robens Report), HMSO.

Hansard (2008), 4 July 2008: Column 473, www.publications.parliament.uk/pa/ld200708/ldhansrd/text/80704-0001.htm.

Health and Safety at Work etc. Act 1974, www.legislation.gov.uk/ukpga/1974/37/contents.

Health and Safety Executive (HSE) (undated), www.hse.gov.uk/.

HSE (2001), *Reducing Risks, Protecting People: HSE's Decision-Making Process*, HSE Books.

HSE (2004), 'Thirty years on and looking forward: The development and future of the health and safety system in Great Britain'.

HSE (2011), 'Preventing catastrophic events in construction', Research Report RR834, prepared by CIRIA and Loughborough University for the Health and Safety Executive.

HSE (2016), 'Helping Great Britain work well', www.hse.gov.uk/aboutus/strategiesandplans/helping-great-britain-work-well-strategy.html.

Institution of Structural Engineers (2010), *Practical Guide to Structural Robustness and Disproportionate Collapse in Buildings*, IStructE Ltd.

Institution of Structural Engineers (2013), *Manual for the Systematic Risk Assessment of High-Risk Structures against Disproportionate Collapse*, IStructE Ltd.

Office for National Statistics (undated), www.ons.gov.uk/aboutus/whatwedo/statistics.

Further reading

Carpenter, J. (2001), 'Health and safety: turning concern into action', *The Structural Engineer*, **79**(12).

Carpenter, J. (2004), 'Competency in structural engineering design', *The Structural Engineer*, **82**(15).

Carpenter, J. (2007), 'Making the most of an opportunity: CDM2007', *The Structural Engineer*, **85**(15).

Carpenter, J. (2008), 'Safety, risk and failure', *The Structural Engineer*, **86**(14).

HSE, 'ALARP "at a glance"' (undated), www.hse.gov.uk/managing/theory/alarpglance.htm.

HSE (2014), 'Risk assessment: A brief guide to controlling risks in the workplace', INDG163.

Institution of Structural Engineers (2012–2017): the Institution of Structural Engineers has published a series of more than 60 short articles, 'Managing Health & Safety Risks', targeted at informing young engineers as to what they should know about health and safety.

Soane, A. (2019), 'Building a safer future: UK government proposals for reform of the building safety regulatory system', *The Structural Engineer*, 97(7).

13 Avoiding failure

13.1 Introduction

It can be quite depressing reading about failures, their multiple causes and terrible consequences. On the other hand, when failures are examined retrospectively, it is rare for there to have been a totally unavoidable cause. Hence an optimist would argue that most are preventable. For structures in normal service, an overview suggests that the vast majority perform satisfactorily; serious failures do occur, but rarely. The reality is that enough is known about structural behaviour to ensure structures are safe if that knowledge is properly applied. Perhaps natural disaster causes are an exception; these are always going to occur with a degree of uncertainty and are often going to be expensive in terms of life and infrastructure loss. Yet even for those, death rates in similar events as between developed and undeveloped countries differ markedly since proper engineering and preparedness can do much to mitigate the worst effects.

There is a safety danger in overreaction. The fact that things occasionally go wrong is no reason to be excessively risk averse. Designing and building infrastructure is not optional, we must have it, and at affordable cost: not having proper infrastructure poses severe societal danger in itself.

It is a platitude to write '*safety is paramount*'. Safety comes with an attached price and it must be accepted that there is an affordable limit to what can be spent to achieve a given reliability. For day-to-day structures this is of little significance because compliance with codes and good practice will achieve adequate levels of safety. But where there are uncertain hazards or where the consequences of failure are disproportionate, what we can achieve is finite and limited by how much we want to spend. Moreover, the level of safety/reliability is often just an informed judgement, especially when trying to cope with very low levels of hazard occurrence. Although hard, designers have obligations to be open with clients, making them aware that nothing is risk free. Philosophically this is a very difficult area, particularly when working with hazardous processes, but it remains an unavoidable challenge for engineers.

To avoid failures, the best strategy is to recognise from the start that there is risk in all projects. As Sir Michael Latham wrote in his important report 'Constructing the Team': (Latham, 1994) 'No construction project is risk free. Risk can be managed, minimised, shared, transferred or accepted. It cannot be ignored'. What might go wrong is varied and the best chance of spotting what might go wrong is to have experience of what has gone wrong elsewhere and why. A root cause of many failures is simply human error; a person or organisation making a mistake. Everyone aspires to be competent, but we are all, corporately and individually, fallible. Hence projects and organisations need to be arranged in recognition of the likelihood of human error and procedures managed to minimise the chances of a mistake progressing far enough to turn into a disaster.

This simple notion has profound implications for the way civil/structural engineering business is conducted. It implies staged checks and reviews. There is a need to be clear and confident about design intent and to have a competent team in place. Once underway, the safety of designs should be checked periodically and formally before issue for construction. Part of the initial design process should be ensuring that projects are 'buildable'. Thereafter, proper dialogue with the construction teams is essential to minimise the risks of failures during construction, and there should be validation during and after construction that what was intended to be built was actually built. All structures degrade in service and periodic maintenance and inspection is a key demand to ensure ongoing safety.

Self-preservation management and regulation do nothing to advance the cause of safety overall. Within the design process, excessive caution and over-specifying does little to help. Anyone in current practice will be familiar with volumes of documentation whose objective is not really to get the project right so much as to protect the document's author if the project goes wrong. However, recognising that all projects included multiple interfaces, what is certainly required is a clear allocation of responsibilities between the parties involved and proper information flow management across those interfaces

13.2 Golden opportunity

The political philosopher Edmund Burke (1729–1797) famously wrote: 'Those who don't know history are destined to repeat it.' Thus failures, their investigation and the dissemination of findings represents a golden opportunity for learning and trying not to repeat the mistakes (albeit, somewhat paradoxically Burke is also quoted as saying 'You can never plan the future by the past'). Every time there is some disaster in public life (not just in engineering) political statements are made that 'this must not be allowed to happen again' and 'lessons must be learned'. And yet in engineering we have no systematic way of teaching the experience of failure nor of disseminating the lessons, either amongst the contemporary workforce or down the generations. Furthermore, the litigation process actively hinders those in the know from sharing their knowledge. This is not in the best interests of society. Rather, it is in the best interest of all to learn from mistakes, and preferably from someone else's. After the Ronan Point failure in 1968, the professional institutions set up SCOSS (Standing Committee on Structural Safety) and their annual reports on current issues have been issued since that time. More recently, a complementary confidential reporting system, CROSS, has been set up (CROSS-UK, undated) to try and collate and disseminate information on things going wrong including 'near misses'. Other industries (such as airlines) have such schemes and they have proven beneficial in promoting safety. The CROSS process was endorsed by the inquiry which followed the Grenfell Tower disaster of 2017 (Hackitt, 2018). The chapters in this book describe many failures grouped under broad categories and then under much more specific sub-headings. It should be observed that there is repetition, commonality and thus opportunities for lessons to be learned and all engineers have a responsibility to learn and a strong self-interest in doing so.

13.3 Risk generally

13.3.1 Introduction

Over recent years the study of 'risk' has become fashionable and that change is positive. Hitherto the word risk just had nasty connotations as something wholly undesirable. Current thinking is to just accept we live in an uncertain world where negative events happen. Put simply, if we look at what might happen, what might go wrong, assess the consequences and weigh up the probability of those events occurring, we can combine those factors to assess the overall likelihood of such and such an event and decide (or not) to take certain precautions. We all do this instinctively in our day-to-day lives, noting that individuals are more, or less, risk averse by inclination. In many circumstances, where the probabilities are low but the consequences high, such as our own accidental death, we might take out personal insurance, which is just a financial tool for spreading risk. We sometimes do this in our domestic lives too, via financial insurance, to mitigate the consequences of fire or flood. On large projects, promoters grapple with the problems of assessing how likely it will be that their project will be completed to time and budget, and if they know where the biggest threats lie, it is often possible to take precautions.

In the abstract, 'risk' is a difficult subject. Governments openly discuss effects on the population in terms of the environment, genetically modified crops, smoking, general health hazards and much else. So, there is some wider recognition that any lifestyle or any activity carries risk. And there is usually a recognition that no progress can be made without taking some risk. But debate about risk brings pros and cons: some significant sections of the population are seriously uncomfortable about accepting any risk at all. The public want 'safety', absolute safety, even if that is not on offer. They do not want to believe train systems may be 'unsafe' or that bridges may fall. But it is very hard to get across the balance of risk and to get across that in truth all engineers can offer is low, not zero, risk. In the Latham Report, Constructing the Team (Latham, 1994), Sir Michael

Latham wrote: 'No construction project is risk free. Risk can be managed, minimised, shared, transferred or accepted. It cannot be ignored.' Nevertheless, in promoting projects, engineers must recognise the feelings of society in these matters and help address them. But they must also be confident enough to argue for acceptability based on something being 'safe enough'.

People's personal response to risk is varied and much has been written about it. Some risks are dreaded more than others, and this is discussed in (HSE, 2001) by the UK Health and Safety Executive. The public reacts in differing ways to risk and in the standards they expect. Multiple deaths in a single incident always generate more reaction than the same number of single deaths in separate incidents. As an example, there was a very severe public reaction to 52 deaths in a Riga supermarket collapse in 2013 (Chapter 8, Box 8.32). The Prime Minister of Latvia even resigned. In the UK there was widespread horror and indignation at the 2017 Grenfell Tower fire tragedy (72 deaths) and the fallout from that tragedy is still underway. In Italy and across the world there was horror at the instantaneous collapse of the large bridge in Genoa in 2018, whilst traffic was crossing. The tolerability of risk argument is an important concept for low probability / high consequence risks. The topic of catastrophic failures (and how to avoid them) which are often low probability / high consequence is covered in the CIRIA publication, 'Guidance on Catastrophic Events in Construction' (CIRIA/ HSE (2011). It also leads onto the issue of who should decide what is or what is not tolerable. Generally, people will accept more risk if they feel in control. The idea of the precautionary principle is also useful: this shifts the burden of proof for demonstrating the acceptance of risk to those responsible for creating the risk. It encompasses the concept that precautionary action should be taken to avoid or mitigate a 'perceived risk' if the outcome would be overwhelmingly adverse; even though analysis suggests the probability is small, it creates the impetus to take action (the precautionary principle is explained in 'The Precautionary Principle: Policy and Application' (Interdepartmental Liaison Group on Risk Assessment, 2002)). Much risk is about balance. Whatever we do in design, we frequently exchange one risk for another. In buildings we enclose people to protect them from the weather, but we then introduce the risk of fire and need for escape. Dealing with balance is a particularly difficult problem at low levels of risk. In nuclear plant, the benefits to society of protecting against a hazardous event (which might not happen) may have to be weighed against the certainty of exposing workers to radiation risk if the preventative measure is implemented. We could think of parallels in normal construction with the benefits of one form of construction over another. Generally, we must all be careful that in reducing one risk we do not inadvertently introduce a greater risk, especially to another party who may have no choice but to accept. One might also think of the balance in the Bradford fire (Chapter 9): crowd control via barriers versus hindrance of escape. Around the year 2000, the UK suffered a spate of bad train crashes with significant loss of life. The reaction was to cut services and slow down trains so as 'to be safe'. This reaction caused severe disruption at huge cost to national life whilst in reality, passengers temporarily exchanged the low risks of train travel for the increased risks of car travel.

A continuing concern is the issue of liability. Whilst everyone must accept the consequences of their actions, irrational fear over liability leads to a risk-averse culture to the detriment of all. A sensible approach is to be aware of the liabilities and take those into account and, in some ways, this maps onto the issue of consequences. If the consequences (or liabilities) linked to failure are excessive, that points to increased effort to ensure safety and reliability.

None of these themes are new. They have certainly been debated for a long time, but new generations need to reflect on the ideas afresh for their own education. Basically, the simple act of looking coldly but rationally at a project and asking oneself what might go wrong is a significant forward step: it seems not to be done widely enough.

13.3.2 Hazard and risk assessments

The terminology of risk assessment occurs often in modern engineering and modern life more generally. Many site activities are required to be 'risk assessed'. How this is done can vary significantly from a simple look at what might go wrong through to a fully documented safety case. The word documented is a key one. It is often essential to produce a written risk assessment which is commonly set out in tabular format under such headings as hazards, probability, risks.

Here it should be pointed out that the word 'risk' can be used in confusing ways. Strictly in engineering: a hazard is an event (say 'dropping something') with potential to cause harm and this can be assigned a numerical value, say 1 to mark a severe event, perhaps down to 5 to mark a trivial event. Probability is a judgement about how likely it is that such an event will occur, again assigned a numerical value which might again be on a scale of, say, 1 to 5 (but it could be any set of numbers). Risk is the numerical product of hazard × probability. Using the scale above, a product of 1 × 1 (severe hazard × high likelihood of occurrence) would suggest some precautionary action to minimise the risk whilst a product of 5 × 5 might suggest the risks were low enough to ignore. The tabular format of a risk assessment might also include a column for 'residual risk' assigned, say, high, medium or low, or a column for mitigating action or a column for responsibility. The document is not the end in itself. In all circumstances, benefit derives from the very process of carrying out the risk assessment collectively and acting on the findings. Liability is controlled if management can be seen to have carried out the process and acted accordingly.

The starting point in any risk assessment is to define the hazards. Whilst some of these are common sense, drawn from everyday experience, a good risk assessment should draw on practical informed experience and the range of hazards described in this book provides a sound starting point (other information may be sourced from the 'Hazards Forum'

(Hazards Forum, undated). Young people starting out in construction would be unlikely to identify all the hazards that might apply due to their limited personal experience, hence the need for learning and guidance. Occasionally a hazard might crop up that no one has previously considered. An example might be the severe disruption caused to air traffic flow over western Europe due to the volcanic eruption in Iceland in 2010 (Chapter 10). Another might be the helicopter crashing into a tower crane in London in 2013 (Chapter 11) or even the terrorism attack of 11 September 2001 (Chapter 11). Terrorism had been known as a threat for a long time, but the rise of suicide bombers provides a new dimension. New threats arise, such as a contemporary concern about cyber-attacks closing down national infrastructure supplies; even solar storms are suggested as possible disruptive elements for our vital communication systems. Some of the hazards mentioned in the previous paragraph might seem rather esoteric, perhaps exposing the process to ridicule.[1] However, the hazards mentioned have all occurred, albeit they will not all be appropriate to all projects: there is no point considering the hazard of cyber-attack on a simple job like digging a trench. But what engineers do require is a good grasp of standard hazards, and in digging a trench, the hazard of side wall collapse is one to be considered and one that has caused many deaths. There are a range of hazards that always occur: slips, trips, falls; falling from height; hot working, etc., and these are ones that should be in the mind of anyone carrying out routine risk assessments. Other hazards are project-specific. During the 2018 Ryder Cup (golf), a spectator was blinded in one eye by a long shot. The hazard was of being hit by a golf ball, which has a certain probability. Mitigation is hardly possible by shouted warning (spectators are too far away to hear) and there is little incentive for staying away from the target green area. A search will reveal that the risk of severe injury from golf ball strike is not that unlikely and, given the number of match activities and the numbers of balls hit, perhaps that should come as no surprise. Overall, the best way of learning is to study what has been written, then keep abreast of news reports of accidents (the CROSS-UK website is a good starting point: CROSS-UK (undated)). Chapter 12 discusses action that needs to be taken after hazard identification. The objective of any risk assessment is to give clarity on what the relevant risks are. Thereafter there are obligations on design teams to manage and reduce these wherever possible (Chapter 12 discusses the approach especially in Section 12.16.2). Chapter 5 describes a more complex approach in Section 5.10, which sets out ways of targeting the defined hazards. One definitive output of any risk assessment should be clarity in respect of who owns the residual risk.

13.3.3 Engineering risk

By and large, engineers are happiest dealing with deterministic events. They feel comfortable with mathematical structural analyses, predictions of member stress and

1 In May 2019, the Japanese high-speed train was affected by a power outage; 26 trains were cancelled and 12,000 passengers were affected. The cause was eventually traced to a slug that had entered and short-circuited some switch gear. More commonly, rodents chewing through power cables cause fires.

definition of acceptance against codified criteria. In the routine of day-to-day design that makes for an efficient process, giving us all an impression of certainty and predictability. But the structural world is not quite like that. In truth things are much less certain. The forces we design structures to resist can defy precise prediction, the complexity of projects and the interrelationship between design and construction are such that it's hard to see at the outset what the totality of 'design' means, and so failures occur. At a more mundane level, project uncertainty and inappropriate risk management result in much cost and programme overrun. 'Risk in Structural Engineering' (Institution of Structural Engineers, 2013b) outline the issues. For routine projects, SCOSS has promoting thinking about risk in terms of 3Ps: People; Processes and Products. Figure 13.1 is reproduced from Carpenter (2008), and the reference offers considerable insight into the wider aspects of risk thinking. The largest single risk remains as 'just getting it wrong'. Certain matters are just pure risk. Are the intensity and frequency of flooding and storms seen over recent years a reflection of increased weather volatility consequent on the climate emergency? How big will future storms be? How do we respond to the threats of terrorism after 9/11? And December 2018 was the 30th anniversary of the Lockerbie tragedy (December 1988): who would account for the risks and consequences of a bombed-out passenger plane landing on a village or a key piece of infrastructure? How do we face up to that risk? What can we, as engineers, do to minimise the impact of these hazards, protect our fellow citizens and protect their investments? These are very difficult topics that cannot be left to politicians to solve. They are our problems in the worldwide engineering community and we should play a major role in solving them. This is part of the topic of accepting that all we do is risky, of accepting that all projects carry technical and financial risks and of educating ourselves to predict,

The '3 P's' categorisation of risk	
Influencing category	Example of risk element
People	Competence (encompassing education, training and experience) Culture Supervision Team resource
Process	Software Procurement Time Use of unfamiliar codes of practice Limitations of codes Checks and reviews
Product	Forgotten problems and shortcomings of some established products New un-tied products Mis-use of products

Figure 13.1
Categorisation of risk

manage and insulate ourselves as much as we can from credible risks. Allied to all this is our responsibility to move the profession forward and educate the next generation. Chapter 10 discusses the risk of natural disasters whilst Chapter 11 highlights some risks in man-made disasters. If nothing else, these chapters show that risks are real and will recur.

The risk of structural failure, at least, includes concepts of likely material strength and likely loading intensity; both are variables. Material properties show as a scatter of test results. All loadings are probabilities, highlighted most obviously for hazard loadings. Some might not be quantified, as, for example, vehicle impact loading to motorway bridges, which has been taken as a much more serious threat over recent years, with measures implemented to increase bridge pier robustness. Judgementally, in initial design, acknowledgement of an impact threat might lead to the repositioning of bridge piers from within waterways onto protected banks out of harm's way. It might lead to the placing of protection barriers to prevent errant vehicles crashing off bridges or into other carriageways. All such responses are just acknowledgements that something might go wrong. Other hazard loadings might be quantified, but with uncertainty. In the UK over the last few years, engineers have become acutely aware of the risks of flooding and storm damage causing power outages, and how to handle these and deal with them in a rational manner; effects are based on concepts and consequences such as the 1 in 100-year storm, and so on. The critical North Sea tidal surge that occurred in 2013 (described in Chapter 10) was the worst since the devastating flood of 1953. To protect against such surges, the Thames Barrier was designed for a 1 in 1,000-year event, whilst the Dutch dykes are supposed to contain floods with a recurrence probability of between 1 in 4,000 and 1 in 10,000 years. The risk of earthquake damage is low in the UK (and so is ignored). But in other parts of the world the effects of earthquakes may dominate the structural design albeit earthquake intensities are so varied, they can only be defined statistically. One practical value of defining hazards in terms of numerical probabilities is that it allows governments, owners and engineers to assess the cost benefits of protecting against extreme events.

13.4 Clarity of purpose

It is important to have an overall target and generally there is one: all structures and projects should 'be safe'. But that, in itself, is not a particularly helpful target as it lacks definition of what safety is and what is 'safe enough'. Hence, we strive to be more specific. At national level this might be achieved by setting standards and appointing regulators. Progress has been made here in inherently hazardous industries where there is now a formal demand for project safety cases and a culture within those specific industries as to what standards should apply. In a similar fashion, major infrastructure such as transport and power structures have specific safety regulations and an active study of safety issues. Likewise, construction generally is covered by the CDM Regulations (HSE, 2015), which have an aim of promoting safety (see also Chapter 12).

The first objective of any project designer must be to have clarity of purpose. In relation to safety, that must include definition of the project's safety function (if any) and a full appraisal of all the applicable hazards. For many structures there might be nothing specific, though certain hazards, such as those linked to fire, are generic. Equally, an appraisal of likely overload or misuse in service might be appropriate. It will always be relevant to define and understand degradation processes and ensure the structure remains safe against these throughout the structure's designated life. No design should start until all the hazards have been defined along with corresponding acceptance criteria.

For certain special structures, specific hazards might be relevant: flooding, impact on bridges, car crashes, terrorism and so on. In some special structures there will be a formal requirement for a hazard/risk appraisal in connection with ensuring robustness (Institution of Structural Engineers, 2013a).

It is not at all unusual for assumptions to be made at the beginning of a project. If that is so, those assumptions should be rational and a strategy for verifying them needs to be put in place, since assumptions add to uncertainty. An obvious assumption relates to below-ground conditions, which will probably be investigated at some stage, perhaps initially by desktop study and then by later site investigation. Nevertheless, any site investigation will be limited in scope and surprises occur when sites are opened up. A good example would be the need to respond to changing ground conditions discovered during tunnelling or, more commonly, assumptions about where underground services might be buried yet turn out to be in 'the wrong place'. In refurbishment projects, assumptions commonly need to be made about the size and condition of the existing structure and such assumptions need some sort of investigation by way of verification and an organisational procedure for response.

One key generic assumption always needs to be challenged; the presumption in operating plant that because nothing has gone wrong so far, it never will go wrong. This line of argument is just false, and a key plank of safety policy is the need for periodic reviews of existing safety cases. Allied to this is the attitude to 'near misses'. Near misses should not merely be treated with relief at having got away with it, but should be dealt with in a proactive manner to make sure a similar event will not cause future disaster. A sobering observation from the Grenfell Tower fire is that several near-miss incidents had previously occurred on high-rise buildings, all suggesting that cladding fires could occur and could be catastrophic. Dealing with near misses requires openness and a no-blame culture.

13.5 Organisation

A root cause of many failures is some form of human error, or failure of management, or organisational weakness. Such causes might vary enormously (Chapter 6). They might, for example, stem from a failure to define to interested parties what is expected of them; they might be a failure to carry out maintenance in service or they might be lack of supervision or even an inability to provide support to inexperienced staff. In certain cases, management pressure to preserve 'production' or imposition of 'cost cutting' has been a direct cause of shortcuts and thence disaster. All of these might be embraced within an umbrella of organisations lacking a safety culture or a management process to ensure safety. After the dramatic failure of the space shuttle Columbia (2003), the inquiry found many causes lay within the organisation of NASA itself; NASA had not learned the lessons of precursors to this disaster. The horrific story of the Chernobyl disaster should be studied by all those who seek to learn the worst form of a lack of safety culture.

Many projects are large or complex or both. Therein lies danger, because the work is distributed across many fronts with many potential interfaces and opportunities for confusion. An essential project management activity ought to be defining how a whole project will be procured, who will do what and when, and making sure that proper instructions and information have been given across the interfaces. In modern construction practice, where packages may be mostly outsourced (like concrete detailing), it is more important than ever to police and check the outsourced work to be sure that the delivered product is satisfactory and compliant with overall design intent. Delegation does not eliminate the problem.

In any project execution plan it should be clear who is responsible for what and when. This is not an abrogation of responsibility; it is merely a matter of proper delegation. But if there is confusion or fudge over instructions along with the hope that someone else will 'sort it out' then safety is not helped. Whatever activity or risk is delegated should be clearly defined and the deliverables should be to agreed standards and subject to overall project lead verification. If the deliverable cannot be defined, this ought to be openly recognised and processes put in place for periodic re-evaluation. A key interface underlying many problems is that between design and construction.

A simple observation which will resonate with many readers is that the risk of mistake increases if staff are harassed, overloaded and juggling too many balls at once. It may look heroic, and all engineers find themselves in such positions now and again, but the commercial risks and the risks posed to public safety are high. Working on a staff base that is too lean is foolish. Not providing adequate technical support or supervision to junior staff is foolish. On the other hand, creating a hugely complex bureaucracy for oversight can serve little purpose. That applies especially when supervision degenerates into excessive paperwork with a 'tick box' mentality.

There is of course no substitute for wise technically competent staff. Within any organisation, competency encompasses more than formal qualification. Ideally it includes a considerable range and depth of experience across the staff base. Many mistakes have occurred which would have been obvious to experienced staff if only they had been offered the opportunity to look.

There are sometimes criticisms of committees overseeing projects and safety with fears of responsibility dilution. On the other hand, for complex problems where judgement is required, it can be enormously beneficial to harness the diverse skills of a competent group rather than rely on the skill and judgement of just one individual; as an example from the financial world, the Governor of the Bank of England looks for advice from the Monetary Policy Committee as a whole.

Overall, if any organisation is asked, 'How is it to be ensured that this project is safe, and the risks contained?', they ought to have an answer and one that is not trite.

13.6 Managing change

All design information and all design outputs (analysis, calculations and drawings) are subject to change. Even finished structures are subject to change throughout life. Risks exist of constructors not working on up-to-date information, so a strict management control procedure is required during the design and construction phases. In service, dangers exist when modifications are made without the original designer's knowledge or input. So, in certain cases a risk mitigation measure is to ensure that one party who has the overall perspective is retained as the design authority and no changes should be made without their say so. The Hackitt enquiry (Hackitt, 2018) after the Grenfell Tower fire highlighted the importance of change management (Chapter 6).

13.7 Structural/civil engineering rules

13.7.1 Starting

At the start of any job, it should be ensured that the team has the right competency and range of experience. If they do not have this, then they may just be unaware of what they don't know. Thereafter, open discussion on the project on all aspects of 'design' should be encouraged. Design is not just about sizing members and doing calculations. It includes everything from ensuring the structural concept is correct to making sure the performance characteristics are adequately understood and defined. All these should be resolved and approved before moving on to more detailed design. There should be absolute clarity on the client's brief (challenged where necessary): bear in mind the chapter headings in this book and the various modes of failure that are possible. Understanding the mode of failure is a powerful 'tool' in risk management.

There are strong recommendations that one designer (or team) should maintain overall control, particularly of issues like stability and robustness. This might also extend to fire protection which demands a mix of architectural, civil, mechanical and fire engineering expertise.

13 Avoiding failure

Civil/structural 'design' is a complex interactive process. In relation to safety, good design must recognise compliance with the objectives of achieving:

- strength
- stability
- stiffness
- robustness
- insensitivity
- durability (including 'inspectability')
- structural fire resistance
- constructability
- etc.

It should be established at feasibility stage that all these objectives are capable of being met. Examples of failures caused by not respecting these objectives can be found throughout this book. Chapters 4 and 5 give an overview of attributes that contribute to structural safety.

It also has to be established that any additional functional demands imposed by the client can be met. These might, for example, be something to do with operability.

Most projects involve a stage of information gathering and designers should be proactive in this. There should not be a later failure predicated by a 'didn't ask and didn't tell' set of circumstances. Designers should probe to assess information they know is required and if that information is unavailable (say precise crane loading or flow rates), rational assumptions can be made and recorded as such and a management process deployed later to check verification before the design goes into service. It is frequently the case that design information is coarse at first but refined with time. A key point on starting and throughout the project, is to be definitive about the status of any information: is it absolute, or subject to confirmation, etc. Gathering information about the site itself is clearly very important.

Once all these matters have been sorted out, the strategy for achieving/ demonstrating them should be recorded in a Basis of Design document, not as a bureaucratic exercise, but as a vehicle so that all parties involved in the project can be informed. The basis of design should certainly record all the overall load paths for the transmission of vertical and horizontal loads from source to foundation and it should record the basis of the overall stability system and the design standards. There ought to be no ambiguity about what part holds what up at any stage of the project. The Basis of Design should schedule all the numerical design targets and any design strategies, such as those for ensuring durability, or fire protection, and any interfaces with other designers.

Any project involves the transfer of information within both the design group and to external bodies. Opportunities for confusion abound. Hence the status of any issued information: ‘preliminary’, ‘for costing’, ‘for construction’, etc. should be formal and definitive.

13.7.2 Analysis

Most structures will require some sort of analysis. This may be a global loading analysis, or a wind loading analysis, or a dynamic analysis, or a flood analysis. In modern design methodology, these are often complex, and computer-based; this may introduce a danger in that analyses may be sublet to a specialist who may not fully comprehend their purpose. Likewise, the receiver may not fully comprehend the significance of the output or might just get lost in a mass of printouts.

The overall designer should maintain control of the work, defining key assumptions. The main designer should also confirm the type and detail of the analysis, for example, whether a non-linear analysis is required or whether finite element analysis is required or whether routine linear elastic analysis will suffice. The overall designer should confirm the boundary assumptions.

Getting analysis wrong has occurred (see Chapter 6). Since many modern designs are totally reliant on the accuracy and validity of the output, it is essential to have a stage of analysis verification before the output is used in later calculation work on details. This is not only for safety but because otherwise a great deal of subsequent work can be abortive. All software used ought to be validated to ensure that it does what is intended to be done and output should be verified at least by simplified hand checks to ensure that predictions are of the right order and make sense (MacLeod, 2005; Mann and May, 2006).

In terms of safety, fully understanding the stress states in a structure is essential. However, ever more elaborate analyses and fretting about ‘stress’ to fine limits is not a fruitful way to proceed. It is a particularly misplaced approach to seek risk reduction by conducting ever-more elaborate structural analyses constrained to artificial stress limits to gain numerical comfort. Safety would be better served by engineers standing back, looking at the real risks, and producing a balanced approach which in part involves producing designs less sensitive to localised effects (see Chapter 3).

13.7.3 Calculations

Most civil/structural projects generate voluminous calculations. Given that volume, it is almost inevitable that there will be mistakes. Consequently, calculation work needs arranging such that it can be broken down into stages that minimise the chances of mistakes being carried forward. The Basis of Design should set out all the key loadings, and all the methods of calculation that will be employed, be they hand methods or dedicated computer programs. And that Basis of Design document should be checked before it is used. Any information passed across interfaces (i.e., where parts of the design are sublet to specialists) should be set out formally so there is no chance of ambiguity.

All calculations should be verified and checked both in stages and before final use. Checking should be for 'sanity' and arithmetic. Crude 'order of magnitude checks' are valuable for ensuring the absence of gross error. Self-evidently, a structured way of setting out the calculations should be followed so that the input and output are easy to extract and the calculations are checkable by an independent person. There is modern tendency to rely on automated software calculations, which often seem to print out in an indecipherable format. If calculations cannot be readily understood or checked, they are dangerous.

13.7.4 Drawings

Drawings remain the main vehicle of communication within the construction industry and considerable skill is required in being able to prepare and read them. That skill is based on experience and a deep knowledge of detailing methodology conforming with accepted practice. It is in the nature of the design/construction process that drawings are periodically modified so a real danger exists in only changing part of a drawing, or changing a detail on one drawing and not others, or not carrying changes through or causing confusion. This risk can only be controlled by strict drawing office management to ensure that those preparing drawings have correct, unambiguous, information and ensuring that the status of any drawing is definitive (i.e 'for comment', 'for costing'. 'for construction' etc). As for calculations, all drawings should be checked and approved before issue for construction. Incalculable benefit is derived from the casting of experienced eyes over drawings during an approval process. What should be checked is what is there plus looking for what is not there. Most structures have a weak link, often a key critical steel or concrete connection (such as column head shear in flat slabs). Experienced eyes should seek these out and verify them.

13.7.5 Specifications

There are two broad groups of specifications: performance specifications and quality specifications. The former require drafting by engineers who know what is required and such documents are in effect a design brief to third parties. Quality specifications, as we now have them, have evolved from years of experience and certain industry-standard ones exist, such as the National Structural Steelwork Standard. This includes options for procurement plus options for the division of responsibilities between designers and constructors. However, many designers do not fully understand the details and fail to understand that there are options to be exercised within the specification clauses. A common dispute revolves around a failure to meet expected tolerances and many designers fail to realise that specifications are no substitute for an overall strategy of understanding what tolerances are essential; what adjustment measures are required and how these will be achieved (Chapter 3). Designers need to be alert for when assumptions about loading paths are sensitive to fit-up tolerances. See box girder bridges; Camden School and Ronan Point, all in Chapter 8.

13.7.6 Procurement

Procurement might be of products, services or materials. In terms of whole product, a definition of requirement might be via a performance specification which requires great skill to produce. The supplier will most likely only supply what is asked for, so a significant responsibility lies with the specifier to define exactly what is required and to be confident that it is achievable. Likewise, if services are procured (such as concrete or steelwork detailing), definition of standards is required along with verification of the finished package. It should also be anticipated that there will be confusion over interfaces. Procurement of materials is very largely covered by appropriate standard specifications, but the specifier must know what is required since options exist. Steel has been known to be procured on strength alone without the options of required toughness being invoked. Specifying concrete by strength alone can be done, but other properties, such as an ability to be placed and compacted, might be more important. Specifiers will generally get only what they ask for. It is not unknown for specifiers to try and cover themselves by citing a plethora of standards, requiring the most onerous interpretation of each. This simply causes confusion and does not serve the cause of safety.

13.7.7 Construction/practicability

There are many examples of structures failing in their partially complete state and this includes major structures such as bridges. The reasons are many. Construction can be highly specialised, and its attributes not fully appreciated by office-based design teams who either might not have the requisite skill set or who might not have the foresight to envisage the details of exactly how a structure will be constructed. On the other hand, many failures occur through inadequate temporary works which are solely the responsibility of the contractor. As a generality, when partially erected, a structure may well be less stable and less strong than when fully complete. Additionally, critical loading cases can exist in the partially complete stage (Chapter 8). Referring to Section 13.7.5 above, designers may not appreciate the effects of tolerances or it might be that constructors fail to achieve tolerances that have been set. The only way to mitigate these many effects is to have full and proper dialogue between the parties as early as possible and to optimise the design of the permanent works to ensure that they are safely constructible. It ought to be acknowledged that both sides of the project (designers and constructors) have a part to play and safety is only fully ensured if there is proper information flow and dialogue between the parties. What can happen if this does not occur is well illustrated by the failure of the Westgate Bridge in Melbourne in 1970 (Chapter 8).

A risk mitigation measure on all projects, large and small, is designer involvement to ensure that what is being built is what was intended to be built. There are risks linked to keeping design teams in a 'design only service'; periodic inspection by the design team during construction is highly desirable. There are distinct advantages to the design team identifying key details dependent on workmanship and imposing QA Hold / Witness points.

13.7.8 Verification of assumptions

The need to verify assumptions about ground conditions or existing building conditions has been described above and many precautions are advised in 'Appraisal of Existing Structures' (Institution of Structural Engineers, 2010) and ISO 13822:2010. Verification of assumptions on stability and load paths might equally be appropriate when assessing the needs of demolition (Thoday, 2004 and BS 6187:2011). Another need arises when safety is linked to complex analyses. Examples might be in geotechnics (as in the contribution of the geotechnics work in the Nicoll Highway Collapse, 2004, Chapter 8). In such circumstances and where, for example, tunnel lining implementation is linked to observations of performance, there is a strong role for monitoring to ensure that what was theoretically predicted is truly what is happening. A failure to properly react to live monitoring was a feature of the Heathrow Tunnel collapse in 1994 (Chapter 8). Structural monitoring and survey can play a major role in the safety of complex frame construction during erection by checking that deformations in temporary/ intermediate states are in line with theoretical predictions. This is an indirect way of assessing that stresses remain tolerable in temporary conditions and equally ensuing that the performance of the frame is progressing as predicted by the analysis.

13.7.9 In service

Structural capacities might degrade in service (Chapter 7). The 2018 collapse of the Morandi motorway bridge in Genoa is prime example. To guard against such failures, all structures should be periodically inspected, looking for signs of deterioration. This is especially important when life-limiting mechanisms such as fatigue / corrosion exist. As a follow-on from that, by design, all structures should be 'inspectable'. And again Chapter 8 contains examples of catastrophic failures promoted by an inability to inspect. A simple inspection gantry once failed. This device was delivered in parts and then secured by pins with retaining latch keys. But a key vibrated out and the gantry came apart, killing the occupant. One criticism of the design was that the latch keys were so hidden that casual observation could never have detected their loss.

13.7.10 Records

Buildings and infrastructure are long lasting; 50–100 years is common. Changes might easily be made throughout life, either in usage and/or physically. To promote safety, access to records of original 'as-built' construction is immeasurably valuable. Nowadays, with electronic data systems, record retention is simple. Indeed, it is a statutory requirement.

13.8 Checking

13.8.1 Introduction

Everyone makes mistakes. The assumption should always be that a designer has made a mistake, that those preparing the analysis or the calculations have made a mistake and that the drawings contain mistakes. It is thus foolish not to check and it is no disrespect to the originator to require checking to be carried out. Of course, there are different degrees of checking and the process can degrade into petty nit picking. But there are equally plenty of examples of glaring errors that ought to have been spotted, at least by experienced eyes. It is especially easy to concentrate on faulting what is there and not seeing what is not there (and should be!). Furthermore, it is surprisingly easy to make gross errors.

Within the design process it is sensible for there to be both high-level checks by experienced staff whose role is to look for the gross error and for there to be routine arithmetic checks on production work.

Commercial risk is minimised if checks are staged. There is reduced value in carrying out a high-level review of the basis of design or analysis when all the calculations and drawings derived from them have already been completed.

On key or unusual projects, it is ideal if a third-party review is carried out. Normally this means details of the project are provided to a truly independent party whose role is to oversee, probe and ask what is not there in addition to assessing what has been provided. This process is only valuable if the third party has an appropriate attitude, is prepared to exercise judgement and is trusted by the originators. If the third-party reports to an extent of detail that is only set out to protect their own liability, little is served. What the originator really needs, and wants, is someone who will be thorough and give the necessary comfort (from experience) that nothing critical has been overlooked. Third party reviews can be carried out on designs and on construction proposals. SCOSS believes they should be mandatory for all Class 3 structures as a protection for public safety. (Class 3 structures are higher-risk structures as defined in HSE (2015)). In some cases, a design can only be properly verified when it is more or less complete. Many companies have internal procedures that require formal design reviews. These might be done by having all the drawings 'on the table' and having a fresh pair of experienced eyes critically review them before issue for construction. At the same time, a formal process may verify that any defined objectives set out in the brief are achievable. Those reviewing need to be competent and from the appropriate disciplines.

13.8.2 Regulation

In higher-risk industries, i.e., nuclear and offshore, the checking process is more formal with the need to submit proposals and designs to a regulator before construction. In bridge design it is normal for designs to be independently checked. Submissions may be staged so that outline intent and safety demands are approved before detailed work is started. The regulator should be completely independent for their prime duty is to ensure safety and not to be diverted by the perhaps pressing needs of cost and programme. That said, designers are entitled to expect standards from the regulatory staff that are achievable, plus an attitude of realism.

As of 2020, the UK government is preparing to introduce a new Building Safety Bill. This legislation has come about as a direct consequence of the Grenfell Tower fire disaster. Details are not finalised, but the objective is to appoint a new regulator tasked with overseeing the safety of higher-risk buildings. There will be a demand for more safety cases (see Section 13.11).

13.8.3 Quality assurance (QA) / quality control (QC)

In all design and construction work there are roles for quality assurance and quality control. QA is an initial plan that oversees what is to be done and defines procedures that (if followed) should ensure the final product will be correct. QC merely checks at the end that the product is correct but, while QC may be essential, without a QA plan to start with, much abortive work can occur. A late check only picks up errors: the ideal is not to have errors in the first place. Conversely, QA also brings danger in that attention can be concentrated on the QA process itself to the detriment of putting effort into the actual work. Thence it can be falsely presumed that because the process paperwork is correct, the output must also be correct. QA is ideally suited to ensuring the reliability of mass-produced products; it is perhaps not so well suited to design work, because that is so varied.

QC activities remain essential to validate that materials and products used in construction have the properties that designers assume. For materials this is most often demonstrated by manufacturer certification, though periodic sampling may also be prudent. Materials that are formed, such as concrete, will always need to be tested to verify their properties. Activities such as welding will also be subject to a regime of QA and QC testing to ensure the adequacy of workmanship. Such routine testing should also be considered for any other components / activities whose capacities may be affected by workmanship: anchor testing may be one. QC is especially important for components that cannot be inspected during life; foundations (once covered up) are an obvious candidate. Foundations are thus also an obvious target for the imposition of 'hold points' within the QA plan. Main designers intermittently visiting sites will not see everything but should consider the importance of particular features within their project and define sensible hold points to verify the quality of key structural items.

13.9 Checking and review of complex systems

13.9.1 Introduction

Civil/structural projects often form part of larger engineering projects with a specific role to play within that project, perhaps as a support, barrier, or containment. In other circumstances, for example, in industrial plant, failure of the structure, perhaps by degradation, may have far-reaching consequences (cf. the Stockline disaster, Chapter 11). Moving structures represent a special case, being a combination of civil, structural, mechanical, electrical and control engineering, where failures in any one discipline's component might affect the safety of the whole. As an example, loss of power in a moving bridge or jamming of some part on a moving structure would represent hazards to consider, with consequences to assess. Such projects can best be thought of as systems consisting of a string of interacting components, any of which might fault and perhaps in more ways than one; valves can stick open or stay closed. Many water projects consist of interlinking pipes, reservoirs, pumps, and valves, etc., and an objective would be to assess whether that system is 'safe' (as a whole), noting that any one of the components might fault, and where there are consequences attached to each identified failure. Human error is always a potential failure cause and the complexity and consequences of the Piper Alpha tragedy or the Buncefield fire (Chapter 9) exemplify the interactive nature of multiple causes and the chain of events that can follow.

Other engineering disciplines, notably in mechanical and chemical engineering, have evolved techniques to examine whole systems in an orderly fashion, and civil/structural engineers can gain insight from an awareness of these techniques (see Section 13.9.2 below).

13.9.2 Terminology

Common techniques are termed: risk assessments; failure modes and effects analysis (FMEA); HAZOPs; and fault tree analysis (all defined later). They all have much in common. One commonality is that the techniques should really be deployed on a project which is highly developed and a second is that for best results, the work needs to be done within an organised team environment. In all that follows, the structure or civil parts may be one system within the whole, playing a role within the overall engineering.

Risk Assessments: these have been referred to above, but here the meaning is slightly different in that when examining a proposed project, the first step is to define all the hazards that might be appropriate and then ensure by design intent that they are not going to become critical. Generic hazards might be 'fire' or 'loss of control'. And there might be a target of design intent associated with each. Thereafter an orderly approach can be taken of notionally faulting each system or component in turn and assessing the consequences. What happens if control is lost? Does the system default to a safe state? Loss of power might be caused by loss of grid power, but mains supply might be backed up by a diesel generator. How is that diesel generator to kick in? How is it known that the diesel generator will work on demand when it is rarely used? (This suggests periodic testing is required.) Such questions might be applicable to a water system where power is required to operate pumps in an emergency. The pumps might fail by loss of power (which might itself occur as part of the same emergency), or by failure of the pump itself, which might point to a need for a standby pump. To ensure this will work, both pumps might need to be connected to both the grid and the standby generator. In examining circuits like this, the reviewer searches for anything that might constitute a single point failure or a common cause failure (the two terms are explained in Chapter 5). For example, if both mains power and backup power went through a single switch gear and that switch gear failed, then both the mains and backup power would be lost. During the 2013 UK winter storms, much grid power was lost regionally, so it would not have been available for emergency flood relief which was needed at the same time. In the Fukushima disaster, the tsunami that knocked out the main power supply also submerged the emergency generators plus the connection points for tertiary backup of brought-in generators (a good example of common cause failure). A risk assessment would normally schedule out every component and would then consider all possible modes of failure and perhaps the likelihood of failure, and then assess the risk. It would be normal to include the possibility of human error. What happens if someone forgets to keep the diesel generator fuel tanks filled? Risk assessments are required for Class 3 structures as a tool for examining whether they are robust enough.

FMEA: (failure modes and effects analysis) is a similar form of failure analysis. All systems in an existing design are scheduled out on a worksheet and in adjoining columns all modes of failure for those systems are defined, along with a judgement of likelihood. A third column then schedules the effects of the failures examined, with conclusions being drawn about tolerability or not. A drawback of FMEA is that it only examines the consequences of failure of the system that exists. It does not inform the design process up front. Thus, the process might decide that a fire risk exists if a certain component malfunctions, but it will not highlight 'fire' as a generic risk to be considered initially. To that extent, a good design might start off with a hazard analysis and risk assessment and the end product might be systematically examined via an FMEA process.

HAZOPs: (hazard and operability studies) are commonly used in the nuclear, chemical and water industries. They were developed for chemical plant which has commonality with water supply / drainage systems. A HAZOP is a team effort and is a structured examination of a system. The first task is to define various components of the overall system. The second is to define a series of key guide words that might be relevant (for structures this might be: overload, blast, degradation, human error, etc.). For plant it might be words like more/less, open/closed, to signify more flow, less flow, etc. The team then examines each of the systems against the key words to see what emerges. In a pipeline system, valves may be examined as jammed open, jammed closed, partially closed, left open by error, etc. The objective is to look at all possible operating scenarios and determine outcomes. This is only successful if the team have the right open-minded attitude and the team should be multidisciplinary to capture all possible scenarios (Kletz, 2018).

Fault tree analysis: (FTA) is a top-down, deductive failure analysis in which an undesired state of a system is analysed in successive steps to combine into a series of lower-level events. A switch might fail 'off' or 'on'. Below 'on', there might be several consequences which are further events in the tree. Setting failures out in a logical treelike pattern allows an appreciation of how individual failures can combine, or not, to create an undesired major failure. It can highlight the combinations of circumstances that might exist to cause failure overall. Note that in serious failures, it is frequently an unusual combination of circumstances that has caused the problem: the 'perfect storm'.

As with all processes, there is a danger of producing too much paperwork and not being able to see the conclusions succinctly. Consequently, an important part of any of these processes is to summarise the key conclusions. These conclusions should be enough to allow the team to create an overall strategy (safety case) which can be implemented.

13.9.3 Structural terminology

Chapter 5 discussed the desirability of structures being insensitive to small variations. Being insensitive is an attribute of robustness. There should be no great change in state linked to some minor change in dimensions or capacity, or to minor failure from whatever cause. The tension system failures discussed in Chapter 3 were cascade-type failures possibly initiated by just one hanger failing. In safety-conscious industries such as the nuclear industry, a positive test is made for excessive sensitivity to loading by 'margin studies' or 'cliff edge studies'. The idea is to examine whether there will be a gross change in state consequent on some variation in any one of the assumed variables: to assess how close the structure is to a 'cliff edge'. For example, earthquake loading is only defined probabilistically; it is certainly not known precisely, so any checks on strength cannot be viewed as accurate. If the assumed load is marginally exceeded, will the structure 'collapse'? Note the examination is not a test of strength directly (although it might be) but more a test of structural performance. Ideally if any loading is marginally above the magnitude adopted for design purposes, the consequence will just be some finite increased deformation rather than disintegration. If the energy of a car hitting a crash barrier is rather more than assumed, the mode of failure should be increased permanent deformation rather than total failure. When governing loading is (dead + live + wind), a marginal increase in any of the three probably would not eat into the safety factor too much, dependent on the ratio of the three load components. But if the loading was entirely wind loading, a minor increase in wind speed (given that force is proportional to (wind speed)^2 can have a dramatic effect on structural performance overall (see Chapter 3). The test of sensitivity should not be limited to loading, it should also encompass a review of the details seeking to identify weaknesses such as those highlighted in the Camden school failure, Box 8.6.

With advanced structural analysis software packages it is possible to perform a 'push over analysis' to examine structural resistance up to collapse. In these, lateral loading is increased incrementally, allowing plastic hinges to develop successively via alternative load paths and to rotate up until the structure effectively transforms from being a structure into a mechanism, whereupon it can take no more load. Examination of the output will inform on the nature of progressive failures and displacements and the demands made on joints, possibly identifying weak links, and will provide information on how much emergency reserve exists in the structure.

Such analyses are far too complex for everyday structures. Nonetheless it is easy to look at the structure and ask the question: what happens if any minor variation in assumptions, details or quality occurs?

13.10 Probabilistic/quantitative risk assessments (PRA/QRA)

13.10.1 Introduction

All the techniques described above are qualitative methodologies making judgements about how likely an adverse effect might be. In the vast majority of cases this is perfectly adequate but there are occasions when a more refined look at probabilities can be justified and there are occasions when looking at probabilities assists in making rational engineering decisions. A common design question is 'what is the worst case' whereas a more useful question is what is the worst probable case (an example is given in Beckett, 2007). Both quantitative and qualitative risk assessment approaches yield different insights into the consequences of applicable hazards. A quantitative risk analysis poses traps, but it can highlight where most risk reduction is achieved against money spent. A word of caution: even if an event is proven 'very unlikely', there is no excuse for not producing robust structures to insure against the unforeseen.

In some branches of engineering, probabilistic methods are used quite widely. If enough similar items are in use, the proportion that fail can be counted and used to calculate the risk of others failing in the future. Thus, for a complex mechanical system, with known failure probabilities for the separate items within that system, it is mathematically possible to define the overall level of safety; or, looking at this the other way around, the overall probability of failure, and this can then be compared with what might be considered an acceptable numerical risk. This would, of course, also need an understanding of the way components interact, the possibility that one failure could trigger another, or that one event could cause multiple failures.

Probabilistic methods provide a rigorous framework for assessing risks and thence their management. When assessing flood risks and the costs of damage if flooding occurs, it can be shown that flood prevention schemes are beneficial up to flood risk returns of a certain return period but not beneficial beyond. Where multiple hazards exist, probabilistic methods also allow the ranking of individual risks and so give insight into the contribution of any one event to the total risk, to the importance of separate risks and thence their management. That will include judgement on how to spend proportionately for risk reduction. In short, there is little point is spending a large amount to reduce a (low) risk which contributes little to the overall risk. Thus PRA/QRA techniques can provide a rational framework for justifying when a risk is ALARP (as low as reasonably practical; or SFARP: so far as reasonably practical) in that the costs of driving a risk down further might be examined. As an analogy, we know that usually double glazing can reduce energy demands significantly and a cost–benefit analysis of window cost versus fuel use can be made to justify double glazing. It would be possible to pose the same question about whether triple glazing would be beneficial and currently in our climate it will be shown that the capital cost of triple glazing offers little benefit gain in terms of comfort or fuel saving.

In areas where populations are at risk from flood, which is an event with a statistical probability, one way of assessing the likelihood of damage, or the required height of a flood protection system, is to assess the risks using statistics and numerical values, against a target acceptance value. That target will be influenced by the losses consequent on the flood (or it could be fire) which might be insured losses, i.e., a direct financial link. In London, some sort of assessment is obviously required to set the height of the Thames Barrier (the consequences of New Orleans' flooding in 2005 illustrate the need). Clearly financial losses would be huge if large swathes of London were flooded. In such investigations, for example, the cost of designing a facility against a return period of 10^{-2} flood as compared to 10^{-4} can be investigated and compared to the mitigation gained on financial loss. In the Netherlands, the consequences of spring floods on the Rhine overtopping the dykes are so great that defence structures are designed for a 10^{-4} event. Those at New Orleans were apparently designed for the 1 in 500-year event (which is arguably too low, given the consequences of failure) but in practice they failed in a 1 in 100-year storm).

Many experienced engineers will carry out exercises like this almost intuitively. For example, given that a large steel beam is to be used (say for deflection), it would not be sensible practice to put a small connection on the end which failed to sustain the strength of the beam in an emergency when the marginal cost between a small connection and a decent connection would be trivial. It is always sensible to put a minimum amount of rebar in a concrete beam since the marginal cost of that above absolute minimum might be trivial. It is frequently sensible to spend more than might be mathematically necessary just for insurance and to achieve desirable robustness.

13.10.2 Wider application

Possibly one of the first rational uses of quantitative risk assessment (QRA) was in the study carried out in 1978 to assess the risks that the chemical plant on Canvey Island posed to London (HSE, 1978). Major accidents to such plants do happen, such as the failure at Flixborough (1974) or at Buncefield (2005) (Chapter 11). How can engineers decide which modes of failure are possible, what the probability of those failures is and hence the risk they pose in terms of their location relative to populated areas? One way of assessing the risks is to express them in numerical terms.

Canvey Island had large number of potentially hazardous industrial facilities including an oil refinery. The study task group sought to 'analyse the kind of accidents which would have the potential to injure people living in the area; to assess the likelihood of their happening; and to assess the chances that if they happened people might be injured by them'. The exercise resulted in a picture of the chances of a serious accident and broad estimates of the risk (couched in numerical terms) which could be put in perspective by the chances of injury for common every day activities. In the nuclear industry, when assessing the safety of facilities, a combination of deterministic (designing for specified events) and probabilistic assessment (the probability of those events not occurring) is used. Neither method is used exclusively, since both give insights into the overall safety of the plant. Probabilistic methods are often used to define the deterministic events. The accepted 'safe' target is that failures leading to significant harm to the public should have a probability of 10^{-7} per year or lower (details are explained in Chapter 4, Section 4.5.13). Similar approaches are used in probabilistic studies of risks to offshore installations or railways. When acceptable risks are expressed in terms of a failure probability per year, care must be exercised when exposure to the risk is only for a short term. As the time interval of exposure is short, the probability of failure at that instant is very low. But if there are many such exposures, the probability that a failure will occur at some stage is much increased. Chapter 2 discusses the probability of failure in the context of the accident at Great Heck in Yorkshire, UK (2001).

13.10.3 Load factors

Structural design generally includes a safety factor / load factor (usually built up from partial factors) which ensures that the probability of overall failure is low enough. The factor value was traditionally subjective, based on collective historical experience, but efforts have been made to justify the values statistically (or occasionally by reliability theory), albeit heavily influenced by what has proven acceptable in the past. There is nothing sacrosanct about the number apart from it providing a level playing field for initial design. Excluding gross error, to get a structural failure normally requires that the combined probability of adverse variation in applicable loading, loading configuration, material quality, workmanship, etc., all become coincidentally too high. For the last 100 years or so codes have kept more or less the same factors, but with a trend towards rationalisation and overall reduction as materials have become more uniform and reliable, and as understandings of structural behaviour have increased (albeit many engineers are dubious about this process). That background knowledge about how the factor is determined can be deployed to useful effect in the assessment of existing structures, not least taking account of the likelihood of any assumed loading or real knowledge of material quality.

It can be accepted that standard design codes serve to deliver products with an acceptably low probability of failure. But occasionally a very high reliability might be sought, and such an objective can be achieved by increasing the load factor. Conversely, in other circumstances a reduced load factor may be acceptable. Sometimes that factor may be reduced by accepting an increased risk. For example, it might be assumed that the time at risk during construction is short, so the wind loading may be changed to the one that has a certain probability of occurrence over, say, a two-year period. Subconsciously, the same exercise is carried out in routine structural engineering design when searching for the 'worst case'. What this usually means is designing for the worst probable load in the worst probable combinations; there can be debate on how far to go in that process.

In some codes, the partial factors that combine to produce the overall factor are varied explicitly to take account of known factors affecting the probabilities. In masonry design codes, the partial factor for materials depends upon the workmanship and quality control of blocks and mortar used. This approach recognises the overall lower probability of failure if workmanship is controlled. In all codes, partial factors are also varied for different load combinations (e.g., dead + live + wind) to reflect the lower overall probability of that combination of circumstances arising with all loads at their peak. Eurocode 0 (BS EN 1990:2002) provides a generalised means of calculating a required safety factor based on knowledge of material reliability, confidence in the accuracy of the analysis and so on. Chapter 4 describes how this methodology was used on the London Eye.

13.10.4 Assessment of existing structures

Calculated probabilities will only be correct if the data used to generate them is correct. For example, it may be assumed that concrete meets the specified characteristic strength. If, as often happens, it is over-strength then the probability of failure may be lower (provided the amount of reinforcement is adequate). It is usual to calculate probabilities using what are known as 'conservative' values, that is, the value is chosen cautiously, so that the final probability of failure is probably an overestimate. Structural engineers are frequently faced with the need to appraise structures that fail to comply exactly with 'modern standards'. This does not necessarily mean they are unsafe, and it may well not be in the best interests of society to spend a disproportionate amount of money strengthening them. With a knowledge of how safety is defined in terms of failure probability, it is possible to assess a structure taking into account the real uncertainty in loading (it may be known highly accurately and may have a low probability of being exceeded or may be controlled) and the real uncertainty in strength (there may be evidence of strength being much better than assumed in design, or lower, or the strength might just be known with high confidence) and then judge the failure probability. It is legitimate, for that judgement, to take benefit from other structural qualities such as redundancy and ductility. It is legitimate to consider how much uncertainty remains overall in the parameters that govern capacity under load (including analysis) since it is those uncertainties which demand a safety margin. Further information is available in Appraisal of Existing Structures (Institution of Structural Engineers, 2010).

13.10.5 Summary

The approaches described above may seem esoteric and distant from the real world but they have practical applications. First, without doing anything complicated, just consciously looking at any structure in probabilistic terms allows rational decisions to be taken on whether a risk is acceptable. Engineers should be comfortable with 'taking a risk'; there is nothing wrong with that provided the risk (i.e., in this case probability) is low enough. Second, the effect of changing any single parameter can be investigated numerically, to understand how much benefit is gained or not and this then allows an assessment of the cost of making such changes and how much 'safety' that expenditure buys or what risk reduction is achieved. Where failure probability is linked to a combination of several hazards, QRA offers insights into the relative importance of each to the total risk. If several modes of failure are possible, there is no point in spending more money to reduce an already low risk linked to one failure mode if another mode is much more likely or more dominant.

13.11 Safety cases

In the aftermath of the Piper Alpha, disaster a recommendation was made that all offshore structures should have a safety case. As in (13.8.2) the forthcoming UK building Safety Bill appears likely to include a demand for safety case documents for all high-rise dwellings. A Safety Case is an evidence-based document that describes the plant (or structure), the hazards from various sources and seeks to justify in a logical manner why the facility is 'safe'. The value of the document is twofold. First, it is a vehicle for the design team to set down their thinking in a structured way and that exercise alone is valuable. Second, it allows for independent peer review. It should be noted that the general UK approach is not to be prescriptive about what makes a structure safe or unsafe since that approach promotes a rote compliance mode of thinking. Rather, many UK standards are non-prescriptive and call for an argument-based approach to justify safety. A safety case should aim to show that specific safety claims are substantiated and that risks are kept as low as reasonably practicable (ALARP).

One benefit of a safety case is that after its completion there should be total clarity on what it is that provides overall safety. Normally a safety case starts out with certain simple principles such as: 'all occupants can easily escape if there is fire'. That principle is easy enough to grasp but is of little value unless all the ramifications are followed through in detail such as warnings / actual escape routes /door systems etc. Alternative principles are possible and an obvious example might be people's safety against fire in high-rise dwellings. Current policy is that occupants are safe for a prescribed period if they stay within their (fire-protected) compartments awaiting later rescue. Hence that knowledge should focus attention on anything that undermines that fundamental presumption, such as inadequate fire stopping, smoke exclusion, and so on. Through life management is assisted when there is a documented 'case'. For instance, it is known that in some high-rise dwellings over time, flat doors have been altered to substandard ones such that dwelling fire compartment standards were rendered inadequate. If a safety case had been available, the facilities management would have had a document guiding them to avoid that occurrence.

13.12 Avoiding failure in service

The most likely causes of in-service failures are degradation (various kinds); fire; or unauthorised changes that may add load or unwittingly undermine the load paths or stability systems. Even major structures sometimes collapse in service, with the 2018 Morandi Bridge disaster being a prime example. In the worst cases, collapses occur without warning. Chapter 5 sets out one safety objective for structures in service as having the highly desirable attribute of showing signs of distress in advance of total collapse. Several failures have been cited where this has not occurred, and the mode of failure has been brittle. The Riga supermarket collapse referred to in Section 13.3.1 was practically instant and the factory tragedy in Dhaka in 2013 (Chapter 11) exemplifies a form of failure that should never be allowed by design. In 2016, a huge part of the old Didcot power station (UK) collapsed almost instantly during demolition (four workers were killed). Chapter 8 also describes some bridge failures that were near enough instant. Chapter 7 describes time-dependent failures linked to material degradation. The key benefit of warning is to offer time to react. That benefit also applies to fires.

Chapter 9 cites examples where fire brigades were on the scene rapidly, but still too late to do anything to rescue occupants or check the progress of the fire.

All structures degrade, with the bulk showing some signs of distress in advance of disaster, which then gives time either for escape or preventative action. To ensure warning for both safety reasons and to protect the asset and prolong its life, periodic inspection is advisable, the objective being to catch the initial weakening before it progresses too far. Uncertainty exists where parts that should be inspected are inaccessible (say fixings on large external panels). Exactly what is inspected is closely linked to the structural form and to an engineer's knowledge of how it might perform. The risk to bridge parapets from de-icing salts causing corrosion is of a completely different order to the risk posed to steel in an enclosed warm environment. It may well be desirable in some buildings to check for foundation displacement periodically. Where there is doubt about degradation and gradual change, the market offers a range of monitoring services that can detect change of shape with time. Such methods can distinguish between movements being one-off or are continuing, and if continuing, whether they are seasonal variations or all trending one way.

Where there are defined degradation mechanisms such as fatigue, then periodic non-destructive examination (NDE) methodologies need to be deployed. Indeed, when there is fatigue, such techniques are probably essential. Where degradation is by corrosion, periodic assessment of extent, supplemented perhaps by thickness surveys, might be applicable. Marine structures may well have an original design that includes a margin of extra material thickness as a corrosion allowance and this is then going to erode away with time.

The theme of this section so far is 'warning' to stave off disaster. Designs can include active warning systems, be they tsunami warnings located in the ocean or smoke detectors in homes. In any complex plant or building, moving structure or civil engineering system, modern control system technology offers plenty of opportunities for incorporating sensors that react to information received and give out warnings of approaching trouble.

In cases of hazardous plant, the norm is to have a safety case which ought to give clarity as to which systems need to be functional to ensure ongoing safety. Most likely there should be a plan to have periodic reviews of those systems as circumstances and understandings change, to make sure that their functionality remains adequate. After Fukushima, nuclear facilities worldwide were reviewed to see what lessons might be learned. After the fire under the M1 bridge in 2011 (Chapter 9), the Highways Agency conducted a review to see if other bridges might be at similar risk. The storms of winter 2013/2014 reignited a debate on just how frequently extreme weather events might happen. In recent years, facility managers at large were alerted to health dangers after members of the public contracted legionnaires' disease from inadequately chlorinated air-conditioning units; a hazard that developed not on installation, but in usage.

Staff competency always needs to be kept under review. Personnel changes are to be expected, so educating recruits is essential. Operational matters change so it pays to reconsider safety periodically against all manner of risks such as fire and security. If the staff at Bradford (Chapter 9) had fully considered the risk of fires spreading through the stands, would management not at least have made sure that underfloor rubbish was cleared and verified that rapid stand evacuation was possible? Periodic review of risk assessments and safety cases is a tool for promoting ongoing safety.

Throughout life, there needs to be management of change (as in Section 13.6).

13.13 Chapter summary

In principle all failures are avoidable; although some caused by natural disasters may not be totally avoidable at affordable cost. For the professions, the challenge is to ensure that failures are avoided.

One key technique is to observe and learn from reported failures, It is in all engineers' interests and in the interests of society at large that they do so. A starting point is to schedule out what has gone wrong before, as in the separate chapters of this book, and learn from history. Rather than being a catalogue of incompetency, there are significant patterns and lessons to be learned in the very complex task of designing and executing projects both large and small. And the lessons can be learned; yet, as with most other aspects of engineering, it is a lifetime's occupation, a process of continual improvement and assimilation: the challenges are enormous but so are the rewards. Having made that point, whatever the potential problem, it's most likely to have happened before.

Thereafter, for most structural designs and for larger systems incorporating civil/ structural engineering, procedural techniques are available to minimise the chances of a latent error remaining undetected. Thus:

- The risks of serious failures maybe thereafter be reduced by striving to incorporate key attributes such as insensitivity and robustness into structural designs as they are being developed.
- Caution needs to be exercised whenever there is change, for fear of undermining any key defences included by preceding parties.
- A very common problem is as built construction differing from Design Intent. Inspection to catch this is sound investment.
- The client should always retain as-built records.
- Engineering systems go into service in one condition and thereafter degrade, no matter how much durability is built in. Throughout life, safety then often depends on periodic

monitoring with the object of detecting degradation before it progresses too far.

- If there are suspected problems, tackle them early. Never ignore warning signs. Putting profit before safety is not a sensible strategy.
- The simple advice set out in Sections 13.6, 13.7 and 13.8 on good practice and checking will save much heartache always bearing in mind that anything produced may contain errors.

It will be found that many of the suggestions made in this book to ensure safety are embraced in the safety assessment principles (SAPs) enshrined in 'Safety Assessment Principles for Nuclear Facilities' (Office for Nuclear Regulation, 2014).

References

Beckett, T. (2007), 'The British Airways London Eye. Part 5: Pier and impact protection system', *The Structural Engineer*, **79**(2).

BS 6187:2011, Code of practice for full and partial demolition.

BS EN 1990:2002, Eurocode 0: Basis of structural design.

Carpenter, J. (2008), 'Safety risk and failure: the management of uncertainty', *The Structural Engineer*, **86**(14).

CIRIA/HSE (2011), *Guidance on Catastrophic Events in Construction*, Report C699, CIRIA.

CROSS-UK (undated), www.cross-safety.org/uk.

Hackitt, Dame Judith (2018), *Building a Safer Future. Independent Review of Building Regulations and Fire Safety: Final Report*, CM 9607.

Hazards Forum (undated), https://hazardsforum.org/.

HSE (1978), *Canvey – An Investigation of the Potential Hazards from Operations in the Canvey Island/Thurrock Area*, HMSO.

HSE (2001), *Reducing Risks, Protecting People: HSE's Decision-Making Process*, HSE Books, www.hse.gov.uk/risk/theory/r2p2.pdf.

HSE (2015), The Construction (Design and Management) Regulations 2015, www.hse.gov.uk/construction/cdm/2015/index.htm.

Institution of Structural Engineers (2010), *Appraisal of Existing Structures*, 3rd edn, IStructE Ltd.

Institution of Structural Engineers (2013a), *Manual for the Systematic Risk Assessment of High-Risk Structures against Disproportionate Collapse*, IStructE Ltd.

Institution of Structural Engineers (2013b) *Risk in Structural Engineering*, IStructE Ltd.

Interdepartmental Liaison Group on Risk Assessment (2002), 'The Precautionary Principle: Policy and Application', https://publications.parliament.uk/pa/ld200304/ldselect/ldsctech/110/110we29.htm.

ISO 13822:2010, Bases for design of structures – Assessment of existing structures.

Kletz, T. (2018), *HAZOP and HAZAN, Identifying and Assessing Process Industry Hazards*, 4th edn, Taylor and Francis Group.

Latham, Sir Michael (1994), *Constructing the Team: Joint Review of Procurement and Contractual Arrangements in the United Kingdom Construction Industry, Final Report (Latham Report)*, HMSO.

MacLeod, I. A. (2005), *Modern Structural Analysis: Modelling Process and Guidance*, ICE Publishing.

Mann, A. P. and I. May (2006), 'The interpretation of computer analysis', *The Structural Engineer*, **84**(7).

Office for Nuclear Regulation (2014), *Safety Assessment Principles for Nuclear Facilities*, Rev 0, Office for Nuclear Regulation.

Thoday, J. (2004), 'Demolition – assessing the risks and planning for safety', *The Structural Engineer*, **82**(15).

Further reading

Blockley, D. (1980), *The Nature of Structural Design and Safety*, Halsted Press.

HSE (1992), *Tolerability of Risk from Nuclear Power Stations*, HSE Books.

Kletz, T., has written numerous books on plant accidents and allied safety topics.

National Audit Office (2000), 'Supporting Innovation: Managing Risk in Government Departments', Report by the Comptroller and Auditor General.

Royal Academy of Engineering (2004), The Risk Debate: 'Trust me I'm an engineer', 16 June 2004. (The RA has numerous publications on engineering risk.)

Tietz, S. B. (1998), 'Risk analysis – uses and abuses', *The Structural Engineer*, **76**(2), which also gives a long list of useful references.

Index

Index

Index